TABLE OF CONTENTS

SECTION ONE
FORMWORK FOR CONCRETE

CHAPTER 1
OBJECTIVES IN FORMWORK BUILDING: FORM LOADS AND PRESSURES[1]

Forms are essential to concrete construction. They control position, alignment, size, and shape. Formwork also serves as a temporary support structure for materials, equipment, and workmen. The objectives of a form builder should be threefold:

Quality—to design and build forms accurately so that the desired size, shape, position, and finish of the cast concrete are attained.

Safety—to build substantially so that formwork is capable of supporting all dead and live loads without collapse or danger to workmen and to the concrete structure.

Economy—to build efficiently, saving time and money for the contractor and owner alike.

Economy is a major concern; formwork costs may range anywhere from 35 to 60% of the cost of the concrete structure. Savings depend on the ingenuity and experience of the contractor. Judgment in selection of

[1] Authorized condensation from *Formwork for Concrete*, Copyright 1973, American Concrete Institute.

3

materials and equipment, in planning fabrication and erection procedures, and in scheduling reuse of forms will expedite a job and cut costs. The architect or engineer can also do much to help reduce formwork cost by keeping the requirements of formwork economy in mind when he designs a structure.

In designing and building formwork, the contractor should aim for maximum economy without sacrificing quality or safety. Shortcuts in design or construction that endanger quality or safety may be false economy. For example, if forms do not produce the specified surface finish, much hand rubbing of the concrete may be required; or if forms deflect excessively, bulges in the concrete may require expensive chip-

Figure 1–1 The contractor must know how to build formwork safely and must not take undue risks for the sake of economy; otherwise, he may not only defeat his purposes, but bear responsibility for loss of life. This formwork collapse caused 15 injuries and one death as well as extensive property damage and construction delays.

ping and grinding. Obviously, economy measures that lead to formwork failure defeat their purpose (Fig. 1-1).

HOW FORMWORK AFFECTS CONCRETE QUALITY

Size, shape, and alignment of slabs, beams, and other concrete structural elements depend on accurate construction of the forms. The forms must be (1) of correct dimensions, (2) sufficiently rigid under construction loads to maintain the designed shape of the concrete, (3) stable and strong enough to maintain large members in alignment, and (4) constructed so that they can withstand handling and reuse without affecting their dimensions. Formwork must remain in place until the concrete is strong enough to carry its own weight, or the finished structure may be damaged.

The quality of surface finish of concrete is affected by the form material used. For example, if a patterned or textured finish is to be secured by use of a textured liner, the liner must be properly supported so that it will not deflect and cause indentions in the concrete surface. A correct combination of form material and oil or other parting compound can contribute to eliminating air holes or other surface imperfections in the cast concrete.

FORMWORK SAFETY—CAUSES OF FAILURES

Premature stripping of forms, premature removal of shores, and careless practices in reshoring have caused numerous failures or deficiencies in completed concrete structures. Improper stripping and reshoring may cause sagging of partially cured concrete and development of fine hairline cracks, which may later create a serious maintenance problem. Inadequate size and spacing of reshores may lead to a formwork collapse during construction as well as to damage of the concrete structure. Proper practices to forestall damage of this type are discussed under stripping and reshoring.

Inadequate Bracing

Inadequate cross bracing and horizontal bracing of shores is one of the factors most frequently involved in formwork accidents. Investigations of cases involving thousands of dollars of damage show that the damage

Figure 1–2 High shoring with a heavy load at the top is vulnerable to eccentric or lateral loadings. Diagonal bracing improves the stability of such a structure, as do guys or struts to solid ground or completed structure.

Figure 1–3 Formwork collapsed at the New York Coliseum when rapid delivery with power buggies introduced lateral forces at the top of high shoring.

could have been prevented or held to a small amount if only a few hundred dollars had been spent on diagonal bracing for the formwork supports. (See Figs. 1-2 and 1-3.)

Unstable Soil Under Mudsills; Shoring Not Plumb

Formwork should be safe if it is adequately braced and constructed so that all loads are transmitted to solid ground through vertical members. But the shores must be set plumb and the ground must be able to carry the load without settling. Shores and mudsills must not rest on frozen ground; moisture and heat from concreting operations, or changing air temperatures, may thaw the soil and allow settlement that overloads or shifts the formwork.

Inadequate Control of Concrete Placement

Temperature and rate of vertical placement of concrete are factors influencing the development of lateral pressures that act on the forms. If temperature drops during construction, rate of concreting often must be slowed down to prevent a buildup of lateral pressure from overloading the forms. If this is not done, formwork failures may result.

Failure to properly regulate the placement of concrete on horizontal surfaces or curved roofs may produce unbalanced loadings and consequent failures of formwork.

When Formwork Is Not at Fault

When forms and slabs collapse during concreting, it is natural to assume that the formwork was at fault. This is not always true; the collapse of one 4-story concrete structure was thought at first to be caused by form-support failure, but later investigation showed that some of the exterior wall columns were not on solid rock as was intended. One of these columns settled, became inoperative, and the slabs collapsed. Other cases have been reported in which slabs collapsed due to weakness caused by duct openings at high-stress points. When lower floor slabs collapse they carry upper floor forms with them, and the collapse sometimes appears to be the result of formwork failure until an investigation is made.

Well designed, strongly constructed formwork can withstand some unusual loads. When a crane boom collapsed during the casting of an upper floor of a concrete apartment building in New York City, the boom fell across the working deck and wrapped itself over both sides of the building. Despite the impact of the falling boom, forms and supporting shores were undamaged, and concreting operations were resumed two days later after minor repairs.

RELATIONSHIP OF ARCHITECT, ENGINEER, AND CONTRACTOR

Generally, design of a concrete structure and specifications for its size, strength, and appearance are the responsibility of the architect-engineer,[2] but planning and design of the formwork, as well as its construction, are the contractor's responsibility. It is desirable to allow the contractor as much freedom as possible to use his ingenuity in planning the formwork and concreting procedures. In practice, however, the architect-engineer usually considers it necessary to include some minimum specifications for forming practices to assure that the structure will be completed to his satisfaction. The contract documents and specifications for any job should clearly indicate the relationships between the contractor and the engineer-architect so that each knows his area of authority and responsibility.

Plans and specifications must first give the contractor a complete description of the structure so that he can develop an efficient plan for formwork. In addition to the structural dimensions, this information may be needed:

1. Inserts, waterstops, built-in frames for openings, holes through concrete, and similar requirements where work of other trades will be attached to or supported by formwork.
2. Descriptions of concrete surface finishes in measurable terms. Tolerances for plumb, level, size, thickness, and location.
3. Number, location, and details of all construction, contraction, and expansion joints.
4. Live load used in design of the structure.
5. Locations and details of architectural concrete.
6. Details of chamfers on beam soffits or column corners; indication of where chamfer is prohibited.
7. Basic geometry of special structural shapes such as free-form shells.
8. Camber requirements for slab soffits or structural members.

If camber is desired for slab soffits or structural members to compen-

---

[2] The terms engineer-architect and architect-engineer are used interchangeably in this section to designate: the architect, the engineer, the architectural firm, the engineering firm, the architectural and engineering firm, or other agency issuing project drawings and specifications and/or administering the work under project specifications and drawings.

sate for elastic deflection and/or for deflection due to creep of the concrete, the contract drawings must so indicate and state the amounts needed. Measurement of camber obtained should be made *after* initial set and *before* decentering.

Where architectural features, embedded items, or the work of other trades will change the location of structural members, such as joists in one-way or two-way joist systems, such changes or conditions should be indicated on the structural drawings.

In addition to a full description of the required structure, criteria for and some details of the forming practices will be specified, depending in part on the type of structure that is being erected. Minimum requirements should be stated to assure the owner and his architect or engineer that the formwork will provide adequate support during concreting and until the concrete has gained sufficient strength to permit form removal.

These items should be clearly covered in the engineer-architect specifications and drawings: (a) by whom the formwork will be designed; (b) by whom, when, and for what features formwork will be inspected; and (c) what approvals will be required for formwork drawings, for the forms before concreting and during concreting, and for form removal and reshoring.

Some of the details of forming and construction practices that may have to be specified are the following:

1. Location and order of erection and removal of shoring for composite construction, complex structures, and permanent forms.
2. Stripping time (in terms of strength of field-cured concrete) and reshoring requirements; decentering sequence for shells and other complex structures.
3. Formwork materials and accessories where these are critical to appearance or quality of finished structure. Sequence of concrete placement for structures where this is critical.

Design, Inspection, and Approval of Formwork

In most instances the contractor will plan and design the formwork. Except for unusual or complex structures, this is desirable since the contractor can best evaluate men, materials, equipment, and procedures and arrive at a design that is both structurally sound and adapted to efficient erection and concreting.

Although formwork safety is the responsibility of the contractor, the engineer or architect may require that the form design be subject to his review or approval or both. Architect-engineer approval may be advisable for unusually complicated structures, for structures where the design is predicated on a particular method of construction, for certain post-tensioned structures, and for structures in which the forms impart a desired architectural finish.

American Concrete Institute's (ACI) Committee 347, Formwork for Concrete, has also called attention to the legal implications of specifying in any set of contract documents, including plans and specifications, both the method by which the work is to be performed and the results to be accomplished. If the method is specified in detail, then provisions regarding the final results may not be legally binding.

Permanent Forms. Where metal deck or other material used as a permanent form is also to have permanent structural value, its shape, depth, gage, dimensions, and properties, as well as shoring requirements, are to be indicated by the architect-engineer in the contract drawings and specifications. The contractor, nevertheless, may be asked to submit fully detailed shop drawings of all permanent deck forms to the architect-engineer for approval.

Composite Construction. The architect-engineer will specify shoring for composite beam and slab construction wherever his design for composite action requires it. He should supervise field-cured cylinder tests of concrete strength, and shores should be removed only after these tests and curing operations indicate to his satisfaction that the concrete has attained the strength required for the composite action. The procedure for shore removal should be specified in such cases.

Stripping and Decentering

Formwork must remain in place long enough to assure that the concrete is self-supporting and stiff enough to carry its own weight without undue deflection or damage. This is especially important for long span members in flexure. To achieve the necessary strength, either the forms are left in place for a specified period of time, particularly on small projects where tests are not practicable or where form reuse is not planned; or, preferably for all important projects, the time of removal is determined by strength of test specimens. In the former instance, the architect-engineer will include curing-time requirements in the specifications or will refer to applicable codes. If form removal is to be based on strength tests, the architect-engineer should include instructions for the preparation and curing of test cylinders in the specifications and should supervise the testing and determine when it is safe to remove forms.

METHODS OF PAYMENT FOR FORMWORK

Payment for concrete formwork may be by any one of three methods:

1. It may be included in the lump sum price for the entire job.
2. It may be included in the unit price paid per cubic yard of concrete in place.
3. A separate unit price per square foot of formed area is established.

Regardless of the contractual basis of payment, most contractors keep separate records of formwork cost, particularly for their larger jobs. This serves the dual purpose of maintaining closer control on formwork costs during the job and of accumulating cost experience data that could be valuable for negotiating any extra work or in bidding on future jobs.

When formwork is paid for at a separate unit price, the specifications should clearly state what areas of the formwork, if any, are excluded as pay areas. Some specifications exclude openings such as doors, windows, and pipe blockouts from the area of formwork to be paid for. This represents a problem for the contractor, who usually will have to build his form panel continuous across the opening and then attach forms for the opening. Other specifications exclude only openings in excess of a designated size, such as 30 sq ft. This latter practice permits normal door openings to be paid for when door bucks are nailed to the main form panels.

Some specifications exclude payment for bulkhead areas; this places the contractor at a disadvantage for two reasons.

1. Construction joint layout is subject to the owner's approval, and at the bidding stage the contractor cannot always be sure of the owner's intent on joint location.
2. The pay basis for formed area is usually used for pricing extra work, which may have a ratio of bulkhead area to formed area that is substantially different from the main job.

If the contract specifications exclude payment for bulkheads, then they should be fully supported by comprehensive construction joint drawings. Contracts that specify payment for all formed areas in actual contact with concrete surfaces may prove cumbersome to administer if the structure has numerous keyways, recesses, chamfers, or moldings. A less exacting method of measurement can be achieved by excluding these items when they are either 3 in. or less in depth or 6 in. or less in width.

Where some formed surfaces are likely to be substantially more expensive than others, and their ratio to the total form area is likely to vary during the job, separate pay rates for the different classes of formwork should be established (Fig. 1-4).

FORMWORK LOADS—DESIGN CONSIDERATIONS

Formwork for concrete must support all vertical and lateral loads that may be applied until these loads can be carried by the concrete structure itself. Loads on the forms include the weight of reinforcing steel and fresh concrete, the weight of the forms themselves, and various live loads imposed during the construction process. Dumping of concrete, movement of construction equipment, and action of the wind may produce lateral forces that must be resisted by the formwork to prevent lateral failure.

Form design must consider conditions such as unsymmetrical place-

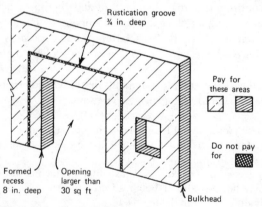

Figure 1–4 Suggested basis of measurement of form area for payment purposes: Pay for all formed areas including bulkheads; include area of recesses formed, but deduct areas of openings larger than 30 sq. ft. on one contact surface. Include formed grooves, keyways, etc., when they are more than 3 in. deep and 6 in. wide.

Figure 1–5 Live load including power buggy and concrete crew. A minimum value of 50 psf for design is usually suggested, but where power buggies are used, 75 psf is more common.

ment of concrete, impact from machine-delivered concrete, uplift, and concentrated loads produced by storing supplies on the freshly placed slab. Rarely will there be precise information as to loads that will come on the forms, and the designer must make safe assumptions that will hold good for conditions generally encountered. The following sections are a guide to the designer for determining the loadings on which to base form design for ordinary conditions normally applicable to structural concrete. (See Fig. 1-5.)

VERTICAL LOADS

Vertical loads on formwork include the weight of reinforced concrete together with the weight of forms themselves, regarded as dead load, and the live loads imposed during construction (such as workmen and equipment). Although concrete may weigh anywhere from 40 to 600 lb per cu ft, the majority of all formwork involves concrete weighing 145 to 150 lb per cu ft. Minor variations in this weight are not significant, and 150 lb per cu ft, including weight of reinforcing steel, is generally assumed for design. Formwork weights vary from as little as 3 or 4 lb per sq ft to 10 to 15 lb per sq ft. When the formwork weight is small in relation to the weight of the concrete plus live load, it is frequently neglected.

If concrete weighs 150 lb per cu ft, it will place a load on the forms of 12.5 lb per sq ft (psf) for each inch of slab thickness. Thus a 6-in. slab would produce a dead load of 12.5×6, or 75 psf, neglecting the weight of forms. If the same slab were of lightweight concrete weighing 100 lb per cu ft, dead load would be (100/12) ×6, or 50 psf.

ACI Committee 347 recommends a minimum construction live load of 50 psf of horizontal projection to provide for weight of workmen, equipment, runways, and impact. Unusual conditions may justify a smaller allowance, but many designers use 75 psf or more for construction with powered concrete buggies.

SHORING LOADS IN MULTISTORY STRUCTURES

In multistory work, the shoring that supports freshly placed concrete is necessarily supported by lower floors that may not yet have attained their full strength, and that may not have been designed to carry loads as great as those imposed during construction, even if they were at full design strength. Therefore, shoring must be provided for enough floors to develop the needed capacity to support the imposed loads without excessive stress or deflection. Whether permanent shores or reshores are used at the several required lower floor levels depends on job plans for reuse of materials, but loads to be carried are the same on either permanent shores or reshore once the lowest floor of shores has been removed.

In any case, shores in the lower stories should be designed to carry the full weight of concrete and formwork posted to them prior to removal of the lowest story of shores supported on the ground or other unyielding support. ACI Committee 347 recommends that shores above the first level be designed for at least 1½ times the weight of a given floor of concrete, forms, and construction loads; often a larger load than this must be supported.

Where selective reduction in the number of reshores required for lower floors is made, the size of the shore should be carefully determined so as to assure its adequacy for the loads posted on it.

In determining the number of floors to be shored to support construction loads above, ACI Committee 347 recommends consideration of these factors:

1. Design load capacity of the slab or member, including live load, partition loads, and other loads for which the engineer designed the slab. Where the engineer-architect included allowances for construction loads, such values should be shown on the structural drawings.
2. Dead weight of the concrete and formwork.
3. Construction live loads involved, such as placing crews or equipment.
4. Design strength of concrete specified.
5. Cycle time between placement of successive floors.
6. Developed strength of the concrete at the time it is required to support new loads above.
7. Span of slab or structural member between structural supports.
8. Type of forming system; that is, span of horizontal forming components, individual shore loads.

LATERAL PRESSURE OF FRESH CONCRETE

Loads imposed by fresh concrete against wall or column forms differ from the gravity load on a horizontal slab form. The freshly placed concrete behaves temporarily like a fluid, producing a hydrostatic pressure that acts laterally on the vertical forms. This lateral pressure is comparable to full liquid head when concrete is placed full height within the period required for its initial set. With slower rates of placing, concrete at the bottom of the form begins to harden, and the lateral pressure is reduced to less than full fluid pressure by the time concreting is completed in the upper parts of the form. The effective lateral pressure— a modified hydrostatic pressure—has been found to be influenced by the weight, rate of placement, and temperature of the concrete mix; the use of retarding admixture, and the effect of vibration or other consolidation methods.

Factors Affecting Lateral Pressure on Forms

Weight of Concrete. The weight of concrete has a direct influence, since hydrostatic pressure at any point in a fluid is created by the weight of superimposed fluid. Liquid (hydrostatic) pressure is the same in all directions at a given depth in the fluid, and it acts at right angles to any surface that confines the fluid. If concrete acted as a true liquid, the pressure would be equal to the density of the fluid—150 lb per cu ft is commonly assumed for concrete—times the depth in feet to the point at which pressure was being considered. However, fresh concrete is a mixture of solids and water whose behavior only approximates that of a liquid, and for a limited time only.

Rate of Placing. The average rate of rise of the concrete in the form is referred to as the rate of placing. As the concrete is being placed, lateral pressure at a given point increases as concrete height above this point increases. Finally by consolidation, stiffening, or by a combination of the two, the concrete at this point tends to support itself, no longer causing lateral pressure on the forms. The rate of placing has a primary effect on lateral pressure, and the maximum lateral pressure is proportional to the rate of placing, up to a limit equal to the full fluid pressure.

Vibration. Internal vibration tends to consolidate concrete. It also results in temporary lateral pressures locally that are at least 10 to 20% greater than those occurring with simple spading, because it causes concrete to behave as a fluid for the full depth of vibration. Since internal vibration is now a common practice, forms should be designed to withstand the greater pressure. They must also be made tighter to prevent leakage. Overvibration or too vigorous vibration are common causes of torm breakouts.

Revibration and external vibration are accepted practices for certain types of construction, producing even higher loads on the forms than normal internal vibration and requiring specially designed forms. External vibration hammers the form against the concrete, causing wide fluctuation in lateral pressure. The frequency and amplitude of external vibration must be adjusted in the field to avoid battering forms to pieces and yet be sufficient to consolidate the concrete. If the maximum vibration that the forms can withstand is inadequate for consolidation, the slump of concrete is usually increased.

During revibration the vibrator is forced down through freshly placed concrete into layers that have stiffened or nearly reached initial set. Local pressures up to 300 psf per ft of head of concrete have been recorded with vigorous revibration. On the whole, however, the effects of revibration and external vibration have not been sufficiently investigated to be expressed in a pressure formula.

Temperature. Temperature of the concrete at the time of placing has an important influence on pressures because it affects the setting time of concrete. At low temperatures the concrete takes longer to stiffen and, therefore, a greater depth can be placed before the lower portion becomes firm enough to be self-supporting. The greater liquid head thus developed results in higher lateral pressures. It is particularly important to keep this in mind when designing forms for concrete to be placed in cold weather or with fly ash or retarding admixtures used in any weather. Tables 1-A and 1-B show maximum lateral pressures for design of wall and column forms, respectively, for different rates of placement at temperatures ranging from 40 to 90° F.

Table 1–A
MAXIMUM LATERAL PRESSURE FOR DESIGN OF WALL FORMS
Based on ACI Committee 622 pressure formula

Note. Do not use design pressures in excess of 2000 psf or 150 × height of fresh concrete in forms, whichever is less.

Rate of placement, R (ft per hr)	p, Maximum Lateral Pressure, psf, for Temperature Indicated					
	90° F	80° F	70° F	60° F	50° F	40° F
1	250	262	278	300	330	375
2	350	375	407	450	510	600
3	450	488	536	600	690	825
4	550	600	664	750	870	1050
5	650	712	793	900	1050	1275
6	750	825	921	1050	1230	1500
7	850	938	1050	1200	1410	1725
8	881	973	1090	1246	1466	1795
9	912	1008	1130	1293	1522	1865
10	943	1043	1170	1340	1578	1935

Table 1–B

MAXIMUM LATERAL PRESSURE FOR DESIGN OF COLUMN FORMS

Based on ACI Committee 622 pressure formula

Note. Do not use design pressure in excess of 3000 psf or 150 × height of fresh concrete in forms, whichever is less.

Rate of placement, R (ft per hr)	p, Maximum Lateral Pressure, psf, for Temperature Indicated					
	90° F	80° F	70° F	60° F	50° F	40° F
1	250	262	278	300	330	375
2	350	375	407	450	510	600
3	450	488	536	600	690	825
4	550	600	664	750	870	1050
5	650	712	793	900	1050	1275
6	750	825	921	1050	1230	1500
7	850	938	1050	1200	1410	1725
8	950	1050	1178	1350	1590	1950
9	1050	1163	1307	1500	1770	2175
10	1150	1275	1435	1650	1950	2400
11	1250	1388	1564	1800	2130	2625
12	1350	1500	1693	1950	2310	2850
13	1450	1613	1822	2100	2490	3000
14	1550	1725	1950	2250	2670	
16	1750	1950	2207	2550	3000	
18	1950	2175	2464	2850		
20	2150	2400	2721	3000		
22	2350	2625	2979			
24	2550	2850	3000			
26	2750	3000	3000 psf maximum governs			
28	2950					
30	3000					

CHAPTER 2
MATERIALS, ACCESSORIES, PROPRIETARY PRODUCTS[1]

LUMBER

Practically all formwork jobs, regardless of the form materials used, require some lumber. Although the species, grades, sizes, and lengths vary geographically, the local supplier will advise what material and sizes are in stock or promptly obtainable, and the designer or builder can proceed accordingly. Frequently the choice of lumber species is a question of local availability and cost; there are usually several kinds that will serve equally well for a given job. Any lumber that is straight and structurally strong and sound may be used for formwork, although the wide distribution and abundance of softwoods make them generally most economical for all types of formwork. The softwoods are usually lighter in weight and are easier to work, although not all species are truly softer than the so-called hardwoods. Hardwood caps and wedges may be introduced where additional strength across grain is needed.

Partially seasoned stock is usually used for formwork, since fully dried

---

[1] Authorized condensation from *Formwork for Concrete*, Copyright 1973, American Concrete Institute.

lumber swells excessively when it becomes wet and green timber will dry out and warp during hot weather, causing difficulties of alignment and uneven surface. Old and new boards should not be used together in the same panel if uniform finish is important. Although final choice of wood for forming will depend on the local market, the following brief description of some of the commonly used woods may be of value.

Kinds of Lumber

Southern yellow pine and Douglas fir are widely used in structural concrete forms and are equally suitable for architectural concrete. They are easily worked and are the strongest in the softwood group. Since both hold nails well and are durable, they are used for sheathing, studs, and wales. Douglas fir, which is appreciably lighter in weight and is a little softer than southern pine, is sometimes used for milled wood forms. Southern pine has moderately large shrinkage but stays in place well when properly seasoned. The choice between the two should be primarily one of cost, since the differences between them are generally small.

California redwood is used to some extent for structural concrete forms and is an excellent material for many uses. It is not recommended for architectural concrete work, however, because of its tendency to stain the concrete. Even for studs and wales, redwood is not suitable as the stain may drip onto an exposed surface when the wood is wet.

Western hemlock is comparable to Douglas fir as form lumber and may generally be used wherever Douglas fir or southern pine is used, although it is not quite as strong. The species of hemlock growing on the Pacific coast should not be confused with eastern hemlock, which is not generally considered suitable for architectural concrete forms.

Woods that are particularly suitable for architectural concrete forms are Northern white, Idaho white, sugar, and ponderosa pines. Since they are not so abundant as Douglas fir and southern pine and are used for purposes for which the latter are not so well suited, they are not generally economical for forms except for special uses. Because the white pines are soft and straight grained, they are especially well suited for run moldings and milled forms for ornamentation. The white pines stay in place well as they are not inclined to warp and twist. This characteristic is especially desirable for forms made up of an assembly of milled pieces since they will remain tight and will insure sharp detailing. Norway pine and eastern spruce have many of the qualities of the white pines and may be used, providing satisfactory grades can be obtained.

PLYWOOD

Plywood has been increasingly used for form sheathing in recent years. Relatively large sheets of plywood save labor in form building, and the correspondingly large areas of joint-free concrete reduce the cost of finishing and rubbing exposed surfaces. Improved production standards, along with waterproof or moisture-resistant glues, have largely overcome early objections that plywood would delaminate when in recurrent or prolonged contact with wet concrete. With proper care and treatment of form surfaces and panel edges, many reuses are possible with plywood sheathing.

Construction: Sizes Available

Plywood is built up of an odd number of thin sheets of wood glued together with the grain of each piece at right angles to the ones adjoining it. The grains of the two outside plies are parallel to provide stability. The layers or plies are dried and joined under pressure with glues that make the joints as strong or stronger than the wood itself. Alternating direction of the grain of adjoining layers equalizes strains and thus minimizes shrinkage and warping of the plywood panels.

Softwood plywood, which is used for concrete forms, is manufactured from several species of woods, of which Douglas fir is the most common. Among other kinds of wood used are western larch, redwood, and various species of pine, spruce, fir, and cedar.

The ¾-in. thickness of plywood in 4×8-ft sheets is most commonly used for formwork except for patent forms that use ½-in. thickness. Other thicknesses are obtainable, but sizes in the range of ¼ to 1 in. are most widely available. In addition to the standard 4×8-ft sheet, 5-ft widths, and lengths ranging from 5 to 12 ft are often available from stock. Some contractors cut the 5-ft width to 2½×10-ft panels, which are easier for one man to handle than the 4-ft width (Fig. 2-1).

Types and Grades

Softwood plywood is made in two types, *interior* and *exterior*; exterior type is bonded with waterproof glue, and interior type is bonded with water-resistant glue. Both types are used for formwork, but the exterior type is chosen where maximum reuse is desired. Within each type, panels are made up of several grades of veneer ranging from A to D depending on the freedom of the surface from knots and other defects.

Figure 2-1 Controlled gluing is important in the manufacture of plywood. Veneer sheets pass through rollers which apply glue evenly. Each sheet of veneer is then placed with the grain at right angles to that of the plies above and below.

Grade B-B, commonly used for formwork, has both faces of B-grade veneer, which is a smoothly sanded solid-surface sheet with circular repair plugs and tight knots permitted.

Plywood grades and types are defined in commercial standards for the manufacture of plywoods established by the U.S. Department of Commerce. These same standards also provide that plywood labeled as concrete form grade shall be edge sealed and mill oiled. Mill oiling does not eliminate the need for oiling or coating on the job, but mill-oiled plywood gives better service than plywood that is job-treated only. Edge sealing protects the glue line from moisture. Some form coatings now on the market require an unoiled base; if the use of such a coating is planned, it is important to specify unoiled plywood.

Various grades of plywood may be found to suit different forming needs, but mill oiling and edge sealing are standard practice only for the grades specifically designated for formwork.

Wherever plywood of known structural qualities is required, as is generally the case for formwork, it is a good practice to specify panels carrying the mark of an approved inspection and testing agency, which indicates type and grade, species of veneer, and conformance with the applicable U.S. Commercial Standard. If there is any doubt as to quality of plywood purchased, a certification of type and grade may be requested. (See Fig. 2-2.)

Overlaid (Plastic Coated) Plywood

Overlaid plywood is exterior type (with waterproof glue) produced the same as other plywood, but with the addition of resin-impregnated fiber faces permanently fused under heat and pressure on one or both sides. Overlaid plywood is also referred to as plastic coated, but it should be carefully distinguished from plywood that is coated or treated by the user with various plastic compounds.

Translucent or opaque overlays blank out the pattern of grain or knots; consequently, these overlaid plywoods are used where smoothest, grain-

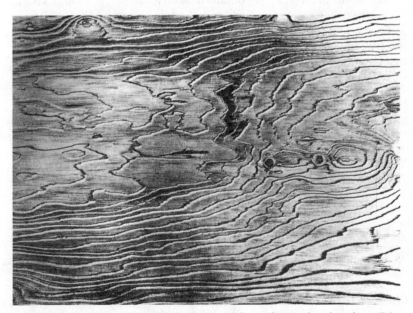

Figure 2-2 Pronounced pattern of grooves in plywood is produced at the mill by removing some of the soft grain growth to accentuate natural swirls and contours.

less surfaces are desirable. Up to 200 reuses in formwork have been reported for overlaid plywood, but performance of this kind depends largely on the care taken in using the forms. The overlay generally improves the abrasion and moisture resistance of plywood, and it decreases requirements for form oiling. Some of the high density overlay surfaces may be used untreated, but light oiling usually prolongs their service life. Since the exact nature of the overlay may vary from manufacturer to manufacturer, the producer's instructions should be followed with regard to oiling or treating overlaid plywood.

Bending Plywood to Curved Surfaces

Simple curves with radii not less than 24 in. can readily be made in plywood form sheathing. Table 2-A shows minimum bending radii for several thicknesses of panel. Note that shorter radius curves can be obtained when plywood is bent across the grain. Shorter radii than those tabulated may often be developed by wetting and steaming, but this should be done only with exterior type plywood made with waterproof glue. Checking and grain rise may be more prominent with this method.

Best results in blending are obtained when a continuous rounded backing is used, but this may not be required for larger radius curves. In applications where there is abrupt curvature, secure the panel to the shorter radius first. In critical bends, two thin panels often work better than one thick one.

Table 2–A
MINIMUM BENDING RADII FOR PLYWOOD PANELS[a]

These radii apply for mill run panels carefully bent. Select pieces with clear, straight grain can be bent to smaller radii.

Panel Thickness, in	Curved Across Grain	Curved Parallel to Grain
¼	24 in.	5 ft
⅜	36 in.	8 ft
½	6 ft	12 ft
⅝	8 ft	16 ft
¾	12 ft	20 ft

[a] Based on information published in "Guide to Plywood for Industry," American Plywood Association, Tacoma, Wash., 1967.

Thick plywood, like solid timber, may also be bent by saw kerfing the inside of the curve. To get a smooth curve the saw cut should not go through the last two plies next to the face. Space kerfs so that the cuts will just close up at the correct radius. The finer the saw, the narrower will be the space on either side of the cut and the smoother the corresponding curve.

OTHER FRAMING AND FACING MATERIALS

Tempered Hardboard
Hardboard is a board material manufactured from refined or partly refined wood fibers, felted into a panel having a density from 50 to 80 lb per cu ft under controlled combinations of consolidating pressure, heat, and moisture. Both standard and tempered hardboard are produced, but the latter is preferred for formwork. Tempering is the supplemental operation of impregnating the hardboard with materials such as drying oils which are stabilized by baking or other heating after impregnation. Tempered hardboard has better strength properties, lower rate of water absorption, and improved abrasion resistance.

Used primarily as a form liner or form facing material, hardboard is not a structural material like plywood and lumber and therefore should be applied to a supporting backing of lumber at least 4-in. wide. Space between supporting members varies from ¾ to 3½ in. Adjoining hardboard sheets should not be tightly butted. To prevent buckling, joints about the thickness of a dime are usually left between abutting edges. (See Figs. 2-3 and 2-4.)

Sheets 4 ft wide are available in lengths from 4 to 16 ft, but the 4×8-ft panel is the standard size. The ¼-in. thickness recommended for formwork weighs 1.4 psf. Large sheets cut down the number of seams and thereby reduce finishing costs.

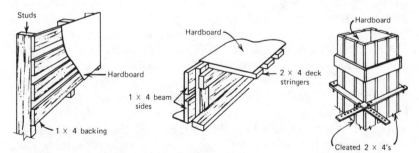

Figure 2–3 Typical assembly of wall, beam, and column forms with hardboard sheathing.

Figure 2–4 Hardboard panels used for deck forming. Stringers (2x4's) cleated together in groups of three or four are placed over horizontal steel shoring. Single 2x4's fill out odd spaces. Hardboard panels, 4x8 ft., are then nailed to the stringers. A gap about the thickness of a dime is left between hardboard panels.

Applications. Hardboard together with 1×4 and 2×4 lumber makes a lining-sheathing surface for wall, deck, column, and beam forms as shown in the several applications sketched. Spacing between the backing boards ranges from ¾ to 3½-in., depending on the loads imposed on the formwork.

The ¼-in. form grade can be bent to a radius of 25 in. without heating or wetting. It can be ordered bent to smaller radii down to 5 in.

Steel

Steel has long been an important material for the fabrication of special-purpose forms: all-steel panel systems for general building construction have been successfully fabricated and used; and steel framing and bracing are important in the construction of many wood and plywood panel systems. Patented steel pan and dome components for slab forming as well as various stay-in-place steel forms are standard construction items, and horizontal and vertical shores of steel are widely used today.

Standard and lightweight structural steel members—channels, angles, I-beams, and others—also are finding increasing use in the framing or supporting of formwork, where they serve much the same purpose as wood members but often permit greater spans or heavier loads than would be possible with sawed timber members.

With reasonable care, steel form framing members will last indefinitely

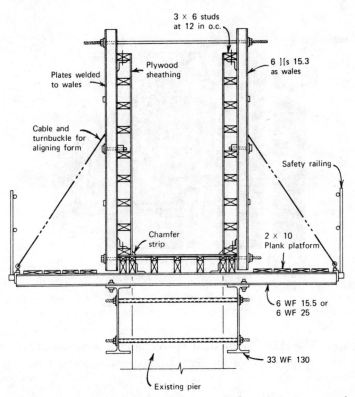

Figure 2-5 Steel channels, angles, and wide flange beams were used with wood and plywood in this form for beams up to 6x10 ft. in cross section. Unsupported form length ranged up to 42 ft. For beam spans over 42 ft., forms were supported at midspan by a heavy wood bent framed of double 2x12 members.

and, hence, will be suitable for many reuses; their selection may be determined on the basis of cost comparisons, or a contractor who has some steel sections on hand may adapt his form design to make use of them. At times, steel members do a job that would be impossible with other materials; their relatively longer spans make possible considerable saving in materials and labor that would otherwise go into intermediate supports. Consider, for example, the large beam form shown in Figure 2-5, which has an unsupported length of 42 ft.

Figure 2-6 shows steel channels in use as wales for a simply framed

Figure 2–6 Double channels held together at the proper spacing by small welded plates serve as wales for this wall form.

wall form. The channels are used double, just as wood members would be, with a space between to allow fitting the ties without drilling the members.

Glass-Fiber-Reinforced Plastic

Glass-fiber-reinforced plastic forms are popularly described both as plastic forms and fiberglass forms, but neither abbreviated term is completely correct. They are truly a combination of materials, up to about one-third being glass fiber. Such forms are finding increasing use in precast concrete construction and in architectural concrete because they produce excellent cast concrete surfaces, minimizing the need for finishing and repair. Only a very light oil coating is required, and some successful use has been reported without oiling the forms at all. A seamless surface with little limitation of size or shape is possible.

Thickness of the fiberglass-reinforced plastic forms has varied from $\frac{1}{8}$ in. for slab material, with no external reinforcing, to $\frac{5}{8}$ in. for column forms, with 3×4's as reinforcement. Dome pans have been used successfully with a wall thickness of $\frac{1}{8}$ in.

Figure 2-7 Fiber-glass-reinforced plastic served as the contact surface for sectional forms used 60 times on the 32-ft.-diameter core of the Marina City apartment building in Chicago. Eight 9-ft.-high sections were joined to make the complete circle. The glass-fiber-reinforced plastic face was ½ in. thick, backed by 4x4's on 1-ft. centers. Additional bracing was provided by curved double-angle wales and double-channel strongbacks.

For complex shapes made in small quantity, the forms are constructed by the conventional hand lay-up method which is used for other reinforced plastic. An alternate method of building up the forms is to use a spray gun to apply the resin to which chopped strands of glass fiber have been added as the reinforcing material. Often a combination of the two methods is used (Fig. 2-7).

With either method of making the forms, it is possible to eliminate all joints or seams. When special conditions dictate the building of a form in sections, it is possible to join the units in such a manner that the several sections may later be sealed together at the job site with additional applications of resin and glass fiber to produce a seamless mold.

These materials generally do not lend themselves to field fabrication, however. Careful temperature and humidity controls must be exercised at all times during the manufacture of the forms. For these reasons all glass-fiber-reinforced plastic forms have been fabricated under factory conditions. Most of these forms have been custom made by a number of firms specializing in this work. However, ready-made column and dome pan forms are now available.

Form Lining Materials

The term "form lining" used here includes any sheet, plate, or layer of material attached directly to the inside face of forms to improve or alter the surface texture and quality of the finished concrete. Substances that are applied by brushing, dipping, mopping, spraying, and the like, to preserve the form material and to make stripping easier are referred to as form coatings and are described later; some of these "coatings" are so effective as to approximate the form liner in function.

Various thicknesses of the several lining materials have been tried and various means of attaching the materials to the forms are used. Some adhesives are suitable for attaching liners to both wood and steel forms, or they may be tacked or nailed to wood forms.

Whether the liner is attached to a horizontal or vertical form surface also makes a difference. Thin sheets that can be used satisfactorily as base mats may wrinkle or sag when attached to vertical forms. Thicker layers of lining material generally have greater rigidity and, consequently, are more adaptable to vertical form surfaces; they also are less subject to accidental damage from vibrators.

Extreme smoothness and relative imperviousness of some of the lining materials give rise to a problem of eliminating air or "bug" holes, particularly when oil or grease is applied. Air bubbles seem to adhere to some of these ultrasmooth surfaces and to resist removal by vibration. Wood, plywood, and hardboard form faces do not present this problem to the same extent because the oil or other parting compound tends to be absorbed into the exterior fibers instead of existing as a film on the form surface.

Highly absorptive materials used as form liners have eliminated voids and air pockets on the surface of the concrete and have improved durability of the surface layer. However, increased cost and difficulties associated with the use of such materials have prevented their widespread acceptance. (See Fig. 2-8.)

Plastic Form Liners. The plastic sheet widely used for concrete molding purposes has a high gloss on one side and a leatherlike texture on the other, giving the user a choice as to degree of sheen he wants in his finished concrete. Various patterns can be imparted to the plastic lining material either by positive pressure or by vacuum forming. The latter method offers an economic advantage, since the side of the plastic that contacts the original pattern is opposite the surface that will contact the

Figure 2–8 Comparison of concrete surfaces cast with internal vibration against different types of material. The surfaces shown measure 4x12 in. and were in a vertical position when cast against: (1) Unoiled steel. Portions of the surface were pulled out on stripping; (2) Polyethylene liner, unoiled. Relatively smooth, uniform surface; (3) Epoxy resin coating on steel. Shows unevenness of coating and a few bug holes: (4) Vaseline coated steel. Shows unevenness of coating and larger number of bug holes.

concrete. Thus, minor imperfections in the pattern are not reproduced, and high quality finish is unnecessary on the model. (See Fig. 2-9.)

Plastic liners clean easily and, except for very thin ones, are suitable for numerous reuses. No form coatings are necessary. The liner should be clean and dry; improved luster may be imparted to the finished concrete surface by polishing the form with a dry towel and by curing concrete in the form, using waterproof membrane to protect any exposed concrete. Thorough vibration is required to break the air bubbles from the formed surface, and external vibration may be helpful. No finishing is needed for surfaces cast against plastic liners. Although the high-gloss concrete surfaces achieved with some plastic liners may be desirable for interior concrete, they are sometimes vulnerable to crazing and uneven weathering when exposed to the elements.

Insulation Used as Form Liner or Form Board. Both thermal and acoustical insulating boards made from such substances as glass fiber, wood fiber, or foamed plastic may be placed in contact with the concrete forms and the concrete cast against them. No oil or parting compound is required; the supporting formwork is easily stripped, and the insulating boards remain in place, either bonded to the concrete or held in place by form

Figure 2–9 Patterned plastic liner, in 17 pieces, each 2 ft. square, placed in column form and held in place by chamfer strips on both sides. Column was cast face down, then lifted into vertical position.

plank clips. The supplier of these products should advise the user of any possible effects on curing of the concrete.

Some of the insulating board materials are strong enough to serve as permanent forms, replacing conventional deck sheathing. A glass fiber form board, for example, is frequently used where the roof deck is made of gypsum concrete or lightweight aggregate concrete. Normal construction with this form board involves placing the boards, finished face down, on a system of subpurlins securely welded to a steel structural frame. The whole area is then covered with wire mesh reinforcement, and the lightweight roof deck concrete is placed (Fig. 2-10).

Wood fiber form boards commercially available are suitable for support on conventional wood framing and carry normal weight concrete slabs up to 8 in. thick. Manufacturers' published recommendations should be followed in setting up spacing of supports for these form boards. The manufacturer should also advise the user of any special curing problems

Figure 2-10 Installation of glass fiber form boards on supporting steel members preparatory to concreting thin gypsum concrete roof deck.

that may arise as a result of the insulating boards remaining in place.

Foamed polystyrene plastic board is another insulating material that serves either as a form liner or as a structural or semistructural form material that stays in place when forms are removed and provides permanent insulation.

HARDWARE AND FASTENERS

Nails

Nails and spikes are the most common mechanical fastenings used in construction of wood formwork and bracing, and their proper use contributes much to the economy and quality of the work. Forms must be substantial, and their component parts should be held together securely;

but the use of too large or too many nails should be avoided. The labor required for fabrication, erection, and stripping of forms will be reduced by holding the number of nails to a safe minimum, and by selecting the best types of nails for differing formwork requirements.

Nails and spikes are available in a wide assortment of lengths, wire diameter (gage), kind of head, and kind of point; shanks may be smooth bright, cement coated, barbed, etched, or galvanized. Wire nails and the shorter spikes have become standardized in "penny" (*d*) lengths through usage, while the longer spikes are measured in inches. The nominal length is customarily measured from under the head to the tip of the point. Figure 2-11 shows the actual size of three 16*d* nails.

Double-Headed and Common Nails. Double-headed nails are a must for nailing kickers, blocks, braces, and reinforcing for wales—any use where considerable holding power is required and, also, where nails must be removed readily when forms are stripped. Double-headed nails can be pulled easily and quickly with a claw hammer or stripping bar without bruising or otherwise damaging the lumber. The size depends on the material to be nailed and the load to be carried. A spike with removable head now available from form hardware manufacturers may be used where loads are heavy.

Common nails are used in assembly of form panels and other components for multiple use, or wherever nails need not be removed in stripping. Their holding power makes them relatively difficult to remove, and

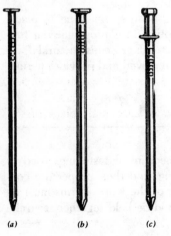

(a)   (b)   (c)

Figure 2-11   Common and double-headed nails.

their heads leave a more noticeable impression in the concrete than some of the special purpose nails (Fig. 2-11).

Nails for Attaching Sheathing or Liners. For attaching sheathing or lining materials to studs, nails whose heads leave the smallest impression on the finished concrete are generally desired. For built-in-place forms, box nails are desirable because the shank is thinner than that of common nails and will pull loose more readily. The size needed depends on the thickness of the sheathing. For nominal 1-in. sheathing or ⅝-in. and thicker plywood, 6d nails are recommended. For panel forms, common nails of this size are better because such forms must withstand considerable racking and abuse.

Where fiberboard or thin plywood liners are to be attached over sheathing, small nails with thin, flat heads such as 3d blue shingle nails are desirable. The heads of these nails leave a very faint impression in the concrete, and the small-diameter shank pulls out of the sheathing material easily without pulling the head through the lining material.

Holding Power of Nails. The diameter, penetration, surface condition, and metal strength are considerations in determining both the lateral and withdrawal strength of nails and spikes. A further variable is the specific gravity of the wood into which the nail or spike is driven.

Nailed joints are generally stronger when the nails are driven into the side grain (perpendicular to the wood fibers) instead of into the end grain of the wood. The joint is also stronger if it uses the lateral resistance instead of direct withdrawal resistance of the nail.

Toenailing. Since withdrawal resistance of nails and spikes driven into end grain is quite low, connections of this type should be avoided whenever possible. This makes toenailing an important technique for form construction. Toenailing, used in place of end nailing, provides a joint that is equivalent to two-thirds of the allowable withdrawal resistance and five-sixths of the lateral load capacity of nails and spikes driven perpendicular and entirely through the side grain of the wood. Best results are obtained when the toenails are started at one-third the length of the nail from the end of the piece, with the nail driven in at an angle of approximately 30° to the face of the piece in which it is started.

Cement Coated Nails. The so-called cement coated nails have some obvious advantages for certain types of formwork construction, since a properly applied coating may double the resistance to withdrawal of nails immediately after they are driven into the softer woods. Different techniques of applying the coating as well as different ingredients in the coating make it impossible to predict the exact improvement in withdrawal resistance. The increase in withdrawal resistance is not permanent but drops off about 50% after a month or so.

Nails in Plywood. Nailing characteristics of plywood are much the same as those of solid wood, except that plywood's greater resistance to splitting when nails are driven near the edge is a definite advantage. The resistance to withdrawal of nails in plywood is 15 to 30% less than that of solid wood of the same thickness, because the fiber distortion is less uniform than in solid wood. Direction of the grain of face ply has little influence on the withdrawal resistance along the end or edge of a piece of plywood.

Ties

A concrete form tie is a tensile unit adapted to holding concrete forms secure against the lateral pressure of unhardened concrete, with or without provision for spacing the forms a definite distance apart, and with or without provision for removal of metal to a specified distance back from the concrete surfaces. Twisted wire and band iron were once the chief tying materials, but because of low strength and the labor of assembly and installation, they are today considered acceptable only for the simplest structures. A wide variety of ready-made ties with safe load ratings ranging from 1000 to about 50,000 lb have gained general acceptance. They consist of an internal tension unit and an external holding device and are manufactured in two basic types:

Continuous single member. The tensile unit is a single piece, and a specially designed holding device is added for engaging the tensile unit against the exterior of the form. These ties may be rod, band, channel, or angle in cross section, and may be cut to length on the job or completely prefabricated. Form-spreading devices are an integral part of some of these ties as Figure 2-12 shows. Some single member ties may be pulled as an entire unit from the concrete; others are broken back a predetermined distance at a section weakened to facilitate "snapping." Some are cut off flush with the concrete surface.

Internal disconnecting type. The tensile unit has an inner part with threaded connections to removable external members that make up the

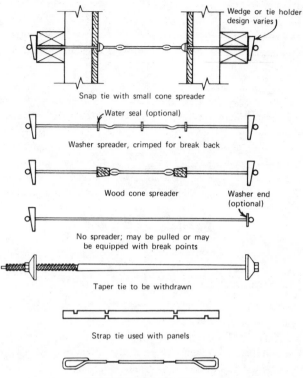

Snap tie with small cone spreader

Wedge or tie holder design varies

Water seal (optional)

Washer spreader, crimped for break back

Wood cone spreader

Washer end (optional)

No spreader; may be pulled or may be equipped with break points

Taper tie to be withdrawn

Strap tie used with panels

Loop end tie used with panels

Figure 2–12 Some typical single member ties. The holding device is shown schematically since a number of different "wedges" or other devices are available. The taper tie is shown with a threaded end holding device, but for single member ties the holder generally slips over the end "button" of the tie and drops down until the assembly is held tight. Working loads vary depending on the kind of steel, diameter of tie, and details of the fastener. A number of ties like these are available with manufacturers' suggested working loads ranging from about 2,500 to 5,000 lb.

rest of the tensile unit and have suitable devices for holding them against the outside of the form. This type of tie (Fig. 2-13) is available with or without spreading devices, and the internal member generally remains in the concrete.

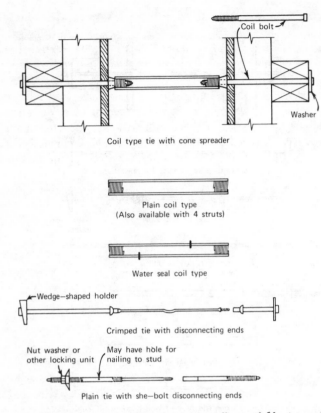

Coil type tie with cone spreader

Plain coil type
(Also available with 4 struts)

Water seal coil type

Crimped tie with disconnecting ends

Plain tie with she—bolt disconnecting ends

Figure 2-13 A few types of commercially available internal disconnecting ties. Either threaded or wedge-style external holding devices may be used, depending on the tie details. Water seal and spreader features are available for many of these ties. Suggested working load for the coil-type units ranges from about 6,000 to 36,000 lb., working load, depending on size and number of struts. The crimped tie is available in a working load range of 3,000 to 9,000 lb., and the plain disconnecting tie is available in range of 3,000 to 48,000 lb. (Data based on manufacturers' statements, but higher or lower strengths may be found since materials and sizes vary.)

These two types of tying devices are identified commercially by various descriptive names, such as form clamps, snap ties, wedge clamps, coil ties, rod clamps, and the like, but no attempt to differentiate them in detail can be made here. The continuous single-member type is generally used for lighter loads, ranging up to about 5000-lb safe load. The internal disconnecting type of tie is available for light or medium loads but finds its greatest application under heavier construction loads.

Since many manufacturers' catalogs show the ultimate load rating for tying devices as well as a suggested working load, a form designer or user can modify suggested loads depending on the safety factor required for a given job. The maximum load on the tie should never exceed the yield point load of the steel, even though the safety factor based on ultimate strength is adequate.

Regardless of the type of tie, the external holding device should have a large enough bearing area so that when the tie is loaded to its maximum safe load, excessive crushing of the wood will not take place.

Effect of Tie on Finished Concrete Surface. Wherever the concrete surface is to be visible and appearance is important, the proper type of form tie or hanger which will not leave exposed metal at the concrete surface is essential. Architectural concrete specifications often require that no metal be left closer than 1½ in. to the surface of the concrete. This requirement can generally be met with either type of tie. However, with the so-called snap tie there is the possibility that break-off will not occur at the specified depth in spite of the weakened tie section.

Where spreader ties are used for exposed work, the spreading device should be of a type that can be easily removed with a minimum of damage to the concrete and that leaves the smallest practicable hole for filling. Spreader cones of wood, plastic, or metal leave a uniform hole for patching. Some wood or plastic spreader cones used with snap ties also permit breaking back ties before the forms are stripped.

A good patch over the tie end, well bonded to adjacent concrete, is essential. If any moisture gets to the tie end, rust stains will gradually appear on the surface of the concrete. Greater depth of breakback or threaded ends of internally disconnected tie units allow a better chance of bonding the patch that covers the tie and also afford a greater factor of safety in case of spalling of the patch. Although the patch remains in place, it may shrink and leave fine cracks through which moisture and rust gradually penetrate. Special admixtures for the patching grout may improve moisture resistance and adhesion of the patch; nonshrink grouts or dry-pack mortar may also be used.

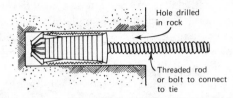

Figure 2–14 One of the several available anchors for off-rock tying. The anchor unit is placed in a hole drilled in the rock. When the bolt or external fastener is inserted, it expands the anchor, thus tightening it in the rock. Some rock anchors are designed to be surrounded by grout in the rock cavity.

Water-Retaining Structures. A tie used for watertight walls must be leakproof. What constitutes a leakproof tie varies with the head of water retained and the resultant pressure on the concrete surface; under some circumstances no ties are permitted at all. If ties are used, they cannot be pulled completely out of the wall because the resultant hole cannot be grouted watertight. The ends of tie metal should be at least 1 in. back from the wall face, and the holes left by tie ends should be carefully plugged with grout.

The process of breaking back or disconnecting tie ends may loosen the tie in the wall, providing a channel for water seepage. To overcome this problem, some manufacturers have crimped or otherwise deformed the tie to improve bond, and have attached round metal or neoprene washers as waterstops at the middle of the ties.

Anchors

Form anchors are devices used to secure formwork to previously placed concrete of adequate strength; they are normally embedded in concrete during placement. Anchors are also used to support forms by off-rock tying (Fig. 2-14). There are two basic parts: the embedded anchoring device, whose design varies with the load to be carried and the strength of concrete in the structure; and the bolt or other external fastener that is removed after use, leaving a set-back hole for grout patching. A number of various types of anchors now available for embedment in fresh concrete are shown in Figure 2-15.

Ultimate strength of anchoring devices in shear and tension is frequently given by manufacturers in sales literature, but their actual hold-

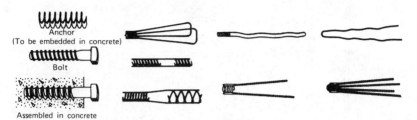

Figure 2–15 Simple screw anchor (left) showing the two parts of the unit assembled in concrete. Embedded anchorage unit only is shown for several other types of form anchors. Some of these require a bolt like that at the left. The crimped anchor requires an internally threaded bolt connection, and the hairpin shape is suitable for looped tie attachment.

Safe working loads for these devices depend on the strength of concrete in which they are embedded as well as the depth of embedment and the area of contact between the anchor and the concrete. Working load data supplied by manufacturers indicate safe loads from 7,000 to 22,000 lb. in 400-psi concrete: 9,000 to 25,000 lb. in 500-psi concrete; and 12,000 to 30,000 lb. in 1,000-psi concrete (pull-out loads on the embedded anchor).

ing power depends on the strength of the concrete in which they are embedded, the area of contact between concrete and anchor, and the depth of embedment.

Many form anchors are set in the relatively low-strength concrete used for massive structures, and development of holding power may be a limiting factor in early reuse of the cantilever forms that the anchors must support. Some manufacturers' strength data for their anchorage products are based on tests in actual concrete specimens at various low strength levels. By using data of this sort together with estimated compressive strength of the concrete mix at early ages (3 to 7 days), tentative selections of anchoring devices needed for a given job can be made. Specific selections for any large job should be field tested under actual summer and winter working conditions.

Hangers

There are a number of ready-made devices for hanging forms from steel or precast concrete structural members. Such forming may be for the construction of a supported slab, for building the fireproofing required for the steel members, or for construction of a slab composite with beam framing. (See Fig. 2-16.)

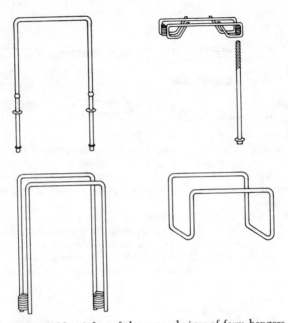

Figure 2–16   A few of the many designs of form hangers used to support forms from existing structural members. Applications of these and similar devices are shown in Chapter 3. Actual working loads vary with diameter and strength of the wire or rods used, as well as hanger design details, but some representative values are: for the snap tie hanger (upper left), 4,000 to 8,000 lb. total load; for the hanger at upper right, 12,000 to 18,000 lb. total load; lower left, 6,000 to 20,000 lb. total load; and for the wire beam saddle (lower right), 2,500 to 6,000 lb. total load. These figures, based on manufacturers' suggestions, show a probable range of strengths for typical commercially available devices; the user must verify the strength of his own specific selection.

Form Jacks. Under some field conditions, use of a jack supported on the lower flange of the steel beam may be preferable to hanging the forms. Several devices available for this purpose may also be adapted for use with precast concrete girders.

A similar jacking device (Fig. 2-17) is used to hold the bottom members of cantilever form panels at the proper distance from a finished wall.

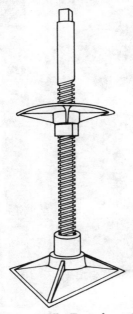

Figure 2-17 Typical cantilever form jack to hold bottom members of cantilever form panels at correct distance from finished wall. A similar jack sometimes is used to support forms from the lower flange of steel beams.

Steel Strapping

Steel strapping, ⅝ to 2 in. wide—the same kind of material that has been widely used in packaging—is used for tying column forms, attaching nailing strips to steel form members, suspending forms for fireproofing structural steel, and tying of low foundation walls. Either standard tensioning tools used for packaging applications or special tools developed for forming applications may be used to install the strapping. When forms are stripped, the straps are cut and discarded.

Steel strapping is held fast by a seal device that is closed after the strap has been placed around the column or other member being tied. Because

Figure 2–18 Workman is tightening steel strapping around column form.

of variations in sealing devices and the tools used to attach them, it is wise to assume that the joint strength may be only about 70% of the strap strength, although some tool-seal combinations do give 100% strap strength in the joint (Fig. 2-18).

Table 2–B
BREAKING STRENGTH OF HEAVY-DUTY STRAPPING

Strap Width (inches)	Strap Thickness (inches)	Feet of Strap per pound	Average Breaking Load (pounds)
¾	0.028	14.0	2300
¾	0.035	11.2	3100
1¼	0.035	6.7	5000
1¼	0.050	4.7	7000
2	0.050	2.9	11,000
2	0.065	2.2	15,500

Representative breaking strength of some of the "heavy-duty" strapping used for concrete formwork is shown in Table 2-B. Both stronger and weaker strappings are available with approximately the same cross sections. For safety purposes, it is suggested that the load on steel strapping be limited to not more than one-half the breaking strength, and a greater factor of safety may be advisable when worn tensioning tools or inexperienced workmen are on the job.

One thing to guard against is the initial tension that an operator may mistakenly put on the strapping before the concrete is placed. High initial tension introduces excessive stress in the strapping and the truss and in the form itself. Ideally, only enough tension is applied initially to cause the strapping to conform closely at form corners or joints. Placing of the concrete then loads the strapping.

When steel strap ties are to be used, rectangular or square column forms are usually provided with some external supporting members that give extra stability under load and keep the steel strapping from making sharp or right-angle bends. For columns 24 in. and larger, this "trussing" is arranged so that the strapping assumes a nearly circular shape around the outside of the form. Most manufacturers of steel strapping have developed their own spacing and truss designs, and these recommendations should be consulted.

Column Clamps

Column clamps or yokes encircle column forms and hold them together securely, withstanding the lateral pressure of the freshly placed concrete. Individual parts of the column clamp may be loaded in bending or tension or a combination of the two, depending on details of assembly. Several designs are available in ready-made reusable column clamps that are adjustable to a range of column cross-section dimensions. Certain clamping devices, like some of those shown in Figure 2-19, are best adapted to shorter columns of small cross section, while other heavier devices are suited to larger cross sections and the greater loads that may be imposed in concreting taller columns.

Figure 2-20 shows a hinged bar column clamp typical of the heavier ones produced by several manufacturers. The two hinged units making up each clamp have all fastening hardware permanently attached; the two parts are laid in place around the column, lock castings are slid together, and the wedge is dropped into the proper slot and tightened with a hammer. Where column forms are in tight quarters, a single bar type, made up of four straight members with attached hardware is used.

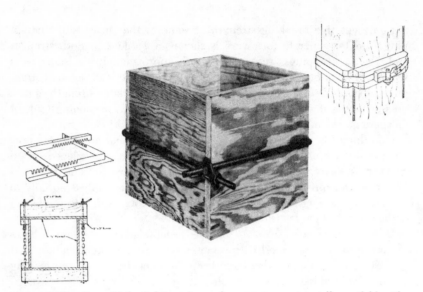

Figure 2–19 Some of the lighter column clamps now commercially available. The clamp made of steel strap may be used for heavier loads when appropriate trussing of the colmun form is provided.

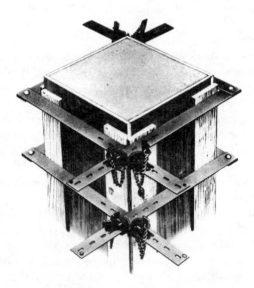

Figure 2–20 Hinged-bar column clamp typical of heavier clamps produced by several manufacturers.

Clamps of the type shown in Figure 2-20, or the single bar variation thereof, are available for column sides ranging from about 8 in. to 8 ft. The longer clamps have a thickened cross section at midspan or are reinforced with attached channels.

It is advisable to examine carefully all of the manufacturer's suggestions and recommendations for the use of these clamps. Aside from dimensional limitations, there may be suggested limits on height and rate of placing concrete when a given type of clamp is to be used. Clamps may be suggested as suitable for only a limited size of column cross section or a special type of framing.

PREFABRICATED FORMS

Prefabricated forms and forming systems have become increasingly important to the concrete contractor as a means of reducing both material and labor cost through the efficiency of mass production. The forms may be purchased outright, rented, or sometimes rented with an option to purchase. Manufacturers of the forming systems frequently provide engineering layouts at no additional cost or for a nominal charge. Supervision for erecting the forms may also be provided.

Prefabricated forms are most advantageous where numerous reuses are expected, and they are generally built with added durability for this purpose. However, some prefabricated forms, such as the fiber tubes and void forms intended for a single use, are destroyed or lost in construction. Other prefabricated forms are built to stay in place as part of the finished structure.

Prefabricated forms may be considered in two groups:

1. *Ready-made forms.* This includes the modular panel systems that are made in relatively small, easily handled units adaptable to forming different structural members. Standardized forms suitable for only one kind of structural member, such as column forms or pans for concrete joist construction, are also in this group.
2. *Custom-made forms.* This group includes special purpose forms built to order, such as tunnel forms, cantilever forms for dams, special forms for bridges, and many others. Although these forms may be made for a single job, they are often reused many times on that job.

Panel Forms and Forming Systems
Prefabricated panel forms are generally manufactured in modular sizes, 2- and 4-ft widths being the most common with heights ranging from 2

to 8 ft. Smaller filler and corner units are of varying sizes. Hardware and ties supplied with form panels vary with different manufacturers. Ties are similar to those already described but are usually fastened against the panel frame, eliminating wales in the typical installation.

Although the many panel-forming systems on the market today vary widely in details, there are four basic types:

1. Unframed plywood panels. Locking and tying hardware are supplied; some panels have attached metal bracing on the back.
2. Plywood panels framed in wood and braced with metal members.

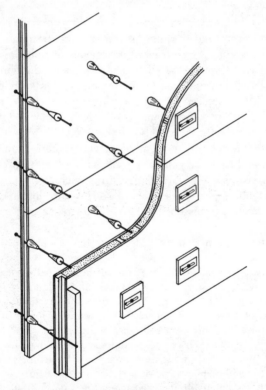

Figure 2–21 Cut-away view of form panel of stressed-skin plywood construction with a foamed plastic insulating material between the two plywood faces. Plywood "donuts" serve as bearing plates for tie holders when wales are not used.

3. Plywood panels set in metal frame, with or without metal bracing on the back.

4. All-metal panels—metal plates welded or bolted to a metal frame.

In addition to these basic types, the relatively new panel form shown in Figure 2-21 uses stressed-skin construction and sandwiches a foamed plastic insulating material between two plywood faces (Fig. 2-21).

Although these panels were developed primarily for wall forming, many of them are adaptable to slab and column forming, and smaller ones have been successfully used in forming beams. The form user should carefully read the manufacturers' data on these forms and select a system that is suited to his job conditions. Some are obviously for light construction, while other form panels are so sturdily built that they can be adapted for a variety of forming needs.

Prefabricated panels are most often used for general and light construction where walls range in height from 2 to 24 ft, particularly on walls ranging from 4 to 10 ft. "Prefab" forms are less frequently used on complicated walls such as curved or battered ones with corbels, but interesting applications for such jobs have been developed (Fig. 2-22).

On large jobs where cranes can work to advantage in handling forms, panel forms are "ganged," that is, joined together with special hardware

Figure 2–22   Newly stripped 50-ft.-diameter sewage treatment tank built in seven days with prefabricated form panels. Small panels made it possible to form the round shape.

provided by the form manufacturer, to form a relatively large wall section. A crane erects, strips, and then moves this large unit to another location for reuse. This is advantageous for high walls where much labor is saved compared with repeated assembly and disassembly of individual panel components. Size of ganged form units commonly ranges up to 30×50 ft.

Pans and Domes for Concrete Joist Construction

Concrete joist construction is a monolithic combination of regularly spaced joists and a thin slab cast in place to form an integral unit with beams and columns. The joists may all be arranged in one direction within a column bay, or they may be arranged in two directions to create a wafflelike pattern. This type of construction has been conventionally formed with ready-made steel "pans" or "domes" of standard size (see Fig. 2-23). Recently, other materials, including hardboard, fiberboard, glass-reinforced plastic, and corrugated cardboard, have been adapted to do the same forming job. Some of the firms that supply these forms also subcontract complete erection and removal of the forms, with or without the supporting centering. Other companies only rent or sell the forms.

The design of steel pans and domes has recently been duplicated or approximated in other materials. For example, the hardboard pans shown in Figure 2-24 are braced with metal supports set in place on wood supporting members. Constructed of 1/4-in. hardboard coated with a waterproof plastic, they require a coating of oil or other bond breaker before concrete is placed. The end pan in each row is trimmed to fill out the row to the exact dimension required. Tapered end pans and end closures are provided. A collapsible wood pan-type form has also been used very successfully.

Reusable special molded fiber pans (Fig. 2-25) are available in widths from 10 to 30 in. and depths from 6 to 14 in. for use on either solid or open deck centering. The fiber is molded under heat and pressure and is treated to withstand the effects of weather and wet concrete, but a release agent or form oil applied before each use facilitates stripping and improves durability. Support chairs and stiffeners are required for most applications of these pan forms. They are suitable for either the nail-down or slip-in systems similar to those shown for steel pans and can also be adapted to forming two-way joist construction.

Single-use corrugated cardboard box-type special pan or dome form members are also available (Fig. 2-26). The corrugated board is asphalt- or wax-impregnated, and these forms require no oiling. An interior core

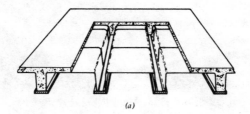

(a)

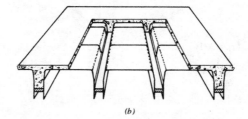

(b)

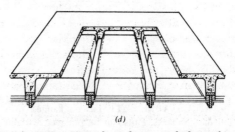

(c)

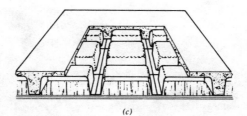

(d)

Figure 2–23 Typical steel pan and dome forms. One other type, the long pan, is shown in Chapter 3 along with erection details for all types.

(a)

(*a*) NAIL-DOWN FLANGE TYPE can be placed, aligned, and nailed from the top side. Simplest to use, but does not generally provide architectural concrete surface.

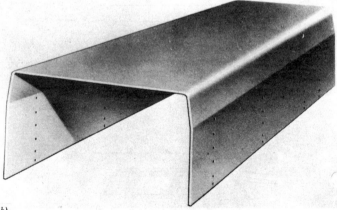

(b)

(*b*) ADJUSTABLE TYPE has no flanges, provides smooth joist bottoms without flange marks. Pans can be removed without disturbing soffits and shore supports.

(c)

(*c*) DOME TYPE used for two-way waffle slabs. This kind of slab design may offer certani structural economies. Flanges are butted at joints, not lapped.

(d)

(*d*) SLIP-IN TYPE is based on the same form as the nail-down type, but uses soffit board between pans to form a smooth joist bottom. Like adjustables, pans can be removed without disturbing soffit supports.

Figure 2-23A Typical steel pan and dome forms. One other type, the long pan, is shown in Chapter 3 along with erection details for all types.

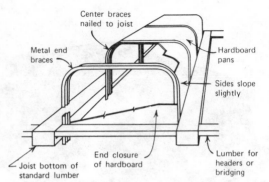

Figure 2–24 Assembly detail for hardboard pans made in widths of 20, 30, and 40 in. with depth adjustable from 6 to 24 in.

Figure 2–25 Molded fiber pans being installed for slab forming.

of egg-crate style or other design provides bracing. The boxes are shipped flat, preslotted and scored, for assembly on the job with tape or stapling machines. They may be either stapled or taped to deck or centering.

Figure 2–26 Assembly of single-use cardboard dome-type forms on a solid plywood deck.

Glass-reinforced plastic dome pans are now available as a ready-made stock item, either for sale or for rent. Range of sizes for these reusable forms is considerably greater than that for steel domes. Clay tile or lightweight concrete filler blocks that remain permanently in the slab structure serve much the same purpose as removable pan and dome forms.

Void and Duct Forms

A number of products have been developed for forming voids and ducts in both slab and beam construction. Laminated fiber tubes equipped with end closures so that concrete will not flow into them are placed near the neutral axis to displace low working concrete, and are tied down with wire to prevent floating or lateral movement when the concrete is being placed. These laminated tubes are available in two types: one with a weather resistant ply designed for use in cast-in-place construction in exposed areas; and the other for use where the tubes will be completely protected from weather during storage and use.

Corrugated cardboard members similar in construction to the cardboard pans and domes described above are also available for forming voids, and inflatable rubber tubing has been used for this purpose too. The patented rubber forms, 3/4 to 12 in. in diameter, are laid in position and then inflated. A loose wire coil is used to hold the tube in place. Reinforcing mesh laid on top of forms also helps keep them from floating. Forms may be tapped with a hammer during placement of slab to vibrate concrete and produce smoother ducts if desired. When the concrete has hardened, the form is deflated and withdrawn through the bulkhead or edge form of the slab. Specially woven nylon or cotton reinforcement governs the way the tube contracts on deflation and thereby aids in breaking bond with the concrete. No form coatings are needed, and little cleanup is required.

Rubber void forms are available up to 60 ft long and can be joined for greater lengths. A total length greater than 330 ft causes excessive friction on withdrawal. Typical inflating pressures range up to 85 psi for smaller sizes, 25 to 33 psi for medium sizes, and up to 25 psi for larger sizes.

For large-diameter (12 to 87 in.) ducts in sewer and industrial work, formers are made of several plies of rubberized fabric. The hoselike form is open at both ends to contain airtight bulkheads through which air pressure is applied for necessary strength of the form. These large forms are inflated at a low pressure sufficient to maintain the shape and carry loads imposed on the form, but without stretching the form material.

Column Forms

Standard forms for round columns are available in both fiber and metal as complete units with no extra fastenings. The fiber tubes are single-use items and require only a minimum of external bracing to keep them plumb. Also available is a square column form assembled from panels of galvanized sheet steel backed by 3/4-in. plywood and braced against lateral pressure with column clamps. Ready-made panels are also adaptable for some column forming.

The tubular fiber column form shown in Figure 2-27 is available for columns ranging from 6 to 48 in. in diameter. Standard lengths are 12 and 18 ft, but greater lengths may be ordered. One manufacturer reports that such forms have been successfully used for continuous placement of columns 40 ft high. Fiber forms can be cut by saw to the exact length desired, and cut sections can be adapted to forming half-round, quarter-round, and obround columns and pilasters. They are also suitable for encasing steel pipe and other members with concrete. Where the form cannot be slipped over existing piling, piers, or posts that are to be encased, a special double-walled type can be cut for installation.

The laminated fiber plies are spirally constructed and are available with wax-impregnated inner and outer surfaces for weather and moisture protection. Where the columns are to be exposed, the inner surface is coated with polyethylene. If the appearance of the columns is critical, a so-called seamless type is available in which fiber plies nearest the inner surface are deckled or scarfed and overlapped to minimize the spiral gaps or seams on the column surface.

No clamps or ties are needed and no form oil is required; since there is no reuse, there is no cleanup. If these forms are left in place, they aid proper hydration of concrete without additional curing procedures. However, it is advisable to strip the forms before placing any critical loads on the columns, to detect any voids that may have formed during placement and may have weakened the member.

The fiber tube form can be stripped by making two vertical cuts with a power saw and pulling the form off, or by making a 12-in. vertical cut in the tube with a linoleum knife and peeling spirally, using a broad bladed tool. With either method, care must be taken to prevent marring the column surface. (See Fig. 2-27.)

Adjustable steel forms for column capitals are made for use with these fiber forms (Figs. 2-28 *a* and *b*). They are similar to the capital portion of the all-steel column forms and are generally a rental item.

Ready-made steel column forms are assembled in sections, the necessary hardware being provided with the form. Bracing is built into the form so that no column clamps are required, and external bracing is needed only to keep the forms plumb during concreting. The capital is a standard part of the form and can be adjusted to different depths and diameters. These forms are manufactured in diameters ranging from 14 in. to 10 ft, with vertical panels 1 to 10 ft high. Some have an access ladder built into the form.

Stay-in-Place Forms

A number of materials used for formwork are left permanently in place, and some become an integral part of the completed structure. Generally described as permanent forms, these may be of the rigid type such as metal deck, precast concrete, wood, plastics, and the various types of fiberboard; or they may be of the flexible type such as reinforced water-repellent corrugated paper or wire mesh with waterproof paper backing.

Where the permanent form serves for deck construction, it is generally

Figure 2–27 Fiber tube column forms require only bracing to keep them plumb and a template at the base for accurate positioning.

supported from the main structural members; however, it may require additional temporary intermediate supports. The permanent form material may be covered in the engineer-architect specifications for the job, but where it is a contractor's optional item, the manufacturers' specifications and recommendations for use are generally relied on.

Metal Deck. Ribbed or corrugated steel sheet is used both as a permanent form for cast concrete and as a combined form and reinforcement; in roofs it may be the permanent supporting member for lightweight insulating concrete fills. Where the steel is to provide continuing support or reinforcement, it should be galvanized. When it acts only as a stay-in-place form, it is generally "black" or uncoated.

(a)   (b)

Figure 2-28 (a) One of the several commercially available steel column forms, showing (b) the general appearance of the column after stripping.

The forms are used for concrete floor and roof slabs cast over steel joists or beams and for bridge decks above high-traffic areas; they may also be used to form the top slab over pipe trenches or other inaccessible places where it is impractical and expensive to remove wood forms. Metal deck forms can also be set in short lengths between precast joists where shear connectors project from the top of the precast members (Fig. 2-29), or over several spans just as with steel members if there are no stirrups.

When the forming material is installed, it is secured in various ways; for example, by clips attached to the top of the joist or by welding to inserts cast in the concrete members. Varying gages and corrugation styles make these forms usable on clear spans ranging up to 8 ft for 20-gage steel.

Some galvanized steel-deck forming materials combine form and positive reinforcing in one piece; this is achieved by deeper corrugations and raised lugs on the corrugations. Transverse wires welded across the corrugations in one such product take the place of temperature steel reinforcing, whereas others require added temperature mesh to prevent shrinkage cracks. Regardless of manufacturers' claims, these products

Figure 2–29 (a) and (b).

cannot take the place of negative reinforcing bars (top steel) required where slabs are continuous over beam, girder, or wall support (Fig. 2-30).

Wherever metal deck forms are to become a working part of the permanent structure, the architect-engineer should take full design responsibility and should specify gage, thickness, depth, physical dimensions, and properties, as well as special shoring requirements, if any.

Precast Concrete. Precast concrete panels or units serve as stay-in-place or permanent forms for various structural elements. In some applications these are exposed aggregate panels that produce a specified architectural surface; another type of precast panel provides an abrasion-resistant surface for mass concrete. Lightweight concrete dome pans have been used as stay-in-place forms for two-way concrete joist systems. Construction applications vary; the precast units may be used as self-supporting form panels or as form liners to produce a special surface on the concrete.

SHORING, SCAFFOLDING

A number of patented shoring systems have been developed with adjustable legs that eliminate cutting, close fitting, and wedging. Miscellaneous hardware is available for joining components, attaching T-heads, braces, stringers, and the like. There is also the so-called horizontal shoring that

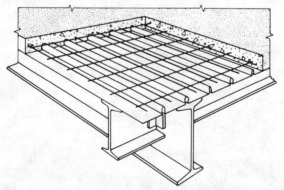

Figure 2–30 One type of combined form and reinforcement. Temperature steel is laid on top of this form, but some combinations have temperature steel welded in place as part of the form. Top steel must be provided where slabs are continuous over supports.

provides a relatively long adjustable-span horizontal support for forms. Most of these items are available either for rental or purchase.

When patented shores or methods of shoring are used, manufacturers' recommendations as to load-carrying capacity may be followed, but they should be supported by test reports from a qualified and recognized testing laboratory. ACI Committee 347 has advised reduced values of allowable loads after materials such as these shoring components have experienced substantial reuse. The user must carefully follow the manufacturers' recommendations for bracing and working loads for unsupported lengths. It is well to remember that bracing of shores may be needed to give safety and stability to the entire formwork assembly, even though it may not be needed to increase load-carrying capacity.

Vertical Shores

There are several types of adjustable individual shores. The simplest of these, shown in Figure 2-31, is based on a clamping device that permits the overlapping of two 4×4 members. A portable jacking tool is carried by the workman from one shore to another to make vertical adjustments.

Figure 2–31   Shores made of two pieces of dimension lumber are joined by a patented clamping device which permits length adjustment.

Figure 2–32 Typical adjustable shore of wood and metal combined. An extension head is shown at the top, but various other head pieces are available.

Special hardware for the top of these shores facilitates joining them to stringers with a minimum of nailing. Typical working load capacity is about 3000 lb. Exact load ratings will depend on condition and quality of lumber as well as the unbraced length.

The shore shown in Figure 2-32 is typical of those made of dimension lumber combined with a steel column and adjusting device; working loads range up to 6000 lb. The steel member may be tubular, T-section, or other shape. If tubular, it may be filled with concrete. Setting is considered a one-man operation. There is a special jacking device for these shores too, and various fittings and extensions for the top are available including T-heads, L-heads, and U-heads. Worn-out wood members often can be replaced with stock lumber to extend the service life. Another advantage of the wood-metal combination is that the shores can be easily braced by nailing to the wood members.

An all-metal individual adjustable shore, sometimes described as a jack shore or simply as a jack, is also available from a number of manufacturers in adjustable heights from about 4 to 16 ft. Figure 2-33 shows a typical line of these shores. There are the usual individual variations in fittings and adjusting devices from one product to another. Safe load ratings range from 1500 to 9000 lb, depending on type of shore and the length to which it is extended. Metal bracing brackets with holes for nailing wood bracing are provided, and heads of different types are available.

For heavy duty at heights up to 50 ft, tripod shoring towers (Fig. 2-34) are available. They consist of a base section, intermediate sections which can be assembled to the desired height, and a bearing member that is attached as the top section. Tripod shoring towers carry up to 33 tons of total concentric load at unbraced lengths up to 45 ft.

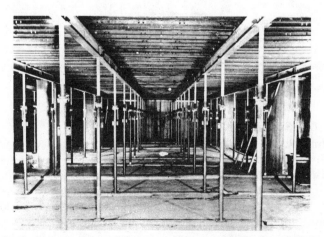

Figure 2–33 One of the many designs of all-metal adjustable shores. Note nailing brackets for attachment of bracing.

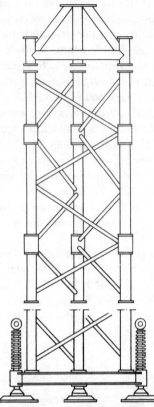

Figure 2–34 Tripod shoring towers are available for heavy loading at heights up to 50 ft.

Scaffold-Type Shoring

When tubular steel frame scaffolding was first introduced, it was designed to support the relatively light loads involved in getting men to the work area. Later contractors began to try out the scaffolding as a support for formwork because of the apparent advantages of its modular assembly and system of jacks for leveling and adjusting elevations. As this shoring application became more popular, it was necessary to develop more accurate data on the carrying capacity of scaffold frames, since the loads of concrete and forming materials were much greater than the loads previously supported as straight scaffolding. As a result of this concern for carrying capacity, heavy-duty frames were designed and placed on the market specifically for shoring installations.

The Scaffolding and Shoring Institute, working to develop more consistent and reliable data on load capacities of scaffold-type shoring, has developed a standard load test procedure and made load tests of scaffold-frame components assembled in relatively high towers. Recommendations for erection of scaffold shoring have also been developed.

The basic components of a scaffold shoring installation are end frames of various designs and dimensions, which are assembled with diagonal bracing, lock clamps, and adjustable bases in shoring towers that may have flat top plates, U-heads, or other upper members for attaching or fitting to forms (Fig. 2-35).

Based on load-carrying capacity, there are three types of tubular steel scaffold shoring:

1. *Standard.* Panel frames available in 2- to 5-ft widths, varying from 2 to 6½ ft high, with various panel designs and bracing details. Average safe working load is 2500 to 6000 lb per leg.

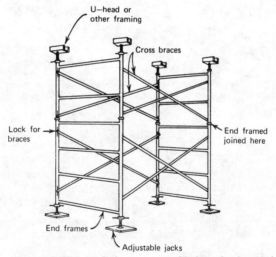

Figure 2–35 End frames assembled with diagonal braces to form typical shoring "tower." A number of different designs of end frames, coupling, and bracing details are available from the various manufacturers.

2. *Heavy-duty.* Tubular panel frames available in 3- to 4-ft widths, 3, 4, 5, and 6 ft high, with various designs and bracing and connection details. Average safe working load is 10,000 lb per leg.
3. *Extra-heavy-duty.* Has heavier tubular members and more rigid bracing, with components that assemble in 6½-ft-high sections. Supports up to 40,000 lb safe load per leg (Fig. 2-36).

Full flights of free-standing sectional shoring, with or without the supported forms, can be moved from one bay to another by skidding, by attaching casters for rolling, or by lifting.

Horizontal Shoring

Adjustable beam, truss, or beam and truss combination members that support formwork over a clear span and eliminate intermediate vertical supports are referred to as *horizontal shoring.* Fixed-length metal support beams (either I- or box-type) that span lengths of about 2 to 8 ft and

Figure 2-36 Extra-heavy-duty scaffold shoring supporting 125-ft. concrete girders. Working loads up to 20 tons per leg are permitted with this type of shoring. Note steel beam sills.

Figure 2-37 These telescoping horizontal shoring members are made of aluminum. An I-section fits within the hollow box member.

that replace timber ledgers are sometimes included in this category too. However, the two major adjustable types are:

1. *Telescoping horizontal shores.* which are made up of lattice, plate, or box members, or a combination of both. Units support forms for spans of 6 to 30 ft and have built-in adjustable camber. Allowable loads range from about 80 to 800 lb per ft of length, depending on span.
2. *Heavy-duty horizontal shores.* which are truss assembly horizontal shores for long-span formwork, capable of carrying 500 lb per lineal ft across spans of 15 to 83 ft.

One disadvantage of such shoring is that the high end loads may cause the member to cut into supporting timbers unless special bearing plates are provided. (See Fig. 2-37.)

CHAPTER 3
BUILDING AND ERECTING FORMWORK[1]

How formwork is built and erected depends on many factors: the materials available or required, the type and cost of local labor supply, and the demands of the job for accuracy and perfection of finish. For most formwork jobs there is no single correct building procedure; instead, there are several workable alternatives, and choice depends on cost comparisons as well as local preferences and customs. At times it may be preferable to handle the work in a way familiar to the available crews, rather than try to train them for a new system.

The techniques presented in this chapter are not the only way to do a given job, but they do represent ideas that have wide current acceptance; and some have proved particularly valuable to experienced form builders.

This discussion of formwork building and erection is divided into six sections according to the following formwork components:

1. Footings.
2. Wall forms.
3. Column forms.

4. Beam and girder forms.
5. Slab forms.
6. Shoring and scaffolding.

[1] Authorized condensation from *Formwork for Concrete*, Copyright 1973, American Concrete Institute.

Sizes and spacing of form members shown in many of the examples are typical of good forming practices, but they cannot be "copied" without making a careful analysis to determine conditions peculiar to the individual job.

FOOTINGS

The principal construction requirements for footings are sound concrete and correct position to match column and wall plans. Tolerances for footing construction suggested by ACI Committee 347 are given in Table 3-A. Since appearance is rarely important because the footing is below

Table 3–A
RECOMMENDED TOLERANCES FOR FOOTINGS

Variation in plan dimensions	Minus ½ in., plus 2 in.[a]
Misplacement or eccentricity	2% of the footing width in the direction of misplacement, but not more than 2 in.[a]
Reduction in thickness	Minus 5% of specified thickness

[a] Applies to concrete only, not to reinforcing bars or dowels.

grade, any old or used material that is sound may be used to build the forms. Sometimes fabricated forms are omitted entirely, and concrete is cast directly against the excavation. When casting concrete against earth, it is frequently desirable to form the top 4 in. of the footing. In case of rain or water buildup, this makes it easier to keep water and slime out of the bottom of a completed or partially completed earth form.

To set the building line from the surveyor's stakes, batter boards are set up outside the limits of the excavation at a convenient location (Fig. 3-1). A string or wire supported by the batter boards is then put up in the plane of the building line as is shown in Figure 3-2. Points below the string are located with a plumb bob. A convenient height should be selected for the batter boards (an even number of feet above or below some horizontal feature of the proposed structure, such as the finished footing elevation) so that they can be used for temporary bench marks.

Wall Footings
In good, cohesive soils that stand up well, all possible excavation down to the top of the footing is done with trencher, backhoe, or other digger,

Figure 3-1 Thick boards used in this grade beam form are braced on outside against the excavation; inside form is tied to outside one and wood spreaders maintain proper spacing.

and the footing outline is cut by hand or by specialized mechanical equipment to the exact size. If the soil is porous or noncohesive, the rough excavation proceeds to the bottom of the footing, and used planks, built-up panels, prefabricated metal forms, or other available material can be used to form the sides of the footing. Since footings are generally shallow, lateral pressure from the fresh concrete is relatively small, and the required bracing is simple (Fig. 3-3).

The panels or planks for one side of the form are adjusted to line and grade and staked into position. After one side is set, the other side is aligned by spreaders and held in position by stakes placed about every

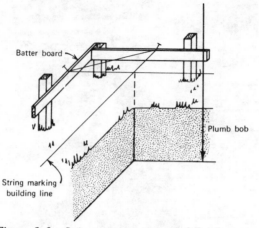

Figure 3–2 String or wire supported by batter boards is set in the plane of the building line.

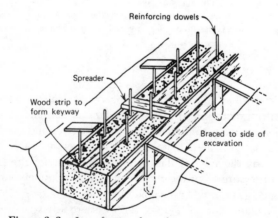

Figure 3–3 Low footing form for wall can be stake braced. No ties are required, and wood spreaders across the top hold sides at correct spacing.

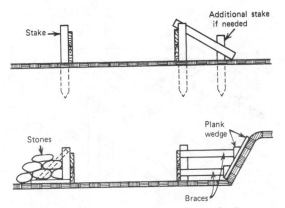

Figure 3–4 Alternate bracing systems for flat footings. Footings of greater depth may require ties.

6 ft. If the holding power of the stakes is poor because of ground conditions, the forms can be braced as is shown in Figure 3-4. Sometimes earth can be used as a brace by simply backfilling around the forms, although this is a disadvantage when forms are stripped. No ties are needed for shallow forms.

Deeper wall footings or grade beams (in noncohesive soil) require more bracing or ties to withstand concrete pressure and are constructed much like wall forms.

If the supporting ground for a wall footing slopes, the footing may be stepped longitudinally with formwork as is shown in Figure 3-5.

Column Footings

The forms for simple, rectangular column footings are bottomless boxes. They are constructed in four pieces, two end sections and two side sections. The end sections are built to the exact dimensions of the footing and the sides are built somewhat longer with vertical cleats to hold the ends as is shown in Figure 3-6. Ties prevent the sides of the box from bulging under concrete pressure, and no external bracing is needed. Since the correct concrete elevation can be marked with a nail, available planks are used, and no time should be wasted in ripping them to exact height. Stakes should be set to maintain the correct position of the form. A template for positioning dowels can be attached to the form box (Fig. 3-6).

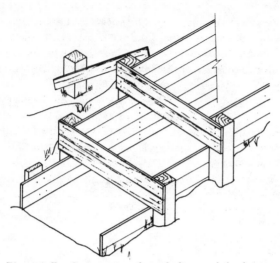

Figure 3–5   Construction of simple formwork for footing stepped down longitudinally because of sloping subgrade.

For small, shallow column footings, the ties are sometimes omitted, and diagonal wood braces are nailed across the top of the form box. Round footings may be formed with short lengths of the fiber tubing used to form round columns.

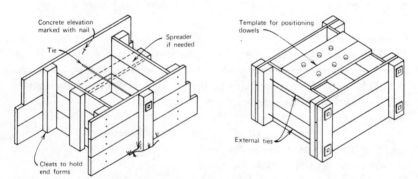

Figure 3–6   Two methods of forming column footings, one with internal ties, one with tie bolts outside the form box, similar to simple column forms.

Combined Footing Forms. Combined or strap footings are used to transfer part of the load of one footing to another or to support two columns on a single footing. If two columns bear on one footing, the combined footing form is built in the same way as for flat or stepped footings. If the load of one footing is to be partly transferred to another, a strap or beam must be constructed between the two footings or pads.

If the beam or strap is to be cast after the pads, the beam forms are supported by the previous placement. If the footing is to be cast monolithically, its construction is much like that of the stepped footing. The boxes for the pad forms are built with a cutout for the beam if necessary. The beam forms are supported on timbers or blocks, and the flat footing portions are covered and tied down like forms for stepped footings. The beam forms can be braced to the supporting timbers. A suggested method of forming a strap footing is shown in Figure 3-7.

WALL FORMS

Built-in-Place Forms

Wall forms are constructed from five basic parts: (1) sheathing to retain the concrete until it hardens; (2) studs that support the sheathing; (3)

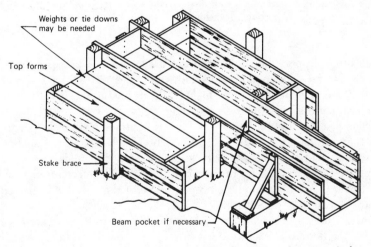

Figure 3–7   One type of strap footing form, where entire footing is to be cast monolithically.

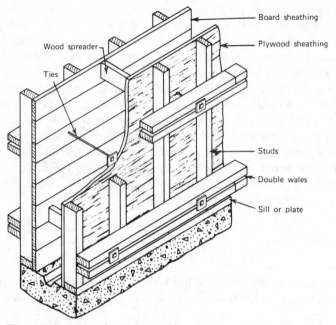

Figure 3–8   Typical wall form with components identified. Alternate sheathing materials are indicated. Wood spreaders are shown, but frequently the spreader device is part of the prefabricated tie.

wales that support the studs and align the forms; (4) braces to hold the forms against construction and wind loads; and (5) separate spreaders and ties or tie-spreader units to hold the forms at the correct spacing under the pressure of the fresh concrete. (See Fig. 3-8.)

The first step in building a wall form is to attach a sill or ledger to the footing as a base for studs. The sill may be attached to preset anchors or stub-nailed to the concrete with specially hardened nails. Special care must be taken in aligning this strip because it will determine the line of the wall. It is set out from the proposed line of the wall a distance equal to the thickness of the sheathing.

After the starting plate or sill is in place, the studs are erected (Fig. 3-9). The bottom of each stud is toenailed to the sill and held vertical by temporary braces of 1×6 or other suitable stock. The stud spacing at the top of the form is held by temporary ribbons until the sheathing is placed.

Figure 3–9 Starting a wall form. In this case, a rustication strip located at the water table served also as the plate on which studs for the outside form were erected. One row of sheathing boards placed at the bottom served as a ribbon to hold the studs in line. Frame was temporarily braced to the ground and brought to final alignment after all sheathing was applied.

To save time and material, studs may be left extending above the top of the form rather than cut off evenly.

The next step is to attach sheathing to the studs. The bottom edge on the first sheathing board or plywood panel is set on the highest point on the foundation and leveled accurately to establish alignment for succeeding boards or panels. If the footing is uneven or stepped and appearance is important, the space between the first sheathing unit placed and the footing may be filled out with specially cut pieces to make a tight joint.

The sheathing for the remainder of the wall is nailed to the inside of the studs after the first unit has been aligned and attached. When nominal 1-in. sheathing lumber is used, it is usually nailed from the inside face with two 6d or 8d nails at each stud. Plywood sheets ⅝ in. and thicker are nailed with 6d nails at 12- to 16-in. intervals. When thin plywood and hardboard liners are nailed over the sheathing, small nails with thin flat heads, such as 3d blue shingle nails, are desirable.

Contractors who install ties as the sheathing is being placed can notch the edge of a sheathing board that is at the required height so that holes through the sheathing are formed at the joint between sheathing boards. Otherwise, holes must be drilled in the sheathing after it is in place. Tie holes must be located at the proper elevation so that the ties will thread through the wales when they are installed.

Wales may be installed as soon as the sheathing is in place or after the other face of the form has been built. Wales are attached to the outside of the studs and held in place by nails, clips, brackets, or patented devices. Wales are generally constructed of two members, which saves drilling for ties. In long wales the joints between members should be staggered. If double wales are used, there need not be a transverse joint completely through the wale.

Permanent bracing can be installed after the wales are in place or as the form is plumbed. These braces extend from the wale to the ground or other solid support. If the brace is designed to act in compression only, wedges may be used at the end to help in plumbing the forms.

If the braces are designed to act in tension only, they must be securely attached to the forms and tightly anchored at the other end. A cable is well adapted to this type of bracing. The cable length can be adjusted during the plumbing of the forms by a turnbuckle or other device.

When the forms are braced on only one side, the braces must take both compression and tension. Therefore, special attention should be given to fastenings at each end of the brace. Proprietary devices are available for attaching the brace and adjusting its length after it has been installed.

Plumbing is the final operation in building a wall form. Ties are usually installed and all permanent bracing is placed before plumbing begins so that the form will not need to be disturbed after it has been plumbed. Most bracing systems have devices for adjusting the braces to facilitate the plumbing operation. If the braces are nonadjustable, the form must be plumbed as the braces are installed. No time should be spent in plumbing both sides of wall forms because the second form built will be automatically plumbed from the first by spreaders.

After the form for one side of the wall has been built and braced, work is begun on the second form. In many operations, steel is placed before the second form is built. If a man can work inside the forms, the second form is constructed like the first. If the wall is too thin for men to work inside the forms but is wide enough for a hammer to be swung, the carpenters can work from the outside over the top of the sheathing as it is placed. For thin walls, the second form can be constructed in an inclined position (or flat if there is enough space) and then can be tilted into position.

When both forms have been erected, and reinforcement has been placed, the ties are tightened. If ties with spreading devices are not used, the forms must be held apart at the correct spacing with spreaders that

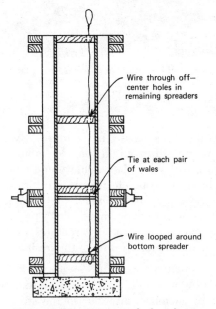

Figure 3–10 Wire attached to bottom spreader and passing through upper ones is used to withdraw spreaders as concrete rises in the form. This system prevents loss of spreaders in the concrete.

Figure 3–11 Heavy external bracing of wall forms resists lateral pressure of freshly placed concrete when ties through the wall cannot be used.

can be removed as the concrete rises in the forms. A wire attached to the bottom spreader, running through all spreaders above, and anchored to the top of the forms, will aid in their removal and will avoid the leaving of any spreaders in as the concrete rises in the forms (Fig. 3-10). Ties and spreading devices are discussed in more detail on p. 109.

Where through-the-wall tying is impossible or prohibited by specifications, as is true for some water-retaining structures, external bracing must be provided to resist the lateral pressure of the fresh concrete (Fig. 3-11).

Low Walls. In building low wall forms, wales are sometimes omitted and the forms braced by members attached to the studs. The general sequence of wall form construction just described applies to this light construction except for the attachment of the wales. Another consideration in this type of construction is the anchorage of the form ties. Since wales are not used, the ties must be anchored to the studs.

High Walls. When high walls are formed full height, greater form rigidity is obtained by using stronger members and adding vertical wales or strongbacks (Fig. 3-12). These vertical wales are designed to support and align the horizontal wales. The method of construction for high wall forms is similar to that for normal one-story walls, but the design may specify more elaborate bracing or other details. High walls are more frequently built by raising the forms vertically in panel units.

Prefabricated Panel Systems

Use of ready-made or contractor-built prefabricated panels for wall forming has been increasing in recent years. Such panels are durable enough for many reuses and simplify and reduce the labor required at the job site. Studs and sheathing are, in effect, preassembled in units small enough to be conveniently handled; the panels are set into position and tied together. Braces, wales, and ties are then attached or inserted as needed to complete the wall assembly. For light construction, wales are sometimes omitted.

There are a number of ways to build the panels, and various materials are used—wood, metal, plywood with metal bracing, plywood with wood bracing, or glass fiber and plastic over wood or plywood. A number of ready-made systems are available usually on either a purchase or rental basis.

Figure 3-12 For walls of great height or where precise alignment is extremely important, vertical wales (strongbacks) are added outside of the regular system of horizontal wales.

Building the Panels. Some contractors find that they can build their own panel systems for less money than they can buy or rent ready-made ones. Manufacturers of accessories sell hardware designed specially for this purpose. The frame is assembled first, often in a jig designed to speed up production. Special care is required to get accurate layout of frame members to assure square corners and straight edges. If any panel is not square and true to dimensions, it will throw other panels out of alignment when the forms are assembled.

The sheathing for panels, whether plywood or metal, must conform to the outside line of the frame. Otherwise, shallow fins will be formed in the concrete at the joints between panels. Since this type of form is usually designed for much reuse, the sheathing should be attached to the frame much more securely than for normal wall form sheathing. Typical construction of such panels is shown in Figure 3-13.

Provision must be made for attaching panels to one another. A variety of clamps and ties are available for this purpose. If they require holes through the frame, the holes must be accurately located so that they will match up when the panels are erected.

Many kinds of hardware and ties suitable for form panel systems are available to the contractor who builds his own panels. Each proprietary panel system usually has its own special ties and other accessories.

Erecting the Panels. Panels—either contractor-made or ready-made—are erected in basically the same way as built-in-place wall forms except that the separate erection of studs and sheathing is eliminated. Some systems with rigid panel frames and connections eliminate the use of wales. An accurate line must be marked on the footing for the alignment of the panels. They are then set into position, and adjacent panels are joined and braced. Other details, such as setting the wales, permanent braces and ties, and plumbing the form, are similar to those for built-in-place forms.

When proprietary panel systems are purchased or rented, the supplier frequently prepares layout drawings and provides erection instructions; sometimes trained supervisors' familiar with a given system are sent out to aid the contractor in proper erection (Fig. 3-14).

"Giant" Forms and Ganged Panel Forms

With greater demand for massive placements, and with the progress in development of cranes and other mechanical methods of transporting forms, the use of giant panels and ganged prefabricated forms for high

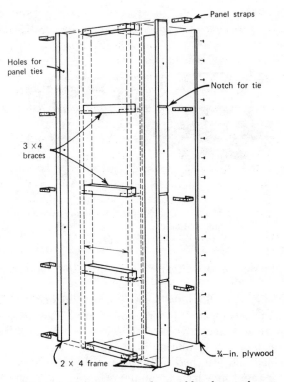

Figure 3–13   Framing and assembly of typical contractor-built panel for wall forming. Metal panel straps are shown, but similar panels can be built without such ready-made hardware.

walls is becoming more common. Ranging up to 30×50 ft, their size is limited only by the mechanics of handling. Large panels can be assembled flat on the ground where it is easier to work. Tools and materials are ready at hand and do not have to be hoisted to the workman. The same kind of delay and wasted motion are avoided in stripping because the giant forms are stripped as a unit.

Materials for the large form units are much the same as for conventional built-in-place walls, except for extra bracing required to withstand the handling stresses. In some instances, a prefabricated cage of reinforcing steel is attached to the large form to be lifted with the unit. These large form units are frequently used as "climbing" forms in high wall

Figure 3–14   Metal framed 2x8 prefabricated panels assembled as a wall form with horizontal and vertical wales and diagonal braces. Note panels at top placed horizontally to obtain desired wall height.

construction, being raised vertically for one lift after another. (See Fig. 3-15.)

The same benefits of the large forms can be obtained by ganging or grouping small panels together to form larger units. They are joined with special hardware and braced with strongbacks or special steel frames that maintain the stability of the unit during handling. After the small panels have been joined and attached to the frame, they are lifted into place as a unit. Such a unit may have a number of reuses and can then be dismantled for use as individual panels on other jobs. The ganged panels are also frequently used as climbing forms, moving up for successive lifts of a high wall.

Square Corners

Handling of wall forms at corners is a critical operation. Continuity of sheathing and wales is broken, making this a potential weak point in the formwork. Corners must be drawn tight to prevent concrete leakage. Leakage here is particularly objectionable as patching to a square edge is most difficult, and chipped, weathered, poorly patched corners are particularly noticeable. Lateral pressure of the concrete, which tends to open a corner joint, must be resisted by ties or suitable connections between the wales of the wall forms on either side of the corner.

One method of making a tight corner is shown in Figure 3-16. The wales overlap and two vertical kick strips are provided at the intersec-

Figure 3–15 Steel-faced giant form for curved wall being positioned by crane. Benefits of building forms in large panels for mechanical hoisting are particularly evident on high walls like this one. Such panels are frequently used as "climbing" forms, being raised vertically for successive lifts.

tion, against which the wales are wedged to tighten the corner. Another method of securing corners in formwork (Fig. 3-17) uses a diagonal tie across the corner. Special corner clamps and ties are available from formwork hardware suppliers.

"Log cabin" corners are sometimes used. Alternate sheathing boards are brought beyond the corner, and vertical kick strips behind the interlaced sheathing boards prevent movement. This type of corner is much more difficult to build and strip than the types illustrated.

Curved Walls and Round Corners

Rounded corners or curved walls of radius greater than 4 ft can be formed with plywood attached directly to studs. Horizontal sheathing boards can be used for curves with a radius of 18 to 20 ft or longer, and

Figure 3–16 Typical formwork corner using overlapping wales held by a vertical kick strip. Stripping has begun at the lower level revealing a sharp, accurate corner line in the board-marked concrete.

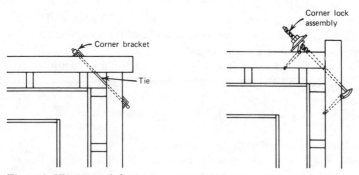

Figure 3–17 Two of the various corner tying devices. Hardware fits between double-member wales so that cutting or drilling is not required.

vertical board sheathing is sometimes used for shorter-radius curves. Rectangular prefabricated panels, with narrow or flexible fillers added, are often used on long-radius curves. Short-radius curves frequently require a smooth lining material backed up by narrow vertical members. Thickness of sheathing and details of construction vary with the radius of the curve. Several of the many forming methods are described in the following paragraphs.

Long-Radius Curves. Long-radius curves are easily formed with prefabricated panels, and special filler panels and accessories, as well as suggestions for erection, are available from panel manufacturers.

A typical detail of a form for a long-radius (not less than 20 ft) curve, using plywood sheathing is shown in Figure 3-18. The studs are vertical and are blocked out from yokes or frames that are spaced 4 to 6 ft apart vertically. By staggering the frames to break joints, a full circle or any part thereof can be held rigid. For the outside form, it is necessary to tie the studs to the frames at the center of the frames and for a distance of two or three studs in each direction to prevent the spring of the sheathing from pulling it away from the frames. For the inside form, it is necessary to tie the studs at the ends of the frame instead of those at the center.

Long-radius curves may be formed with horizontal board sheathing in much the same way that straight wall forms are erected. After the wall

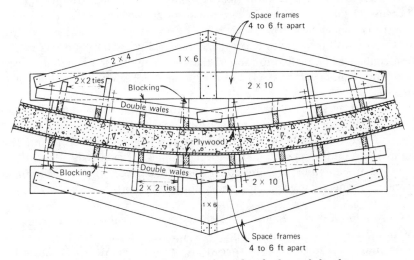

Figure 3–18   Details of long-radius curve formed with plywood sheathing.

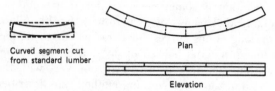

Curved segment cut
from standard lumber

Plan

Elevation

Figure 3–19   Arc segments cut from standard lumber sizes may be laminated to form a rigid curved piece. The built-up curved member may be placed next to the sheathing, with straight vertical wales or strongbacks on the outside, or it may serve as a curved horizontal wale outside of straight vertical studs.

line has been marked on the footing, a sill is nailed down to set the line for the studs. Since wales cannot be bent to the curvature of the wall, short sections are used to span four to six studs according to the curvature of the wall as is shown in Figure 3-18, or curved wales may be constructed from standard lumber as is shown in Figure 3-19. With the former method, care must be taken to assure the correct amount of blocking so that a smooth curve will be formed after the ties are tightened. The installation of ties and external bracing and the plumbing of the form are similar to those for straight wall forms. More external bracing is required for this type of form than is required for normal straight walls because of the short wales. Where pressures are not excessive, a simpler construction may be used.

Curved wall forms with vertical sheathing (Fig. 3-20) require curved horizontal ribs fabricated to support the sheathing. These ribs are usually cut from 8-in. or wider stock and can be made of single or double segments.

Building curved wall forms in place with vertical sheathing is somewhat more difficult than the method of horizontal sheathing using vertical studs. The ribs must be set in place and held by vertical members and braces until the sheathing has been applied. After the sheathing has applied, ties are inserted and vertical wales are placed. Vertical sheathing is more frequently used to support the lining for short-radius curves.

Short-Radius Curves. Figure 3-22 shows a short-radius corner formed with hardboard or plywood lining. Yokes are cut to the required curvature from 2-in. stock and are spaced about 30 in. apart. The yokes are sheathed with vertical 2×2 dressed strips, and the lining is then nailed

Figure 3–20   Completed curved wall form using vertical sheathing boards.

securely to the 2×2 backing. Since the lining material will tend to spring back to its original flat shape, it must be nailed at 6-in. intervals in both directions. (See Figs. 3-21 and 3-22.)

Two thin sheets of form material may be used rather than one thick one because they can be bent to a smaller radius. Figure 3-23 shows one way of using double plywood for corner forming. The corner sheathing is brought out beyond the spring line to form tighter joints where it abuts the wall sheathing. The sheathing is supported by horizontal members cut to the curvature of the corner.

Wall Openings

The forms for window and door openings must be made rigid so that they will not distort under the pressure of the fresh concrete. If metal sash or door frames are to have concrete cast around them, they may be braced and aligned to serve as a part of the formwork for the opening. The box forming the opening should be made of at least 2-in. material, well braced, to give adequate strength and to prevent deflection so that the sash will fit properly.

Any strips required to form the recess to receive the sash are securely nailed to the 2-in. plank forming the frame. Cross braces should be

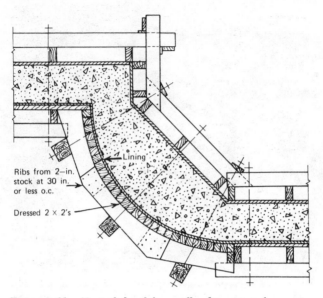

Figure 3–21   Typical detail for small-radius corner form using hardboard c. plywood liner.

located at each cleat horizontally and vertically unless there is an inner frame of 2×4's. The cleats should be not more than 24 in. apart. A closer spacing will be necessary if less than 2-in. material is used. Extra cross bracing should be provided in large window forms. The frame for the opening is supported by being nailed to the form sheathing. (See Figs. 3-24 and 3-25.)

Except for windows with very steep sills, it is necessary to be able to get at the sill to finish it and to work the concrete into place properly. For these reasons, the sill of the form may be omitted altogether, or the sill piece may be made in two sections for easy removal.

Door opening forms require special attention since they are larger and distortion can occur easily without proper bracing. The opening is made in much the same way as the forms for window openings. Wales running across the opening will help prevent twisting of the frame.

Small openings in walls can be formed by attaching a box frame to the forms to block out the concrete. The box should be rigid or braced so that it will not deform from the pressure of the concrete. It should be

Figure 3–22 Workmen bending two thicknesses of ¼-in. plywood to form curve of intermediate radius. This same technique is applicable for curves as small as 15-in. radius without wetting or steaming. Spacing of backing members depends on form strength requirements as well as radius of curvature.

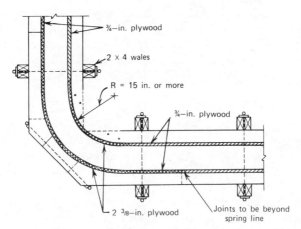

¾—in. plywood

2 x 4 wales

R = 15 in. or more

¾—in. plywood

2 ³⁄₈—in. plywood

Joints to be beyond spring line

Figure 3–23 Small radius corner formed with two thin sheets of plywood supported on curved horizontal members.

Figure 3–24 Cross bracing for large window forms. Note that full sheathing is used on one side of the formed opening to give added stability.

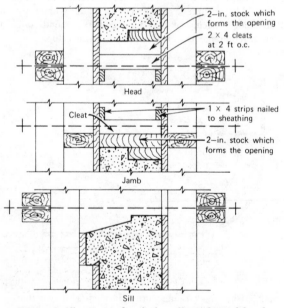

Figure 3–25 Form details for sill, jamb, and head of window.

lightly tied to the face of the form so the form can be stripped leaving the box in place in the hardened concrete. Kickers to resist the uplift action of the fresh concrete may be required.

Joints

Vertical Construction Joints. Vertical construction joints are formed with a bulkhead placed in the forms at the end of the proposed concrete placement. Since the reinforcing steel continues past the joint, some provision must be made to allow it to pass through the bulkhead. The bulkheads may be built up from short pieces placed horizontally and cut so they will fit evenly between the two faces of the wall form. Vertical boards may also be used (Fig. 3-26). Reinforcement is passed through the bulkhead by notching the boards at the desired location.

The bulkhead resists the pressure of the fresh concrete by cleats nailed to the inside face of the forms, outside of the bulkhead sheathing. The short pieces forming the bulkhead are lightly attached to these cleats to hold them in place until the concrete is placed. The pressure of the concrete will hold them against the cleats. Keyways or waterstop materials are frequently attached to the bulkhead forms as required by design specifications.

Horizontal Construction Joints. The technique of forming high walls with lift forms (or climbing forms) is used where it is impractical to place them full height. When the first concrete lift is placed, ties or bolts are placed in the concrete near the top of the lift. After the concrete has

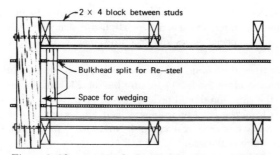

Figure 3–26   One method of bulkheading a wall form, showing strip for keyway attached to bulkhead.

hardened sufficiently, the forms are stripped and raised to the elevation of the next lift and are supported on the previously placed bolts below.

A row of ties should be placed about 6 in. above the joint between the concrete lifts to prevent leakage. The bolts or ties in the previous lift support the weight of the form, but cannot be relied on to prevent a slight spreading of the forms at the joint. The form members connected to the bolts must be an integral part of the form panel. If the wales are securely attached to the form, they may be used to support the panel. Sometimes a timber ledger is attached to bolts cast in the previous lift, and the studs of the form rest on it. Specially constructed strongbacks can also be used. Special brackets, attached to the previous lift by bolts, may also be used to support the forms.

Control Joints. Control joints must be placed in concrete walls as they are shown on the plans. Such joints are formed by a beveled insert of wood, metal, or other material which is tacked to the form. This insert produces a groove in the concrete that will control surface cracking. It is frequently designed to be left in the concrete for some time after the forms have been stripped, and in this case it must be only lightly attached to the form. If this insert or strip is removed too soon, the area around the joint may be damaged. Special care is required if control joint strips are removed along with the main form panel. If wood strips are used, they should be kerfed to prevent swelling that might crack the concrete.

Suggested Tolerances for Walls

The following tolerances suggested by ACI Committee 347 apply to a finished wall. The forms should be constructed to give a finished wall within these limits, unless otherwise specified.

1. *Variation from the plumb* should not be more than $\pm\frac{1}{4}$ in. per 10 ft, but in no case should it exceed 1 in. High-rise structures (over 100 ft tall) or other special types of structures may require special tolerances.
2. *Variation from the plumb for conspicuous lines* such as control joints should not be more than $\pm\frac{1}{4}$ in. any bay or 20-ft maximum. In 40 ft or more, the variation should be less than $\pm\frac{1}{2}$ in.
3. *Variation from the level* or from specified grades should be less than $\pm\frac{1}{4}$ in. in any bay or 20-ft maximum. Not more than $\pm\frac{1}{2}$-in. variation in 40 ft or more is allowed.
4. *Variation of linear building lines* from the established position in the plans, and variation of the related position of walls and partitions,

should be less than ± ½ in. in any bay or 20-ft maximum. In 40 ft or more, the variation must be less than ± 1 in.

5. *Variation in sizes and locations of wall openings* should not be more than ± ¼ in.
6. *Variation in thickness* is limited to − ¼ in. and + ½ in.

COLUMN FORMS

Because of their comparatively small cross section and relatively high rates of placement, column forms are frequently subject to much greater lateral pressures than are walls. Tight joints and adequate anchorage at the base are required. A frame of 2×4's nailed to the completed slab is generally used to position and anchor the column form. Because of the confined space in which concrete is placed, tall columns frequently have pockets or windows at midheight or at other intervals to make placing and consolidating the concrete easier. A cleanout opening must be made at the column base for removal of waste and debris before concreting begins.

Columns may be round, rectangular, L-shaped, or of various irregular cross sections. Irregular shapes are frequently formed by attaching special inserts inside square or rectangular forms, and L-shaped columns may be formed like wall corners (Fig. 3-27). Forms for round columns may be built of wood, but ready-made forms of metal, fiber, or other materials are more commonly used. Prefabricated panel system parts are also used at times to form square or rectangular columns (Fig. 3-28), and special techniques and materials are used for custom fabrication of special column shapes. Column forms of irregular or decorative cross section may also be built in place from wood and plywood.

The following detailed descriptions are limited to some of the more common column forms built by the contractor. The framing and sheath-

L–shaped

Octagon

Cutout corners

Figure 3–27 Irregular shapes of column cross section are frequently formed by using inserts within standard square or rectangular forms.

Figure 3–28 The 16-in.-wide filler panels of a prefabricated panel system were clamped together to make a 24-ft. column form which was erected by crane. Shorter panels (4 ft.) at the 12-ft. level were removed for placing and vibrating concrete at the bottom of the column, then replaced as concreting reached that level.

ing members and yoke or clamp spacings shown in the drawings are representative of actual practice, but should not be used without first considering the proposed application on the basis of design principles. Temperature and rate of placing as well as column size influence the design, and a full investigation should be made to see that sizes and spacings are adequate for actual job conditions.

Erection Practices

Square or rectangular column forms are generally built in four panels, round ones in two or three circular segments. Clamp spacing and code numbers should be marked on the panels as they are built. The sequence and method of erection vary somewhat depending on the total job schedule, lifting equipment available, and plans for setting the reinforcing bars. The column form may be erected in place, panel by panel, or the forms may be assembled into a complete column box and set in place as a complete unit. Reinforcing bars may be assembled in place or prefabricated in cages and may be set in place either before or after the forms are in position, depending on individual conditions. For example, if there are any ties that pass through the form, the cage may be set in place and wired to the column dowels, then the forms may be set in place around the steel and ties threaded through. For heavy bars or large columns, it is common practice to build the reinforcing cage in place.

If the panels are assembled in place, the first one is aligned and temporarily braced, then the others added and full bracing (like that in Fig. 3-32) is set up. The other method of assembling the entire column first is preferred if equipment is available to lift the assembled form because it saves some of the work of temporary bracing.

A template is generally set in place on the floor slab or footing to locate the column form accurately. Interior dimensions of the template should be slightly larger (1/8 in. or so) than the outside measurements of the assembled form panel to make it easier to fit the form into proper position.

Careful erection of column forms is necessary to avoid "twisting" from the square or rectangular cross section. Alternating the direction of bolt members of yokes, or the direction of tensioning for certain types of clamps, helps to overcome this. Line and thickness tolerances will not control twist unless they are referenced to a grid on the floor such as column center lines. Column tolerances for width and thickness should not be checked from a reference point on the column form or from a single point on the floor unless twist is separately checked to about $-2°$ to $+5°$ F.

True height of the column form is the story height less the slab thickness; if the slab is formed with the columns, the slab sheathing thickness must also be subtracted. Because of irregularities in the finished slab at column locations, the made-up panels are frequently cut 1/2 in. or more shorter than true height and then are shimmed up at the bottom or are pieced at the top to exact height in the field. Exact amount of height reduction depends on local field practice and job conditions. Sometimes grade is checked at the base and any necessary correction is indicated by a mark on the template; or elevation may be checked at the top of the form to make final adjustments.

If columns are formed, cast, and stripped independently ahead of the forming of other structural members—a fairly common practice to permit columns to undergo initial shrinkage—the adjusted form should be slightly higher than the true column height to allow for shrinkage and variations that may occur when other members are framed in. Some contractors allow about half an inch of extra concrete at this point to cover small shrinkage and any irregularities, but the exact amount depends on local conditions. Frequently, specifications require the contractor to chisel or sandblast the top $\frac{1}{16}$ in. or more of concrete to remove laitance before casting the next lift.

Square or Rectangular Columns

The method of building form panels depends on the materials being used as well as the means of clamping or yoking the columns. Figure 3-29 shows construction suitable for light column forms up to 12×12 in., held together with a combination wood and bolt yoke. Battens attached to the plywood side panels are a part of the yoke, and ties or bolts with washers form the other two sides of the yoke. This same kind of construction may be used for columns up to 18×18 in. by adding battens

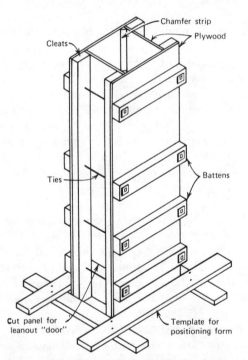

Figure 3–29 One method suitable for forming light columns, up to about 12x12 in. Plywood is backed by battens whcih are a part of the wood and bolt column yoke.

to the end panels. Similar construction can be used with board sheathing as is shown in Figure 3-30. Light column forms can also be built for tying with metal strapping by "trussing" the panels to eliminate right-angle bends in the strap tie.

Heavier column forms are commonly tied with adjustable ready-made column clamps, which are available for column cross sections ranging from about 9-in. square to approximately 5×8 ft. Figure 3-31 shows a widely used method of forming columns using plywood backed by vertical 2×4 stiffening members. To build these panels, two ¾-in. plywood sheets are cut to exact column size, and the 2×4 vertical backup strips are attached to extend 1⅜ in. over each side of the plywood. The plywood sheets for the other two sides are cut 1½ in. wider than that column dimension, and the backing members are nailed flush along the plywood edges. Additional vertical strips are used for wider columns to prevent excessive deflection of the plywood. The same type of assembly

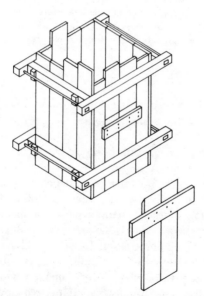

Figure 3–30 Column formed with board sheathing, using wood and bolt yoke wedging between bolt and side panel batten to tighten form.

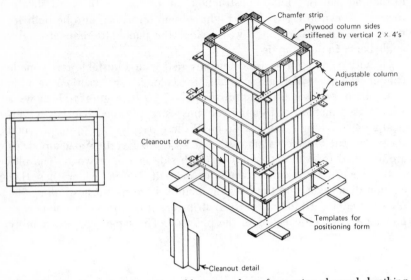

Figure 3–31 Typical construction of heavier column form using plywood sheathing backed by vertical stiffening members and clamped with adjustable metal clamps. Small section shows an alternate method using board sheathing with horizontal cleats.

can be used with 1×4 vertical backing members if a design check indicates strength and stiffness are adequate. Such a column form is shown in Figure 3-32.

The foregoing method of form building produces two matching sides and two matching ends. A useful variation of this scheme for square columns, shown in Figure 3-33, makes all panels identical and, therefore, completely interchangeable (if they are not cut at the top for beam pockets). If columns of this general type are to be tied with strap or band iron, trussing is required as is shown in Figure 3-34.

All square or rectangular column forms should have chamfer strips in the corners (unless architectural limitations prohibit them) because sharp edges are likely to be chipped or damaged while the concrete is green. Chamfer strips are usually nailed to two opposite panel units when the forms are being made up. Strips may be cut from wood for a sharp 45° flat fillet, or ready-made inserts of sheet metal or other materials may be used for a quarter-round fillet.

Figure 3–32 Rectangular column (bridge pier) form built with 1x4's backing the plywood, clamped with adjustable metal clamps. Note four-way bracing for alignment, template at base.

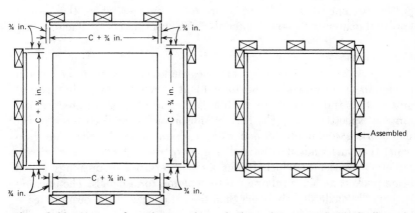

Figure 3–33 Square column form similar to the form of Fig. 3–35, but with all panels identical. Panels are interchangeable unless beam pockets are cut at the top. (Dimensions based on ¾-in. sheathing.)

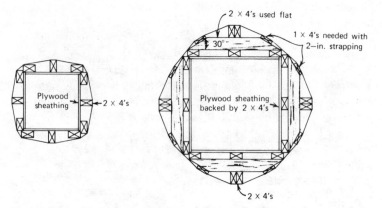

Figure 3-34 Typical arrangement of "trussing" when steel strapping is used to tie column form together. Smaller cross section shows arrangement suggested for columns up to 18x18 in. Method at right is suitable for larger columns, up to 32x32 in.

Round Columns

The ready-made single-use fiber tubes or multiuse steel forms for round columns require only joining the parts, setting in place, and bracing to maintain alignment. No outside yokes are required since lateral pressure is resisted by hoop tension in the forms.

Round wood column forms are expensive to construct, and are rarely justified except where only a few are to be built or where the surface finish attainable with other materials is not considered satisfactory.

Column Heads

The top of the column form is sometimes made separate from the column proper to avoid remaking the form to fit varying sizes of beams and girders that frame in, or to facilitate the use of prefabricated capital forms of special shapes. The flaring "mushroom" capital common in flat slab work is commonly formed with ready-made steel units. The drop panel that surrounds it is formed as part of the slab.

For columns into which beams and girders frame, openings called beam pockets are cut in the top of the column form panels. The opening is generally made slightly larger than the actual size of the beam to leave some room for adjustments when final assembly is made in the field. The beam bottoms and sides may be placed within the pocket or merely brought up flush with it. In the former case, the opening must be cut to allow for the thickness of beam side and bottom forms in addition to the ¼ to ½ in. allowed for ease in assembly. The beam pocket may be reinforced around the edges with 1- or 2-in. material that also serves to support the ends of the beam form.

The beam forms are also cut a little short of true dimension to allow some leeway in assembly. These beam-column intersections are difficult to form, and it is easier to fit a small closure piece or bevel strip in the field than to have to make saw cuts to enlarge openings or shorten panels. One contractor reports success with a field-fitted metal closure tacked on at beam and column intersection lines to cover any gap left by approximate fitting of shop-built panels. Figure 3-35 shows another method of handling this closure, using a wedge-shaped key that also facilitates stripping.

Changing Sizes of Form Panels

If a plan for form reuse requires reducing column widths, it is better to build form panels of cleated boards that can simply be ripped along

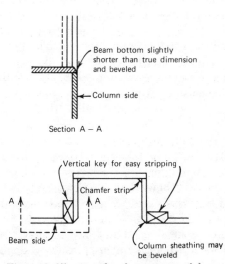

Figure 3–35 Details of one type of beam-column form intersection, showing a wedged vertical key as the closure piece. Removal of the key simplifies stripping the forms later.

one or both edges to the desired new size. Plywood backed by vertical stiffeners would require much more refabrication to change panel widths. Reductions in column height may be made by cutting the bottom short if beam and girder pockets are to remain the same at the top. If column cross section remains the same while sizes of members framing in vary, several different head boxes can be devised to extend one common set of panels. Column boxes added to the top are also suitable for height changes where no inframing members exist.

BEAM AND GIRDER FORMS

Beam and girder formwork consists of a beam or girder bottom and two sides, plus necessary ties or braces. Usually the bottom is made to the exact width of the beam is supported directly on shore heads. Beam sides overlap the bottom form and also rest on the shore heads. For fireproofing structural steel and for some composite construction, beam or girder forms may be supported from an existing structural frame.

Details of the formwork assembly vary, depending on plans for stripping as well as materials to be used, location of the member in the building, and the anticipated loads to be carried. Whether the beam forms are assembled in boxes prior to installation or are handled as separate bottom and sides depends on the lifting equipment available as well as the planned sequence of form stripping. The following examples show some of the many ways the forms may be built.

Figure 3-36 shows a typical interior beam form with slab forming supported on the beam sides. This drawing indicates plywood for beam sides and bottom, supported longitudinally on the beam bottom by 2×4's and with blocking or stiffeners along the beam sides. The vertical side members are sometimes omitted if the beam is less than 20 in. deep and slab loads carried by the beam are not excessive. The ledger supports loads from adjoining slab forms, and the kicker holds beam sides in place at the bottom. Figure 3-37 shows the same type of framing using different materials—cleated boards for the side panels and a plank for the beam bottom.

Figure 3-38 indicates another method, less commonly used, in which the beam bottom is carried on stringers resting on double-post shores. Beam sides are carried on the beam bottom. This design permits wider spacing of the shores and offers more resistance to tipping when loads become unbalanced because of unsymmetrical placing of concrete. Double-post shores may also be desirable when unusually heavy slab

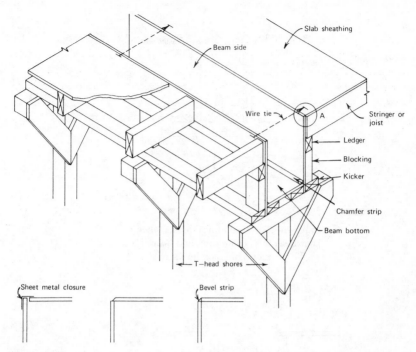

Figure 3–36 Typical components of beam form-work with slab framing in. Plywood is the sheathing material, and the beam bottom panel is backed with dimension lumber. Alternate details show different treatments of the form at beam-slab intersection.

loads are carried on the beam sides. This might occur when horizontal shoring is used to support long slab spans (Fig. 3-39).

Beam Pockets in Girder Sides

There are two ways of framing the beam into the girder, similar to ways of framing a beam or girder into a column. The beam pocket may be made to exact size of inframing beam plus a small allowance for give in the forms, and the beam form then may be butted up against it. Alternatively, the pocket may be made to beam dimension plus allowances for thickness of beam bottom and sides and a small allowance for easy fitting. The beam form is then placed in the pocket with its ends beveled toward the supporting members. The beam pocket should be strengthened by framing with 1×2 or heavier stock around the opening, setting

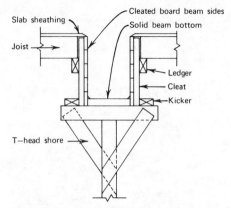

Figure 3–37 Beam form details when cleated boards serve as beam sides and beam bottom is a solid piece of dimension lumber.

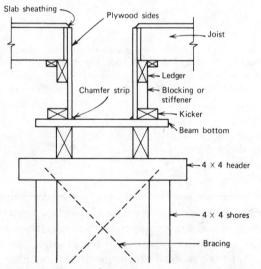

Figure 3–38 Another less common beam forming method. Beam sides rest on beam bottom, which is carried on stringers resting on double-post shores. This design permits wider shore spacing and offers resistance to tipping when loading is unbalanced.

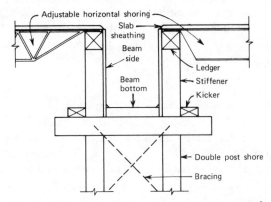

Figure 3–39 Heavy ledgers and stiffeners supported directly on head of double-post shore carry the relatively heavier slab load transferred from long-span horizontal shoring members.

back the bracing as required by the intended framing in and support of sides or bottoms of beams.

Ties and Braces

A simple wire tie across the top of the beam is frequently used; for deep beams, ties like those in wall construction are required. Depth and spacing are determined according to design principles and depend on the rate and method of placing concrete. Knee braces to the shore heads are generally used to hold beam sides against lateral pressure and are particularly important for spandrel beams or interior beams that are formed independently of the slab.

Spandrel Beams

Spandrel beams require careful forming because of their critical location at the outer face of the building where accurate alignment is the key to good appearance, even though they may be covered over with other materials. Spandrel beams are frequently deep enough to require ties through the beam forms, and the ties must be carefully located so as not to interfere with heavy reinforcement or with the numerous inserts that are common in beams of this kind.

Shore heads are commonly somewhat extended on the outside to accomodate knee braces that are required to maintain alignment. The extended shore head frequently supports a catwalk (Fig. 3-40) for work-men placing and vibrating concrete. In addition to the regular bracing of supporting shores, some form of tie back to permanent anchorage is desirable to prevent the entire spandrel beam assembly from being shoved out of line. (See Fig. 3-41.)

Suspended Forms for Fireproofing, Composite Construction

The various commercially available hangers that support beam forms for concrete encasement, often required for structural steel frames, make possible several different methods of forming such encasement (Figs. 3-42, 3-43). Frequently, as with other types of beams and girder form-

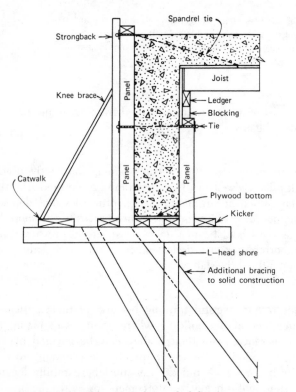

Figure 3–40 Forming of spandrel beam using pre-fabricated panels with ties. External strongbacks and knee bracing are shown. L-head shore, is braced to solid, complete srtucture.

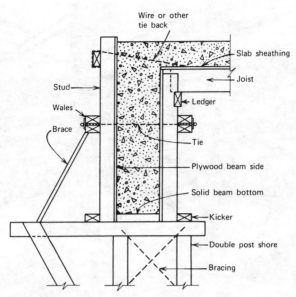

Figure 3-41   Stud and wale forming for spandrel beam, supported on braced double-post shore.

ing, the beam form also supports deck forms. Where steel framing members are fairly closely spaced, this may be the only support required for the entire deck and beam concrete placement. (See Fig. 3-44.)

This same suspension system is adapted to placing concrete decks on steel framing which is not encased. Similar methods are used for cast-in-place work supported on precast concrete members. Shoring may be required by the nature of the structural design for composite action, regardless of the ability of the structural frame to support the weight of forms and freshly placed concrete. This is a decision to be made by the structural designer, and he should clearly state in the contract documents the amount and kind of such shoring required. Otherwise, the contractor will be completely justified in supporting the formwork and its applied loads on the already placed structural frame.

Panel Assembly and Erection
Beam and girder forms are made up in the bottom and side panels previously described. If stripping of sides and bottom at the same time is planned, they may be assembled in a "beam box" for erection if neces-

Figure 3–42 Snap tie hangers on structural steel frame support beam bottom ready for the setting of beam sides.

sary hoisting equipment is available. However, these forms are often handled as separate panels; sides are stripped for early reuse, and beam bottoms remain in place for the longer time required.

For a typical structure where the beam and girder forms are preassembled in boxes, the erection sequence may be as follows: girder boxes are set in place spanning between column forms and supported on them at the ends; a few shore supports are set under the boxes. Beam boxes are then set in place between girder forms with a few of their shores supporting them. Deck panels are placed, and shoring and bracing are completed as required under all members.

If the contractor has chosen to use separate side and bottom panels instead of boxes, girder bottoms are first set in place between column forms and shores placed under them. Then, girder sides are set in place, are lightly nailed to the bottom, and are tied across the top or elsewhere as required. Any ties that are in a position to interfere with the setting of reinforcing bars may have to be added later. Beam bottoms between girders are then set on their shores, and beam sides set in place and

For unfinished work, hanger ends remain at concrete surface

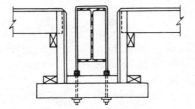

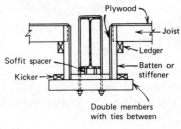

Plywood

Joist

Ledger

Batten or stiffener

Soffit spacer

Kicker

Double members with ties between

Snap tie hanger

Coil—type hangers

Double wales

Fascia hanger

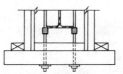

For exposed surfaces, where setback is specified. Recess to be grouted when bolts are removed

Figure 3–43 Typical beam encasement forms, showing both coil and snap types of hangers. Soffit spacers or spacing devices on the hangers hold beam soffit form at required distance from the beam to be encased.

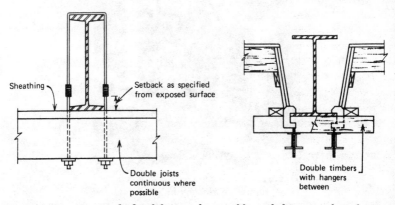

Sheathing

Setback as specified from exposed surface

Double joists continuous where possible

Double timbers with hangers between

Figure 3–44 Two methods of formwork assembly with hangers where bottom flange of steel beam does not require cover.

lightly nailed. Finally, intersection details are taken care of, such as fitting in or attaching any filler pieces required for a neat closure. (See Fig. 3-45.)

After any slab forms that rest on beam or girder sides are placed, kickers are securely nailed to shore heads, and the entire form is brought to desired elevation by wedging or adjusting devices on the shores. When correct adjustment is obtained, cross bracing of the shore system may be completed.

If camber of beam or girder forms is required by structural design or formwork plan, it may be introduced by wedging, blocking, or other adjustment at the shore.

Beam and girder forms should be built to ensure finished work within the specified tolerances for completed construction. In the absence of

Figure 3–45   One stage in the erection of beam forms. Shores are in place and the bottom panel rests on column forms and shores between columns.

other stated tolerances, the recommendations of ACI Committee 347 for reinforced concrete buildings may be followed.

SLAB FORMS

Forming for the following common types of slab construction is discussed in this section:

Beam and slab construction

Flat slab and flat plate construction

Ribbed or waffle slabs (concrete joist construction)

Slabs supported on structural steel frame

Typical methods of construction are shown, but choice of individual members and determination of support spacing must be based on established design principles.

Beam and Slab Construction

Where the slab is cast monolithically with the beam and girder system, ledgers on the beam sides support joists on which the slab sheathing rests. If spans between beams are long enough to require additional support for the joists, stringers perpendicular to the joists are used; additional shoring or scaffolding is provided to support the stringers (Fig. 3-46).

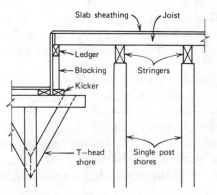

Figure 3–46 Slab form resting on beam ledger and stringers. For short spans between beams, intermediate stringer and shore support may not be required.

The ledgers are generally a part of the beam side, which is erected before the slab forming begins. The ledger is nailed to the beam side at a distance down equal to the exact depth of joists (except in cases where slab sheathing is butted against beam side, in which case the distance down is equal to joist depth plus sheathing thickness). Beam sides must be carefully erected so that ledgers are at the proper elevation to give level support to slab joists. If this is not done, the ends of individual joists may have to be wedged up or cut down to make a level base for sheathing. If the beam form is cambered, slab forming follows the beam camber. Some contractors nail block supports to the beam sides at specified joist spacings instead of, or in addition to, the long ledger.

If stringer supports are needed for the joists, their shores must be set in place, and then the stringers put in position. The stringer elevation should match that of the ledger top so that the slab form will be level. As soon as all joist supports are in position, the joists are dropped into place at the required location.[2] To facilitate stripping, the joists are frequently not nailed to their supports. However, joists that are narrow relative to their depth must be prevented from turning under load; nailing or bridging is often used. Some contractors prefer to use 4×4's, which do not have a tendency to roll, and thus avoid the need of bridging or nailing.

Slab sheathing is applied next. Plywood sheets, individual boards, hardboard, or boards cleated into panels and laid loosely on the joists are used. A few nails at the corners keep the panels in place. To further facilitate slab forming, joists and sheathing can be prefabricated into large panels. The whole area between the beams can then be formed with a few large panels in a single operation without setting the joists separately.

Where a slab load is supported on one side of the beam only, edge beam forms should be carefully planned to prevent tipping of the beam because of unequal loading.

The details of the intersection of slab decking and beam side require special attention so that the panels will not become keyed into the concrete. Keying can be prevented by cutting the sheathing ¼ to ½ in. short of the inside face of the beam side form and beveling the sheathing

---

[2] Since the deck panels between beams or girders are not loaded during erection, they are sometimes placed on the joists before the shoring and stringers are set. Stringer and shore supports can be placed and adjusted for elevation later on during the form erection work.

panel edge. This and alternate methods are shown in Figure 3-36, page 107.

Flat Slabs, Flat Plates

Flat slabs and flat plates are supported directly by columns without the aid of beams and girders. In flat slab construction, columns flare out at the top to form capitals, and the slab may be thickened at the columns to make drop panels. In flat plate construction, capitals and drop panels are always omitted. Forming methods are generally the same for both flat slab and flat plate work, except for the omission of drop panels in the latter.

The slab decking (sheathing) is usually formed of panels, but board sheathing can be used. Since there are no formed beam sides to support the joists, they are carried on stringers or ledgers resting on shores. In some designs the joists may be carried directly by the shores. (See Fig. 3-47).

The first step in construction of formwork for this type of structure is to erect and temporarily brace the shores. Stringers and joists are placed next and are leveled by wedging or adjusting the length of the shores. Joists or stringers that are narrow in relation to their depth may require bridging or other lateral support to prevent their turning or tipping under load. Deck panels or sheathing are then placed over the joists. Deck panels adjoining the drop panel are beveled, and they extend over the top of the drop panel form just as slab sheathing extends over beam side forms.

Drop panels (Fig. 3-48) are built independently of the slab decking. They are usually supported on four shores or a pair of scaffold-type frames that are braced to form a rectangular supporting structure. Two ledgers may be attached to the top of this frame to support joists for the drop panel. Sheathing applied over the joists is cut out to fit the column cap, and the shallow panel side is attached. Drop panel units may also be shop fabricated and set in place on supporting shores (Fig. 3-49).

Some flat slab and flat plate designs have voids in the slab. Rubber or fiber tubes with the ends blocked to prevent concrete entering are commonly used for forming these voids. They are placed between upper and lower layers of reinforcement and are tied to the bottom steel or formwork (Fig. 3-50) to prevent them from floating.

"Permanent" shores are frequently needed in flat slab construction; that is, shores that remain in position while the rest of the formwork is stripped around them. This is accomplished by setting a small piece of

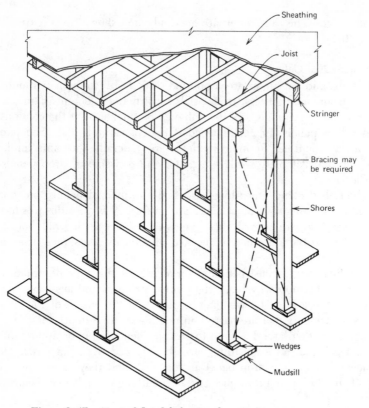

Figure 3–47   Typical flat slab formwork components.

formwork that can be detached from the rest, usually at the corner of four adjoining panels directly over a shore head. This method, however, leaves a conspicuous mark or indentation on the slab soffit and may be undesirable for exposed ceilings.

Concrete Joist Construction

Concrete joist construction is a monolithic combination of regularly spaced joists and thin slab cast in place to form an integral unit with concrete beams, girders, and columns. When the joists are all parallel, this is referred to as a ribbed slab or one-way joist construction. If the joists intersect each other at right angles, it is two-way joist construction or the so-called waffle slab.

Figure 3-48 Drop panel supported on a pair of scaffold-type frames is built independently of slab forming, then framed in. Metal column form has flaring capital that joins drop panel.

Figure 3-49 Flat plate forming with prefabricated panels resting on steel stringers supported by adjustable metal shores. Slab sheathing is pieced around top of already cast column.

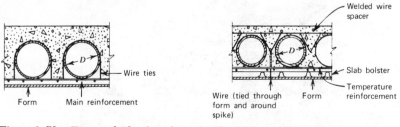

Figure 3–50   Two methods of tie-down for fiber tube void forms.

Proprietary forms made of a number of different materials are available for this type of forming, but this discussion of formwork erection is limited to the widely used metal pans and domes. The forming system is basically the same regardless of the pan material. These forms are available on either a rental or purchase basis, and the supplier frequently provides engineering layouts for the forming. Installation and stripping of the forms, including the erection of necessary shores, stringers, and soffits or decking, is also included as part of the service of a number of suppliers.

One-Way Joist System. Flange-type pans nailed down to supporting soffit forms are used for this system, as well as the adjustable no-flange pans which are nailed to the side of the soffit form with double-headed nails. With the adjustable pans, depth of joists can be varied by using nail holes at different levels on the sides of the pans. The use of adjustable pans nailed to the *sides* of soffit forms permits stripping of the pans while main soffit supports remain in place. Some flanged pans are also available with nail holes in the side and can be used for slip-in construction similar to the no-flange system. Long-span steel pans (Fig. 3-51) or wood forms for joist construction improve the appearance and reduce finishing where ceilings are to be exposed.

Nail-Down Pans. After beam, girder, and column forms are in place and braced, shoring and stringers are erected to approximate elevation. Wood soffit planks are then placed on top of the stringers and are extended over to and on top of the beam side. (In some instances the joist soffits are supported directly on shores.) Headers for bridging joists, beam tees, and "fishtail" wedges to accommodate tapered end forms are placed as

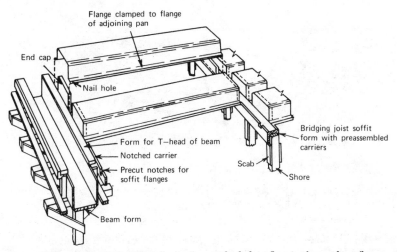

Figure 3–51 One type of "long" pan form which has flange clamped to flange of adjoining pan. This longer pan reduces the number of seams, producing smoother exposed construction.

needed. Shores are then adjusted to proper elevation in relation to supporting beams. (See Fig. 3-52 and 3-53.)

Flanged pans can then be nailed into position. A chalk line on the soffit form can be used to align the pans, so that precise alignment of the soffit boards is not required. End caps are placed first, and work progresses toward the center of the member from both ends, overlapping the pans 1 to 5 in. and setting a filler piece in the middle if required. After the pans are in place, they are oiled before reinforcing steel is set and mechanical trades make installations. A final check of elevation and alignment of the forms should be made before concrete is placed.

When concrete has obtained specified strength, shores and centering are removed. Stripping of the pans is the final step, in contrast to the adjustable pans described below, which may be removed for reuse ahead of supporting shores and joists.

Solid decking may be used to support nail-down pans if desired. It may offer some simplification of assembly and may provide a freer working area for laborers. It is particularly advantageous where all joists, beams, girders, and the like, are designed at the same depth for a uniform floor thickness.

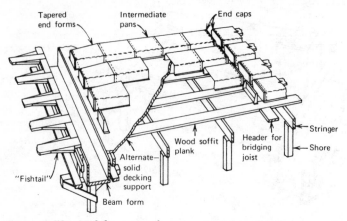

Figure 3–52 Nail-down pans for concrete joist construction may rest on either solid decking or an open system of soffit boards. If the latter are used, "fishtail" piece (left) must be attached to soffit boards where tapered end pans are used.

Figure 3–53 Workman setting nail-down type of pan on joist soffit form supported by heavy stringers.

Adjustable Pans. Shores, stringers, and soffit forms for this type of forming are erected in the same way as for nail-down pans, except that the stringer (if one is required) is parallel to the soffit form and must be of a size to fit between the pan sides. Soffit boards must be cut to exact dimensions unless joist thickness has been designed to use standard lumber sizes. The boards must also be cut to fit between tapered end pans. The pans are then adjusted to height and nailed to the side of the soffit form (Fig. 3-54)). A template (Fig. 3-55) will aid in setting pans to proper height and will hold them temporarily in position at one end for nailing. This type of forming is frequently chosen for exposed joist ceilings because joist soffits are formed with flange marks. The soffit forms must be very carefully positioned and aligned to give true, straight joists.

Two-Way Joist or Waffle Slab. These slabs are formed using dome-style pans. Shores and stringers are placed first, as was done for the one-way system. Soffit boards or solid deck forms are then set on the stringers. The design of a waffle slab usually requires some areas around the

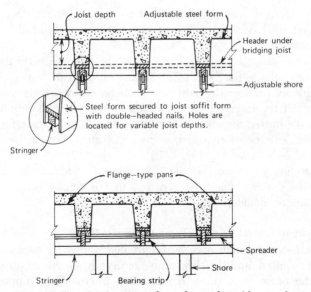

Figure 3-54 Cross section shows how adjustable pan forms are supported. Both flanged and unflanged types are shown.

Figure 3–55 Template used to position adjustable form. One end of the pan is supported on the template, the other rests on a form already nailed in place.

columns to be cast solid to the full depth of the joists. Domes are omitted, and solid deck forms are placed in these areas level with the joist soffit forms.

The steel pans or domes are placed next. They are nailed to the soffit form through holes in the flanges. If holes are not provided, special hook-headed nails are used, or the pans can be anchored by driving a washer-headed nail at the intersection of four pans. Most designs require that the flanges be butted together on the soffit form, which facilitates alignment. If wider joists are required, a chalk line should mark the outside alignment of the pan flanges (Fig. 3-56).

Lightweight concrete or clay tile filler blocks that remain in place are used in some designs. The formwork required is similar to that which supports the steel domes.

Slabs Supported on Steel Beams or Precast Concrete Beams

If structural steel framing is to be fireproofed, forms for encasing steel beams are required, and the slab forms are usually supported, at least in part, on the beam form sides. If additional intermediate supports are required, conventional shoring is used. If the slab is simply on top of the steel or concrete or has only the top flange of the beam embedded in it, slab formwork is frequently supported entirely from the already placed

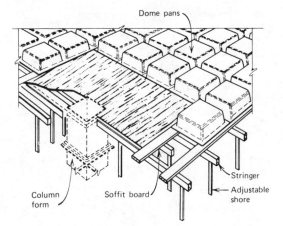

Figure 3–56 Two-way joist forming with prefabricated dome pans. Plywood deck form shown in area where solid slab is required may be extended to provide support for all the pans, replacing soffit boards.

steel or concrete frame. If the structure is designed for composite action of slab with its beam or girder support, the design engineer should indicate whether special shoring is required.

Decking is laid on timber or steel crosspieces hung from the beams by conventional wire beam saddles or various other hanging devices. The general system and materials are similar to other slab forming, except that support is from above rather than from shores below. (See Figs. 3-57 and 3-58.)

Metal Decking. Corrugated metal decking can also be used as formwork for slabs supported on previously placed steel or concrete members. It is laid across the existing structural members; joints may be left loose, welded, or clipped, according to details recommended by the manufacturer of the forming material. Longitudinal joints between decking panels are made by overlapping corrugations or interlocking flanges. The steel form remains in place after the concrete is cast.

Certain types of this decking material are designed to serve also as positive (bottom steel) slab reinforcement. If the decking form becomes part of the slab reinforcement, it should be galvanized and should not be subject to atmospheric corrosion.

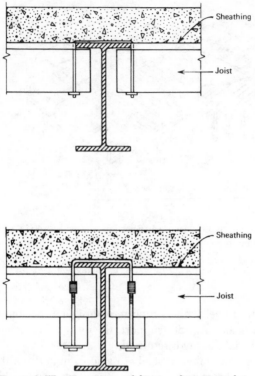

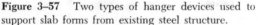

Figure 3–57 Two types of hanger devices used to support slab forms from existing steel structure.

Suggested Tolerances for Slab Work

In the absence of more rigid contract specifications, ACI Committee 347 suggests that formwork be constructed to produce slabs to the following tolerances.

Variation of slab soffit from the level or slope indicated on drawings should not exceed ¼ in. in 10 ft, ⅜ in. in any bay or 20-ft maximum, or ¾ in. in 40 ft or more. Variations in soffit level are to be measured before removal of supporting shores. The contractor is not responsible for variations resulting from deflection except when such deflections indicate inferior concrete quality or curing, in which case only the net variation due to deflection can be considered.

Figure 3–58 A simple method of slab forming when supporting steel joists are closely spaced and slab load is light. Clips attached to steel member support 2x8-ft. panels of plywood.

Variation in sizes and locations of openings should be no more than 1/4 in. *Variation in slab thickness* may be no more than −1/4 in. or +1/2 in.

SHORING AND SCAFFOLDING

Regardless of the type of shoring system—single-post wood shores, adjustable shores, scaffold shoring towers, or horizontal shoring—the layout or plan should be worked out in advance by a qualified individual, with careful consideration given to the possibility of stress reversals in partially cured slabs when construction loads are applied later. This danger exists particularly when long-span horizontal shoring is used on multistory work. A copy of the shoring layout should be kept on the job at all times, and it should be followed closely. If changes are necessary or desirable because of field conditions, the approval of the shoring designer or engineer-architect should be secured. Permissible tower heights (for scaffold-type shoring) and necessary external lateral bracing should be shown on the layout.

Inspection before concreting should include checking the actual layout against plans to see that shoring is correctly placed. Further inspection during and after concreting is also recommended.

Special attention should be given to beam and slab, and one-way and two-way joist construction to prevent local overloading when a heavily loaded shore rests on the thin slab. Shores resting on new construction that has not yet developed sufficient strength for full removal of shores should be placed as nearly as possible above the shores or reshores below to avoid excessive loading of the partially cured slab or beams. When vertical alignment of these shores and reshores is impractical, the shoring location should be approved in advance by the engineer.

Vertical shores for high multitier scaffolds must be set plumb and in alignment with lower tiers so that loads from upper tiers are transferred directly to lower tiers. Particular care must be taken to transfer lateral loads to the ground or to completed construction of adequate strength. This may be done with guy wires, diagonal bracing, struts, or a combination of these, depending on the height and location of the load within the structure. ACI Committee 347 recommends two-way lateral bracing at each splice in the shore unless the entire assembly is designed as a structural framework or truss. Columns cast ahead of the deck are a further aid to lateral stability.

All shoring members should be straight and true without twists or bends. Shores or vertical posts must be erected so they cannot tilt and must have firm bearing. Inclined shores must be braced against slipping or sliding. Shores supporting inclined formwork members should be firmly connected to the formwork after final adjustments of elevation have been made. Splices, couplings, or joints should be secure against bending and buckling. Connections of shore heads to other framing should be adequate to prevent the shores from falling out when reversed bending causes upward deflection of the forms. (See Fig. 3-59.)

The importance of adequate diagonal bracing to the safety and stability of the entire shoring system cannot be overemphasized. Diagonal bracing must be provided in both vertical and horizontal planes to provide stiffness and to prevent buckling of individual members of the formwork. For multiple-tier shoring, high scaffolding towers, or any other high shoring, increased attention is required to provide bracing that will prevent sway or lateral movement of shoring and buckling at splices. If the braces are required only to prevent buckling of individual members, struts may be used if anchored to masonry of adequate strength or to panel points of adjacent braced bays.

Figure 3–59 X-bracing is needed for every tier of high wood shoring towers.

A problem may exist where two kinds of shoring are used in a single bay or section of formwork. Such a combination of different types of shoring is not recommended. Tubular steel shoring does not show a gradual set or buckling as vertical load is applied; instead, it fails abruptly when a certain maximum load is reached. On the other hand, wood shores (particularly adjustable wood shores) undergo an initial set when vertical loads are applied. When the two types are combined for shoring a single bay, if the wood shores are not raised enough to allow for this initial vertical set, part or all of the concrete load can be transferred to the rigid steel shores. This sudden shifting of load overloads these shores and may cause collapse. Where tubular metal shoring is used on top of wood shores, the two assemblies must be individually braced and independently stable, as complete structural connections between the two are not practical.

Mudsills or Shoring Foundations

Another consideration of critical importance to the stability of shoring is a provision for adequate mudsills or other foundation support. A good foundation or sill distributes the shoring load over a suitable ground area. The footing must be firm, solid, or properly planked so that the load is evenly distributed to each leg. Unequal settlement of mudsills changes shore reactions and may cause serious overloading of some shores that do not settle as much as others. Mudsills should not be placed on frozen ground, on recently placed backfill, or where water will flow over them.

If the soil is of low bearing capacity, use of spread mudsills (Fig. 3-60) is suggested. If the soil is, or is likely to become, incapable of supporting the superimposed loads without appreciable settlement, it may be stabilized with cement or lean concrete or by other methods. Another alternative is to use piles or temporary concrete sills.

Shores supported on previously constructed floors may be assumed to have equal and uniform bearing. However, sills may be required to distribute the shoring load on green concrete. Suitable sills are particularly important in concrete joist construction or any other floor system involving voids, where a shore could concentrate an undesirable load on a thin concrete section.

Wood Shores

One-piece wood shores, cut slightly short of the desired elevation and adjusted by driving wooden wedges under the shore or at its top, are generally used for slab and beam shoring where great heights are not involved. Braces can be readily nailed at any desired elevation.

Ends of all such shoring should be square cut and have a tight fit at splices. According to recommendations of ACI Committee 347, field constructed lap splices should not be used more often than for alternate shores under slabs or for every third shore under beams. Wood shores should not be spliced more than once unless diagonal and two-way lateral bracing is provided at every splice point. Such spliced shores should be distributed as evenly as possible throughout the work. To avoid buckling,

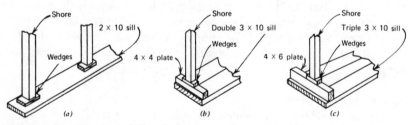

Figure 3–60 Wood mudsills, showing spread types for fair and poor bearing (a), (b), (c).

the splices should not be located near the midheight of shores without lateral support, nor midway between points of lateral support. Splice pieces should be 2 ft or longer, no thinner than 2-in. nominal lumber or ⅝-in. plywood, and as wide as the shore being spliced. Round shores should have three splice pieces; rectangular shores should have one on each face.

Shores formed by overlapping wood members supported by patented splicing devices may be used when their load capacity has been established by tests and is guaranteed by a reputable manufacturer. However, when all shores are of this type, much more rigid lateral bracing is required. Field inspection of each splice is also necessary to insure that guaranteed capacity can, in fact, be developed.

For beam support, wood shores are usually assembled with T-heads (Fig. 3-61) or with L-heads for spandrel beams. Where loads are relatively heavy, a double-post shore is frequently assembled with cross bracing. The double-post shore gives greatly improved lateral stability and resistance to overturning when loads are unbalanced.

Figure 3–61 Heavy T-head wood shores for beam support. Note bracing of heads, and wedges at base.

Adjustable Shores

All-metal and wood-and-metal shores available with jack- or screw-type adjusting devices are used in much the same way as wood shores. They simplify the problem of making fine adjustments of elevation after shores are in place. Manufacturers' load ratings are usually given for a certain extension of shore, and they should be carefully followed unless extra-diagonal bracing is added to the system. Bracing may be nailed to the wood-and-metal shores anywhere on the wood members. All-metal single-post shores usually have nailing brackets at fixed intervals or a movable device for attaching braces.

Adjustable shores have various fittings that can be interchanged at the top, providing extended height, flat bearing plate, U-head, or T-head. For support of stringers or other horizontal timbers in slab forming, the U-head is preferable because it permits nailing laterally into the stringer.

Scaffold-Type Shoring

Scaffold-type shoring is generally assembled in tower structures consisting of a pair of prefabricated frames and the diagonal cross bracing required to make the tower. The manufacturer usually specifies or provides tower bracing, but additional bracing between towers is advisable for stability where large or high structures are involved, and bracing or guying to some solid construction is also needed.

Before scaffold shoring is assembled, the parts should be checked, and those that are heavily rusted, bent or bowed, or have damaged welds should be rejected. Locking devices, coupling pins, and any pivoted cross braces should also be examined to make certain that they are in good condition, and any that are not should be rejected.

Location of each tower, as shown in the shoring layout, should be marked on the floor by a chalk line or other simple method. Sills are placed first, then adjustment screws or base plates are distributed to each tower location. The adjustment screws should be set to their approximate final adjustment before tower assembly begins. The base unit should be leveled after assembly, regardless of how many upper frames are to be added.

Assembly should be planned so that the shoring load is carried on the legs of the frames, not on top horizontal members, unless the frames are specially designed for such a condition. It is important to avoid eccentric loading by centering stringers on the U-heads or top places of the frame legs. Adjusting screw extensions should be kept to a minimum if the shores are being used at maximum rated load capacity.

The following checklist of points to be covered in a final inspection of scaffold-type shoring is based on one prepared by the Scaffolding and Shoring Institute. It is a good indication of approved erection practices.

1. Check to see that there is a sound footing or sill under every leg of every frame on the job. Check also for possible washout due to rain.
2. Check to make certain that all base plates or adjustment screws are in firm contact with the footing or sill. All adjustment screws should be snug against the legs of the frame.
3. Obtain a copy of the shoring layout that was prepared for the job. Make sure that the spacings between towers and the cross brace spacing of the towers do not exceed the spacings shown on the layout. If any deviation is necessary because of field conditions, consult the engineer who prepared the layout for his approval of the actual field setup.
4. Frames should be checked for plumbness in both directions. The maximum allowable tolerance for a frame that is out of plumb is $\frac{1}{8}$ in. in 3 ft. If frames exceed this tolerance, the base should be adjusted until they are within tolerance.
5. If there is a gap between the lower end of one frame and the upper end of another, it indicates that an adjustment screw must be turned to bring the frames in contact. If this does not help, it indicates that the frame is out of square, and it should be removed.
6. All frames must be connected to, at least, one adjacent frame to form a tower. Check to make sure that the towers have all cross braces in place.
7. While checking the cross braces, also check the locking devices to assure that they are all in closed position and are all tight.
8. Check the upper adjustment screw or shore head to assure that it is in full contact with the formwork. If it is not in full contact, it should be adjusted or shimmed until it is.
9. Check to see that the obvious mistakes of omitting joists, using the wrong size ledger, or placing timber flat have not been made. Check to see that the lumber used is the same as that specified on the shoring layout. Check the general formwork scheme to make sure that it follows standard practice for formwork.
10. If the shoring layout shows exterior bracing for lateral stability, check to see that this bracing is in place in the locations specified on the drawing. Check to make sure that the devices which attach this bracing to the equipment are securely fastened to the legs of the shoring equipment. If tubing clamps are used, make sure that they have been properly tightened. If devices for holding timber have been

used, check to see that sufficient nails have been used to hold the bracing securely to the frame legs.

Horizontal Shoring

Adjustable metal members are used to support slab forms over comparatively long spans without intervening vertical shores. This reduction in the number of shores leaves open spaces clear for work and is a major advantage of horizontal shoring, but it frequently results in much heavier loads on the fewer vertical shores required. Consequently, greater care must be taken in lacing and bracing the vertical shores that are used and in providing solid footings or mudsills.

Because of the greater concentration of load at vertical shores, more settlement because of compression of timbers will take place, and ledgers supporting the horizontal shoring should be set at an elevation to allow for this: about $\frac{1}{16}$ in. higher for each wood member concerned is suggested, or a total of $\frac{3}{16}$ in. for mudsill, shore, and ledger combined. Also, the prongs of the shoring member should have a minimum bearing of $1\frac{1}{2}$ in. on the supporting vertical shore or ledger.

If beam sides are to carry the horizontal shoring, they must be built heavier than usual to carry the load. (See Fig. 3-41.) A 4×4 ledger with stud support at each point where the horizontal shoring rests on the ledger is advisable. Extra bracing for stability must be provided because the loading from the horizontal shoring on one side of the beam will occur before the beam is cast. This is even more critical for spandrel beams, and greater attention to diagonal bracing and tying is needed to keep spandrel beams in correct position.

The manufacturer's recommendations for adjusting the length of shoring members should be followed. Usually, these members have a built-in camber of the correct size to produce a level slab as cast when horizontal shores are loaded to the amount recommended by the manufacturer. The designer must provide further camber if he wishes to offset deflection of the reinforced concrete member itself. If for some reason the full loading will not be carried—if, for example, horizontal shores are parallel to beam sides or walls, or a bulkhead occurs at midspan of the horizontal shore—compensating adjustments in camber must be made according to instructions of the manufacturer. Otherwise, the finished slab may have an upward camber at that point.

This initial camber means that screed chairs set by instrument to desired elevation will be too low after concrete is placed in the forms and camber is removed. Screed chairs should be set from the deck sheathing to give correct slab thickness. It is preferable to place the screed bars perpendicular to the span of the horizontal shoring and to load the entire span before final strike off is made.

Adjustment and Jacking

Jacks or wedges should be available to permit alignment and facilitate stripping, and also to provide a positive means of adjustment and realignment if excessive settlement takes place at shores. Wedges may be used at the top or bottom of shores, but not at both ends. The best wedges are made of hardwood, driven in pairs to an even bearing. When final adjustment of shore elevation is complete, wedges are toenailed to the shore. Any wedges at the top of shores under sloping forms, which serve to establish bearing of form on shore, should be securely nailed to the shore head when adjustment is completed.

The various patented shoring devices have their own jack or screw adjustments for elevation. Screw jacks for pipe shores or scaffold-type shoring may be used at both top and bottom so long as they are secured by the shore or scaffold leg against loosening or falling out. A minimum of 8 in. embedment in the pipe leg or sleeve is recommended by ACI Committee 347.

Where major movements of heavy loads are required, such as when a large section of shoring is to be lowered or raised as a unit, or where very careful control of decentering is required, hydraulic or pneumatic jacks are often used.

For decentering only, sand jacks lower heavy loads up to 300,000 lb and offer the advantages of little deflection under load and no danger of failure during placing and curing. Their rate of travel is easily controlled by a slight finger pressure over a plug hole to change the flow of sand. Construction details of such a jack are shown in Figure 3-62.

Permanent Shores

Since reshoring is such a costly and critical process, and early reuse of form panels is so often desired, the installation of permanent shores frequently becomes important. A "permanent" shore is one that remains undisturbed as form panels are stripped around it. Use of permanent shores avoids the special attention required to assure that reshores are placed uniformly tight under the slab. It also provides greater assurance that shores are placed in the same pattern on each floor.

Two basic systems of permanent shoring are the *king stringer system* and the *king shore system*. The former uses ledgers on the sides of the stringer which may be released, permitting the removal of the joist and

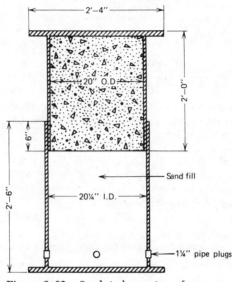

Figure 3–62 Sand jack consists of concrete-filled piston and sand-filled cylinder. Useful for slowly lowering heavy loads; motion can be stopped by replugging holes at the base.

form contact surfaces between the stringers. Figure 3-63 illustrates how this may be done with panel slab forms. In the king shore system, the stringer is attached to the side of the shores so that it may be removed, permitting the release of joists and sections of the form contact surfaces. The shores and a trapped strip of contact surface are all that remain in place.

Shoring for Composite Construction
Shoring required by the structural design for composite action should be specified by the architect-engineer, and his instructions must be carefully followed. Such shoring must generally remain until concrete is fully cured, so permanent shores are desirable.

Shoring of members that will act compositely with the concrete should be done with great care to assure sufficient bearing, rigidity, and tightness to prevent settlement or deflections beyond allowable limits. Wedges, shims, or jacks should be provided to make necessary adjustments before or during concreting as well as to permit removal of shoring without jar-

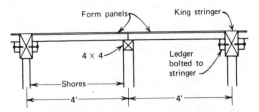

Figure 3–63 One method of leaving shoring in place when slab forms are stripped. Prefabricated panels 4 ft. long used with king stringers 8 ft. apart. Panels supported at one end on 2x4's bolted to king stringers and on 4x4 stringers at the other end. The 4x4's and their shores can be removed, freeing panels and leaving king stringers and their shores to support slabs as long as desired.

ring or impact on the completed construction. Provision should be made for readily checking the accuracy of position and level during concrete placement.

Where camber is required, distinction should be made between that part which is an allowance for settlement or deflection of formwork or shoring and that which is provided for design loadings. The former is generally the responsibility of the contractor who designs the forms and supports unless such camber is stipulated by the engineer-architect. For acceptance purposes, measurement of camber provided for design loadings should be made after hardening of concrete but before removal of supports.

CHAPTER 4
USING THE FORMS[1]

Construction procedures should be carefully planned in advance to achieve the proper balance between safety and economy in producing quality concrete work. This advance planning should be coordinated with the design and construction of the formwork. A knowledge of conditions for which the forms were intended and a willingness to respect these limitations are absolutely necessary. For example, if forms are designed for a rise of concrete of 4 ft per hr, that rate should not be exceeded if forms are to function properly. If slab forms are designed for 125 to 150 psf total load, it should not be expected that heavy bundles of reinforcing steel or other construction materials can be placed without damage or deflection of the formwork. It is also especially important that heavy construction loads be kept off new, partially cured concrete structures.

The purpose of this chapter is to point out the many factors to be considered in using forms, such as cleaning and coating forms before concreting, care in placing reinforcement and inserts, and operation of placing equipment in such a way that forms are not damaged or misaligned. Inspection of the formwork before, during, and after concreting is important, and some suggestions on how this should be done are included.

[1] Authorized condensation from *Formwork for Concrete*, Copyright 1973, American Concrete Institute.

Stripping forms and reshoring require considerable care to protect the concrete, and for this reason they are often under the direct control of the engineer-architect. Some of the factors affecting his decisions are enumerated, and recommended techniques are included. Reconditioning and proper storage and handling of forms benefit the contractor who wants his forms to give maximum reuse, and suggestions for these practices are given. Insulation of formwork for use during cold weather is also discussed.

PLACING REINFORCEMENT AND INSERTS

Detailed recommendations for placing reinforcing bars have been published by the Concrete Reinforcing Steel Institute; the comments made below cover only those aspects of bar setting that have a direct effect on the formwork. The place of bar setting in the construction sequence depends on the member involved as well as the system of form erection and other construction details, including facilities for hoisting and placing reinforcing cages and forms. For walls, the external form is usually set in place and followed by the reinforcing mat; then the interior wall form is erected and secured in place. For columns, the reinforcing bars may be preassembled and set in place, after which the panels of the column form are erected, aligned, and clamped around them; alternatively, the reinforcing cage may be dropped down inside the preassembled column form. Sometimes the entire preassembled column form is set down over the erected reinforcement. (See Fig. 4-1.)

Beam and girder boxes are usually fully assembled to receive the reinforcing, and beam and girder reinforcement is assembled in place in the forms; less frequently, the cage is preassembled and set in place as a unit. Spandrel beams having reglets, anchors, and the like usually have them placed by the form builder ahead of the steel setters. If beam sides are to be tied across the top to prevent spreading, this must be done after the bar setters have finished, and such ties should not interfere with the bars.

Regardless of the exact sequence established, it is extremely important to have the work of the bar setter coordinated with that of other trades so that any inserts, sleeves, conduits, ducts, straps, and anchors that should be placed ahead of the bars are so placed, and any that should properly be placed later are not in place when the bar setter goes to work. Similar coordination with the form builders is necessary so that formwork is in position, braced and aligned ready to receive bars. Pro-

Figure 4–1 Tying steel for the baffle walls of a filtration plant. One side of the wall form is in place with some lateral bracing at the top. All ties are in position, the bulkhead at the end of the wall panel is built, and bars project for the next wall section to be concreted later. Wooden box inserts for wall openings are in position and steel has been set around them.

vision must also be made for formwork parts that cannot be erected until after the bars are set. (See Fig. 4-2.)

Form oil or other coatings should be applied before steel is placed so that the coating material does not get on the steel where it could reduce or destroy the bond between hardened concrete and reinforcement. Most sleeves and inserts should be securely fastened in place before steel is set, and bar setters should take care not to remove them or to kick them out of place. Sheet metal pipe sleeves held in place by a dozen or more nails can have a very damaging effect on forms. When the forms are stripped, these nails tear the surface, and often the form material cannot be reused. A pipe sleeve made of fiber material now available can be held in place by a metal cup and a single nail. Such a sleeve is cheaper for the mechanical trade to use and offers a considerable saving to the form contractor.

Support for Reinforcing Bars

In footings, concrete bricks or precast concrete blocks are commonly used to support mats of bars. For top bars in heavy foundation mats, special heavy bar supports can be obtained with bearing plates to rest on

Figure 4-2 Beam and slab deck form with network of conduits, pipes, circular inserts to form openings, mountings for anchor bolts, and reinforcing steel all in position. Complex installations such as this make it extremely important to coordinate the work of bar setter and form builder with electrical, mechanical, and other trades.

the earth subgrade. For other members such as joists, slabs, beams, and girders, a wide variety of ready-made wire, plastic, and even concrete bar supports are available. Any steel bar supports used at concrete surfaces that are to be exposed to the elements should be cadmium plated, galvanized, or plastic coated to prevent rusting; it is preferable that the entire bar support be made of plastic or other nonrusting material.

For horizontal members, bar supports are commonly spaced 5 ft on centers, but because of possible effects on the formwork, this spacing is often reduced to 4 ft or less. Bar chairs or bolsters should be strong enough and spaced closely enough so that they are not excessively loaded and bite into the form; this can be particularly troublesome in architectural concrete. Heavy reinforcing cages may cause beam bottoms to deflect objectionably at the point where bar chairs or other bar supports are located. This can be overcome by hanging the reinforcing cage from the top of the form, or by closer spacing of bar supports to distribute the load more evenly over the beam bottom.

Bar supports for cantilevered reinforced members should be made extra heavy to maintain reinforcement in proper design position. This requires particular attention because the major reinforcement is near the top of the member where it is more vulnerable to accidental displacement because of construction activities.

Vertical bars in wall mats are usually tied to dowels at the bottom and are attached to the forms in some way at the top. Nails driven into the formwork near the top of the bars, projecting the required amount of cover and matched to the vertical bars, together with wire looped around the nail head make good spacers. Sometimes the mat is wired to the form ties; or wire or stamped metal spacers are used, and the mat is wired to the wall forms and is pulled tight against these spacers. From the point of view of the form builder, the wall reinforcing should be attached to nails in the forms as little as possible. Nails make stripping difficult and may cause damage to form panels. Also they may later cause streaks on the surface of exposed structures.

Another method of spacing vertical reinforcement mats is to rip long strips of 2-in. material to the required dimension and to insert them at intervals between the wall sheathing and the steel mat before concrete is placed. The strips are raised as the form is filled, and care should be taken to make certain that they are entirely removed before completing concrete placement. If there is more than one layer of steel in a wall or other vertical member, the same procedure can be used in both faces, or else spacing can be done from one face only, with the second mat of steel spaced from the first by accurately preformed spacer rods.

For some exposed walls, small concrete cubes with tie wire cast in them may be used as spacers to hold reinforcing bars at the proper distance from the forms. The short lengths of embedded tie wire are used to fasten the block to the reinforcing steel before the outside wall form is positioned. Such spacer blocks can be economically cast in large quantities on the job.

Steel in columns is spaced from the form and is tied to it, at least, near the top. Wherever access permits, column reinforcing is braced away from the forms at three or four points as near to the top, bottom, and midheight as possible. For such spacing, some erectors cast concrete doughnuts that are slid onto the appropriate column verticals during assembly. Others slide wire slab bolsters of suitable height down the erected column cage to bear against the formwork and then wire them to the steel. Any practices that hinder stripping or damage forms should be avoided.

Forms must be thoroughly cleaned of all dirt, mortar, and other matter such as chips, blocks, sawdust, or ice before each use. If the bottom of the form cannot be reached from the inside, access panels should be provided, preferably at the end of the form rather than the side, to permit thorough removal of all foreign matter before placing the concrete. A jet of air, water, or steam may be used effectively to remove debris. All cleanout openings must be carefully closed after the forms are washed out. Some specifications prohibit cutting washout holes in the forms. In such instance, pumps, air-lifts, or siphons may be required to remove washings (Fig. 4-3).

Oil and Other Coatings

Before concrete is placed, form surfaces should be wetted, oiled, or coated with satisfactory materials that will not stain or soften the concrete. The form coating or treatment must serve as a parting compound, or release agent, to prevent sticking of the concrete to the forms, thereby aiding in stripping. It may also act as a sealer or protective coating for

Figure 4–3 Water jets are used to clean out this dam form resting on irregular rock surface.

the form, preventing absorption of water from the concrete into the form material. Numerous available form-coating materials perform one or both of these functions to varying degrees; some offer additional benefits of improved concrete surface finish as well as longer form life through improved weathering and wearing qualities of the forms. A few types of coatings make possible several reuses without recoating the form.

Wood and Plywood Forms. Most of the commercial oils are satisfactory for wood forms; light-colored petroleum oils and oil emulsions of various types have been successfully used, and these oils have also proved satisfactory for concrete molds. The oil should be capable of penetrating the wood to some extent while leaving the surface only slightly greasy to the touch. Linseed oil cut with kerosene has also been reported as satisfactory for plywood forms; if the plywood is mill oiled, it should be used once without oiling, then coated lightly for each reuse.

Occasionally, lumber or hardboard material contains sufficient tannin or other organic substance to cause retarded set of concrete surface. When this condition is recognized, it can be remedied by treating the form surfaces with whitewash or lime water before applying the form oil or coating.

Products other than oil are sometimes used for treating plywood and wood forms to seal and preserve the material as well as to make stripping easier. Plywood may be coated with shellac, lacquer, resin base products, or plastic compounds that offer almost total exclusion of moisture from the plywood. This prevents wet concrete from raising the grain which would later detract from the surface smoothness of the finished concrete. When such coatings are used, generally only a very light oiling prior to use of the form is needed; some coating manufacturers suggest no oiling with their products. (See Fig. 4-4.)

Metal Forms. Sticking of concrete to steel forms may result from (1) abrasive cleaning that exposes bright metal, (2) abrasion opposite openings or other areas where entering streams of concrete are directed against forms, and (3) unsuitable form oil.

Form oils that are satisfactory on wood are not always suitable for steel forms, especially where there is a sliding movement of concrete against forms as in tunnel lining. If one oil does not work as well as desired, others should be tried, and there are a number of satisfactory compounded oils available. Paraffin-base form oils and blended oils consisting of a petroleum base blended with synthetic castor oil, silicones, graphite, or other

Figure 4–4 Dip tank used for treating plywood panels with sealing compound, showing how panels are stacked on edge to drain and dry. Tank is shaded to reduce evaporation of sealer.

substances have been successfully used. Some marine engine oils have also given good results on steel tunnel forms. A heavier form oil should be used in hot weather if difficulty is encountered with concrete sticking to forms when they are stripped early.

Rough surfaces on steel forms where sticking occurs may be conditioned by rubbing in a liquid solution of paraffin in kerosene, or the forms may be cleaned and oiled with a nondrying oil, and then exposed to sunlight for a day or two.

Plaster Molds. When the plaster waste molds used for architectural concrete are thoroughly dry, they should be given two coats of shellac or equivalent waterproofing coating before leaving the shop. After these molds are set in place, and all joints are patched and touched up with shellac or other waterproofing, they should be greased with a light yellow cup grease which may be cut with kerosene if too thick. Grease should be wiped into all crevices and recesses, and all surplus should be carefully removed. Grease and shellac must be kept off any hardened concrete or reinforcing steel in the area.

Application of Coatings. Surfaces to which coatings are applied should be clean and smooth. Application may be by various methods: roller, brush, spray, wiping, and the like, depending on the type of coating being placed on the form. The important consideration is to make coverage complete and uniform, if good stripping and good appearance are to be attained. There should be no excess coating to stain the concrete or leave undesirable residue on the finished surface. If oils or greases are used, the excess can be wiped off; some types must be applied quite carefully because wiping is not permitted. (See Fig. 4-5.)

Whenever possible, form panels or form materials should be coated before erection; this is desirable because it permits dipping of plywood panels and various other techniques for faster coating. It is sometimes necessary because special coatings may require several days drying or curing. Forms subject to continuing reuse are generally coated just after stripping and cleaning. When coatings are applied after forms are erected, the application must precede steel placement so that no form coating gets on reinforcing bars. Construction joint surfaces should also be kept free of form coating.

Figure 4–5 Epoxy-resin plastic coating is applied by brush to a large made-up form panel. Two coats of this material were used to achieve greater smoothness and longer protection of the plywood sheathing.

Use of Water on the Forms

An effective way of keeping untreated board forms tight is to soak them continuously with water for about 12 hr before concreting. This also stops absorption of water from fresh concrete and thus prevents warping and swelling of form members after concrete is placed. If forms are badly dried out, soaking with water, at least, twice daily for three days prior to concreting may be necessary. Untreated wood inserts should also be soaked well in advance of concreting; otherwise, they may swell and cause concrete to split.

Thorough wetting of untreated wood forms just before concreting also facilitates stripping, but increasing use of oils or sealants that protect wood from absorption of moisture tends to reduce the importance of wetting. Wider use of plywood with greater dimensional stability than board sheathing has also often eliminated the need for wetting; protective sealant coatings for the plywood forms and form panels are more generally used because they contribute to extended form life as well as aid in stripping.

Many contractors wet forms just before concreting in addition to using protective coatings or release agents. Cooling of forms and reinforcing steel by sprinkling with cool water inside and outside just before concreting has been recommended to help lower the temperature of freshly placed concrete in hot weather. Shading forms and steel from the hot sun is also suggested.

Where concrete is placed against earth forms of dry sands or other absorbent material, the soil should be wetted to prevent the too-rapid absorption of water from the freshly placed concrete.

INSPECTION AND FORM WATCHING

Before concreting is permitted to progress, forms should be inspected to determine that they are in the correct location and are properly built to produce concrete of the required finish and dimensions with adequate safety for the workmen on the job. Job specifications should state clearly when, by whom, and for what features the owner desires an inspection of formwork, and what approvals, if any, are required by his representative before concreting can begin. Local building codes, like that of New York City, for example, may also require a certification of inspection and approval of the formwork[2] to be filed by the architect or engineer with

[2] The New York code requirement applies to forms over 12 ft clear height.

the proper government official before concreting can begin, or after concreting is completed, for a permanent record.

If the owner's representatives or local building officials do not require inspection of the formwork, the contractor will bear the sole responsibility. It is clearly to the interest of the contractor as well as the owner to make an inspection for accuracy, stability, and satisfactory workmanship before concrete is placed; however, the contractor is not likely to leave all this until the last minute. There is too much detail, and too much of it is hidden by the time forms are ready for concreting. The foreman or superintendent probably will maintain a continuing check, closely watching each phase of form erection as it progresses. The contractor is also concerned with a continuing check of the forms during and after concreting by experienced form watchers.

Before Concreting

A competent inspector must be thoroughly familiar with the entire job and its requirements and also must have a general knowledge of good concrete construction practices. The *ACI Manual of Concrete Inspection* is a valuable pocket reference guide for any inspector. Approaching the problem from the point of view of the owner, this manual suggests a three-stage inspection as the work advances:

Preliminary inspection. This is made when excavation has been completed or forms built. If dimensions and stability are satisfactory, the contractor may then clean the foundation or oil the forms and may install any reinforcement and fixtures.

Semifinal or cleanup inspection. When everything is in place for concreting; this is a detailed inspection of forms, reinforcement, foundations, and all equipment or parts to be embedded in the concrete. If the installations are satisfactory, the work is ready for final cleanup.

Final inspection. This occurs immediately before concreting to make certain that forms, reinforcement, and fixtures have not been displaced. Surfaces must be clean and wetted if so specified.

Some of the points that must be considered, whether the inspection is performed in these three stages by the owner's representative or done by the contractor as the job advances, are listed in the following sections.

Overlapping Inspections. Inspectors for different features—structural, electrical, mechanical, or others—should coordinate their work. If the architect-engineer's representative cannot make the inspection for all trades or all parties concerned, he must make certain that no changes required by others affect structural features that he has already approved. The final inspection should cover structural requirements.

Alignment, Location, and Dimensions. Forms should be checked for accuracy of line and grade as early as possible so that delays for any necessary adjustments can be minimized. Location and dimensions of the forms after they are filled with concrete may not be the same as when they were built, since loading may cause them to settle, sag, or bulge. To insure that line and grade of *finished work* be within the required tolerances, forms should be built to elevation or camber shown on the formwork drawings. If settlement of supports or sagging of spans is to be expected, the form designer generally will have planned for this and will have included compensating allowances in establishing his dimensions. Any indicated elevations and cambers should allow for joint closure, settlement of mudsills, dead load deflection, and elastic shortening of form members as well as any camber specified on structural drawings.

Various means of checking location and alignment are used. Governing points of line and grade will have been set by the engineering staff, but the inspector will need to make additional measurements from and between these points. Transit and level may be used along with direct observation and measurement by the inspector; plumb lines and stretched wires may be necessary in some locations. An accurate straightedge should always be on hand, and in many instances homemade templates will serve as convenient and accurate means of checking dimensions.

In pavement construction, the alignment and crown of screeds should be checked. Screeds for floor and roof slabs should be set to assure the desired thickness of the member; for example, if a slab form is cambered to compensate for dead load deflection after form removal, screeds should also be cambered to give a uniform slab thickness to the member.

After the final check of alignment and location is made, telltale devices should be installed on supported forms and elsewhere to facilitate the detection and measurement of formwork movements during concreting. Wedges or jacks should be secured in position after the final check of alignment, but there should be some positive means of realignment or readjustment of shores if excessive settlement occurs after concrete is placed.

Adequate Strength and Stability. In addition to the obvious verification of position and dimensions, this question must be asked: Are the forms likely to keep the proper position and dimensions during concreting? The strength and stability of the formwork depend in large measure on a properly developed form design, but there are a number of details the inspector can check closely to see that the designer's plan is being properly executed:

1. Are the bracing and tying of the formwork adequate? Are all the necessary tie rods or clamps in the proper location and properly tightened? This point is critical since it is usually impossible to force a form back into position after it has bulged while being filled.
2. Are the shores properly seated and adequately braced? Is the bearing under mudsills adequate? Sills or spread footings should not rest on frozen ground.
3. Are the shores adequately connected to formwork at the top to resist any upward movement or torsion at the joints?
4. Does the concrete-placing crew know the placing rates and sequence planned for the job?

Quality and Cleanness of the Formwork. Joints and seams in the forms should be checked for tightness to prevent accumulation of dirt before concreting or formation of fins of mortar when the concrete is placed. The final inspection should include examination of formwork and construction joints for cleanness and to determine that necessary fittings and reinforcement are attached in the proper locations.

The inspector should check to see that the form sheathing or lining can be reasonably expected to yield the desired or specified finish. He should also be sure that wetting, oiling, or other specified form treatment has been adequately performed before concreting and that there is no form coating on the reinforcing steel.

During and After Concreting

Formwork should be continuously watched during and after concreting by competent men (the number depending on the size of the job) stationed below or alongside the forms being filled. Precautions should be taken to protect the formwork watchers and to maintain an area of safety for them during concreting. Some means of communicating with placing crews in case of emergency should be planned in advance.

The form watchers will use previously installed telltale devices to maintain a constant check of elevations, camber, and plumbness of the formwork system. They should tighten wedges and promptly make appropriate adjustments of elevation by jacking or wedging wherever necessary. All adjustments must be made before the concrete takes its initial set.

If bulging of *vertical* formwork goes beyond tolerated amounts as work progresses, the superintendent should be notified and the filling of the form slowed down, possibly even stopped, until additional bracing or other corrective measures can be taken. As previously noted, it is almost impossible to push back these bulges in the filled forms without removing some of the fresh concrete, in contrast to the considerable leeway possible in adjusting jacks and wedges for horizontal forms.

If any serious weakness develops during concreting, such as would endanger workers or cause undue settlement or distortion, work should be halted while the formwork is strengthened (or concrete removed to permit form adjustments). If the affected construction is permanently damaged, it may be necessary to remove a portion of it, but this requires approval of the engineer-architect, since it may affect the safety and stability of adjoining construction.

Although a most critical stage has passed once the concrete is in the forms, the form watchers should remain on duty until the concrete has been screeded and telltale devices show that deflection has ceased. An impending form failure often gives warning by gradually increasing deflection.

PLACING AND VIBRATING—EFFECT ON FORMWORK

Properly designed formwork of good quality will not be adversely affected by proper internal vibration or normal placement of concrete, although it is advisable to watch for loosening of nut washers and wedges during vibration.

For good form performance, rate of rise of concrete in the forms should not exceed that for which they were designed, and any limits on vibration set by the designer should be followed. Necessary depth of vibration varies somewhat with the depth of layers in which concrete is placed. When not provided for in the design of the form, revibration of previously placed layers should be avoided because vibrator action in the stiffened concrete can cause overloading of normal forms.

If forms are not designed for external vibration, extreme caution should be used in applying or attaching vibrators to the outside of the forms. It has been repeatedly found that external vibration can destroy the strongest form. For this reason, some agencies prohibit its use except when forms are specially designed for such external vibration.

Vibration should be used for the purpose of consolidation only, not for lateral movement of the concrete. Reasonable care by the operator is necessary to avoid scarring or roughening the forms by operating vibrators against them.

Runways for moving equipment should be provided with struts or legs as required and should be supported directly on the formwork or a structural member. They should not bear on or be supported by reinforcing steel unless special bar supports are provided. Formwork must be suitable for support of such runways without intolerable deflection, vibration, or lateral movement.

Abrasion of forms caused by an entering stream of concrete can be prevented by use of protective sheets of metal, plywood, or rubber belting. For forms more than 10 ft high, tremies or chutes should be used to avoid impact on forms as well as segregation of concrete.

REMOVAL OF FORMS AND SHORES

Although the contractor has general responsibility for design, construction, and safety of the formwork, the time of removal of the forms and shores should be specified by the architect-engineer in the contract documents or made subject to his approval, because of the danger of injury to concrete that may not have attained full strength or that may be overloaded in the stripping or subsequent construction operations. Where reuse of forms is planned, it is vital to the interest of the contractor to remove forms and shores as early as possible. In warm weather, early stripping is sometimes desirable because it permits specified curing to begin. Another advantage of early form removal is that the necessary surface repair or treatment can be done while the concrete is green and favorable to good bond. In cold weather, curing requirements and the danger of thermal shock to the concrete make early removal less advantageous.

Stripping Time Based on Concrete Strength

Since early form removal is usually desirable so that forms can be reused, a reliable basis for determining the earliest proper stripping time is necessary. When forms are stripped, there must be no excessive deflection or distortion and no evidence of cracking or other damage to the concrete, as a result of either removal of support or the stripping operation. Supporting forms and shores must not be removed from beams, floors, and walls until these structural units are strong enough to carry their own weight and any approved superimposed load. Such approved load should not exceed the live load for which the member was designed unless provision has been made by the engineer-architect to allow for temporary construction loads as, for example, in multistory work. Generally, forms for vertical members such as columns and piers may be removed before those for beams and slabs.

The strength of concrete necessary before formwork can be stripped and the time required to attain it vary widely with different job conditions; the most reliable basis is furnished by test specimens cured under job conditions. In general, forms and supports for suspended structures can be removed safely when the ratio of cylinder test compressive strength to design strength is equal to or greater than the ratio of total dead load and construction loads to total design load, except that a minimum of 50% of design compressive strength is required. Some agencies specify a definite strength that must be obtained for example, 2500 psi or two-thirds of design strength. However, even when concrete is strong enough to show no immediate distress or deflection under load, it is possible to damage corners and edges during stripping, and excessive creep deflections may occur.

If strength tests are to be the basis for the engineer-architect's instructions to the contractor on form removal, the type of test, method of evaluation, and minimum standards of strength should be stated clearly in specifications. The number of test specimens as well as who should make the specimens and perform the tests should also be specified. Ideally, test beams or cylinders should be job cured under conditions that are similar to those for the portions of the concrete structure that the test specimens represent. (These specimens must not be confused with those cured under laboratory conditions to evaluate 28-day strength of the concrete.) A curing record including time, temperature, and method for both the concrete structure and the test specimens together with weather record will assist both the engineer and contractor in determining when forms can be safely stripped. It should be kept in mind that specimens that are relatively small are more quickly affected by freezing and drying conditions than is concrete in the structure.

On jobs where the engineer has made no provision for approval of shore and form removal based on strength and other considerations peculiar to the job, ACI Committee 347 suggests minimum times forms and supports should remain in place under ordinary conditions, as shown in Table 4-A. These periods represent cumulative number of days or fractions thereof, not necessarily consecutive, during which the temperature of the air surrounding the concrete is above 50° F. If high-early-strength

Table 4–A
MINIMUM TIME FORMS SHOULD REMAIN IN PLACE
(Under Normal Conditions)

	(in days)
Walls[a]	1 to 2
Columns[a]	1 to 2
Sides of beams and girders[a]	1 to 2
Pan joist forms[b]	
30 in. wide or less	3
Over 30 in. wide	4

Where design live load is:		(in days)
Arch centers	14^d	7
Joist, beam, or girder soffits[c]		
Under 10 ft clear span between supports	7^d	4
10 to 20 ft clear span between supports	14^d	7
Over 20 ft clear span between supports	21^d	14
Floor slabs[c]		
Under 10 ft clear span between supports	4^d	3
10 to 20 ft clear span between supports	7^d	4
Over 20 ft clear span between supports	10^d	7
Posttensioned slab system	As soon as full tension has been applied	
Supported slab system	Removal times are contingent on reshores, where required, being placed as soon as practicable, but not later than the end of the working day in which the stripping occurs.	

[a] Where such forms also support formwork for slab or beam soffits, the removal times of the latter should govern.
[b] Of the type which can be removed without disturbing forming or shoring.
[c] Distances between supports refer to structural supports and not to temporary formwork or shores.
[d] Where forms may be removed without disturbing shores, use half of values shown but not less than 3 days.

cement is used, these periods may be reduced as approved by the engineer. When higher cement-content concretes are used, yielding higher early strengths, the engineer may also approve some reduction in the required time before stripping. Conversely, if low temperature concrete or retarding agents are used, these periods may be increased at the discretion of the engineer-architect.

Form Removal Related to Curing Needs

In warm weather, wood forms left in place furnish good protection from the sun but do not keep concrete moist enough to be acceptable as a method of outdoor moist curing. Metal forms, and wood forms that are kept thoroughly wet, provide satisfactory protection against loss of moisture if forms are loosened and the exposed top surfaces are kept wet in such a way that the water finds its way down between the concrete and the forms. Under these conditions only, the forms may be left on the concrete as long as practicable. Otherwise, they should be removed as soon as concrete strength development permits so that prescribed curing may be commenced with the least delay after placing. The U. S. Bureau of Reclamation has found that surfaces of ceilings and walls inside buildings may require no other curing than that provided by leaving the forms in place for four days; this period may vary, however, with the humidity and drying conditions inside the building.

In cold weather, moist curing, although important, is not so urgent, and protection afforded by forms, other than steel ones, is often of greater importance. In heated enclosures, these forms serve to distribute warmth more evenly and to prevent local heating. With suitable insulation the forms, including those of steel, will usually provide adequate protection without supplemental heating. In cold weather, therefore, in view of the reduced period of protection required, it is usually advantageous not to remove forms until the end of the prescribed minimum period of protection.

Stripping Techniques

Forms and shoring should be designed for easy, safe removal in a way that permits the concrete to take its load gradually and uniformly without impact or shock. Stripping was formerly referred to as "wrecking," and this description was often given a literal interpretation when forms were removed—there was little or no salvage of materials. Today, with increasing emphasis on panel systems, modular formwork components, and economy through maximum reuse of forms, much more attention is given to building forms that can be removed intact. (See Fig. 4-6.)

Considerable damage can be done to formwork that has not been planned for orderly dismantling. The sequence of stripping is one consideration when making or assembling forms. For example, column forms should be made so that they can be stripped without disturbing adjacent beam and girder forms. Column panels can be pried out from the bottom so that they can drop down free of beam form. Beam and girder side forms may be made to come out before slab soffits. The designer should provide crush plates or key strips to facilitate removal at difficult form intersections or where there is danger of damage from stripping tools.

Figure 4–6 Proper use of screw-type leg adjustments on shoring members makes it easier to lower forms gradually without shock to structure. Pneumatic and hydraulic jacks, sand jacks, even slowly melting blocks of ice, are also used to facilitate gradual release of forms.

Small form openings to permit introduction of air or water under pressure sometimes simplify stripping (Figs. 4-7 and 4-8).

Time devoted to training the stripping crew in both the order and method of form removal is well spent. Stripping requires considerable care on the part of workmen to avoid damage to the green concrete, which can be marred by scratching and chipping even though it has sufficient structural load-bearing strength. The contractor is concerned both about protecting the concrete and extending the useful life of his forms by careful handling. Not only must the forms hold together, but they must remain dimensionally accurate, and edges should stay in good condition to make accurate alignment and clean joints possible.

Form panels and shoring components should not be dropped but should be handed down or lowered on stretched ropes, cables, or other devices to avoid damage. Various rigs on wheels can be devised for different kinds of stripping jobs to improve the safety and speed of the workers; traveling suspended scaffolds are particularly helpful in bridge work.

Multistory Buildings

Removal of shores from multistory buildings requires special considera-

Figure 4–7 Compressed air inserted through holes at the form center aids in stripping otherwise airtight forms such as these dome pans for waffle slab construction.

tion because a given floor slab may be required to support one or two stories of live and dead loads from the work going on above, depending on the rate of progress being made on the job. The contractor's schedule of shoring and reshoring should be approved by an engineer or architect representing the owner. The engineer's approval should be based on strength of field cured test cylinders as well as consideration of weather, placing conditions, and time and quality of curing. The total load of the upper structure, including freshly placed concrete, formwork, workmen, placing equipment, runways, and motor driven buggies, must not exceed the live load that the lower structure is capable of carrying at the strength then available.

Figure 4–8 Careful stripping and handling pays off by keeping form panel corners and edges straight and true.

For a typical multistory building and proper curing conditions, three stories of shores are usually required for progress at the rate of one story per week. (See Fig. 4-9.) However, for faster rates of construction, or with structures designed for light live loads such as 40 or 50 psf, it may be that more than three floors of shoring are required to support one floor of freshly placed concrete and construction loads. Reshoring, discussed in the following sections, becomes a particularly important consideration in multistory work.

RESHORING

Reshores are shores placed firmly under a concrete slab or structural member that has just been stripped of its original formwork and has to support its own weight and construction loads posted to it. Such reshores are provided to transfer additional construction loads to other slabs or members or to impede deflection due to creep which might otherwise occur. Reshoring is done to facilitate maximum reuse of the formwork, making use of the strength of completed construction below as well as the partial development of strength in the member being reshored.

Premature reshoring and inadquate size and spacing of reshores have been responsible for a number of construction failures. Since reshoring is such a highly critical operation, it is essential that the procedure be planned in advance and be approved by the architect-engineer. Detailed instructions and prohibitions regarding reshoring may be written into the job specifications, and the time and sequence of reshoring may be made subject to engineer approval.

When the study of form removal and reshoring is being made, the live loads for which the completed structure is designed, as well as the actual strength of the partially cured concrete, must be considered. Allowance must also be made for any additional live and dead loads to be imposed as construction continues. The reshoring system must be designed to carry all loads that will be imposed. It is important to remember that partially cured concrete deflects much more than fully aged concrete, even at identical compressive strength levels, for both immediate and sustained loads. Often, the critical limitation on temporary construction loads will not be for prevention of collapse under load, but rather, for avoidance of excessive deflection, cracking, and inadequate bond for splice details of the reinforcement.

Locations, spacings, and type of reshores to be used should be considered carefully. Story heights, speed of construction, type and structural design of slabs, and spacing of columns are influencing factors. Under proper conditions, metal shores, 4×4 wood shores, or 6×6 wood shores can be economically and safely used as reshores.

All reshoring members must be straight and true without twist or warp. Reshores must be plumb and adequate in capacity. When placing reshores, care should be taken not to preload the lower floor and also not to remove the normal deflection of the slab above. The reshore is simply a strut and should be tightened only to the extent that no significant shortening will take place under load. In no case should wedging or jacking be permitted to lift a slab above its formed position to the point of causing cracking. Excessive wedging will change the load distribution on various floors supporting the shores and may easily relieve one and overload another.

Operations should be performed so that at no time will large areas of new construction be required to support combined dead and construction loads in excess of their capability as determined by design load and developed concrete strength at the time of stripping and reshoring. While reshoring is under way, no construction loads should be permitted on the new construction. For high-story levels, adequate provisions should be made for lateral bracing of reshores during this operation. Reshores should be located in the same position on each floor so that they will be continuous in their support from floor to floor. Where slabs are designed for light live loads, or on long spans where the loads on reshores are heavy, care should be taken in placing the reshores so that the loads do not cause excessive punching shear or reversed bending stress in the slab.

Reshoring should never be located where it will significantly alter the pattern of stresses determined in the structural analysis, or where it will induce tensile stresses where reinforcing bars are not provided. Where the number of reshores on a floor is reduced, such reshores shall be placed directly under a shore position on the floor above. When shores above are not directly over reshores, an analysis should be made to determine whether or not detrimental bending stresses are produced in the slab.

Reshoring Beam and Girder Construction

When stripping forms before the structure is strong enough to carry its own weight plus any construction loads above, the forms should be removed from one girder at a time, and the girder should be reshored before any other supports are removed. After the supporting girders of a bay are reshored, the beam forms in the bay should be removed one at a time, and the beam should be reshored before other supports are removed. Long-span beams, over 30 to 35 ft, should have, at least, one substantial shore that remains permanently in position during reshoring operations.

Slabs should not be reshored until supporting beams and girders have been reshored. Each slab with a clear span of 10 ft or more should be reshored along its center line at regular intervals or with continuous shoring. If a line of reshores at midspan is not in line with shores on the floor above, the slab should be checked for its capacity to withstand reversal of stresses or punching shear.

Reshoring Flat Slabs

When flat slab formwork is stripped before the slabs are strong enough to carry their own dead load plus construction loads above, shore removal and reshoring should be planned and located to avoid reversal of stresses or inducement of tension in slabs where reinforcement is not provided for design loads. Reshores should be placed along the intersection line of the column strip and middle strip in both directions. Such reshoring should be completed for each panel as it is stripped and before removing forms for adjacent panels. For flat slabs whose column spacing exceeds 25 ft, it is desirable to plan the form construction so that shores at intersections of column strips with middle strips may remain in place during stripping operations. For average flat work, 6×6 wood reshores are usually most economical and satisfactory.

The recent development and use of expandable horizontal shores with vertical shore lines at greater spacings (perhaps 14 ft instead of 2 or 3 ft) requires careful consideration of location of shores and reshores so that stresses are not reversed, particularly in flat slab areas where no negative (top) reinforcement is provided. Construction loads present a greater potential problem in this type of forming because the large spacing between shore lines gives the contractor work space and loading areas for construction materials not possible with typical vertical shores placed every 2 or 3 ft on centers each way.

Removal of Reshores

Generally, final removal of reshores is based on the same standards as removal of shores. Reshoring should not be removed until the slab or member supported has attained sufficient strength to support all loads posted to it. Removal of reshores should be planned so that the structure supported is not subjected to impact or loading eccentricities.

ACI Committee 347 recommends that in no case should reshores be removed within two days after placing a slab above or within two floors below such a new slab.

Heavy Industrial Structures

For heavy industrial structures a "permanent" shoring system is desirable. Early stripping and subsequent reshoring may allow undesirable deflections or cause development of fine cracks that in later years would create a serious maintenance problem. When quick reuse of the formwork is desired in such structures, forms should be planned so that original shores can remain in place while forms are stripped around them (Fig. 4-9). Shores can then remain in place until strength tests or elapsed time indicate that it is safe to remove them.

CARE AND STORAGE OF FORMS

Prefabricated forms and unframed plywood panels should be thoroughly cleaned and oiled as soon as possible after stripping. Frames and any

Figure 4–9 Three floors of shoring are commonly used in multistory flat slab construction. Formwork for this structure was planned so that original shores remained in place after slab soffit forms were stripped.

metal parts should be thoroughly scraped to remove any accumulated deposits of concrete. The frames should be regularly inspected for wear, and any split or damaged lumber should be replaced. After the frame has been checked, the form face should be cleaned. For wood or plywood surfaces, a hardwood wedge and a stiff fiber brush are good tools. The wedge will remove any odd lumps of concrete, and the brush will remove dust with a minimum of damage to the face. Unless used with great care, scrapers and wire brushes loosen fibers on the form face, and the "wooly" surface becomes progressively harder to strip. A hard scale of concrete may require some light tapping with a hammer, but this can be carefully done to avoid breaking fibers and damaging the form face.

Any open seams in panels should be filled, warped boards planed, metal facings straightened, and joints rematched. Plywood panel faces of prefabricated forms may be reversed on the frame or replaced if badly damaged. Tie holes may be patched with metal plates, corks, or plastic materials.

Cleaning knives, steel scrapers, and wire brushes are satisfactory for use on steel panels, and mechanical cleaning devices (Fig. 4-10) are sometimes used on large projects. Metal forms should not be sandblasted or abraded to a bright surface as this may cause sticking.

As soon as forms are cleaned and any necessary repairs made, they should be coated with a good oil or other preservative. Steel forms should be oiled on the back as well as the face to prevent rusting and sticking of spilled concrete. With some coatings, a curing or drying period is required before the forms can be stacked on one another. (See Fig. 4-11.)

Following any necessary drying period for the coating material, the forms should be stacked off the ground on 2-in. or heavier lumber at a slight pitch for runoff (unless indoor storage is available.) It is preferable to place strips of wood between the forms to promote evaporation of moisture. Panels should be stacked according to sizes and types to simplify handling; they should be arranged so that any code markings can be read without further moving. Old forms may be placed on top to protect newer ones from sun and rain.

Careful techniques recommended for stripping should carry over to handling to prevent chipping or denting edges. Forms should be piled face to face and back to back for hauling; they must not be dumped or thrown from a truck but should be passed from man to man for stacking. On-site stacking procedures for panels and components vary with the superintendent and site conditions. Orderly storage will obviously do

Figure 4–10 Gasoline powered cleaner for metal form panels. Forms are placed on a conveyor belt which pulls them under a rapidly rotating steel wire brush. Form faces are cleaned and oiled, and panels emerge at the far end of the conveyor belt.

Figure 4–11 Large size panels stacked for orderly storage at the job site. Wood strips separate groups of panels to allow for circulation of air and evaporation of moisture.

much to reduce loss and damage and to prevent panels from being used for other purposes.

Dimension lumber, as well as form panels, may be saved for reuse, in which case it should be scraped free of concrete deposits, and all nails removed. It should be sorted by sizes and stored off the ground in a location that will minimize weathering and rotting.

Reusable form hardware requires careful handling or it will become a costly item through losses or deterioration. Boxes or buckets should be kept on hand to store small parts as they are stripped. Parts can be soaked in a special solvent to loosen hardened concrete and afterwards brushed. Some form coatings in concentrated form will soften the buildup of hardened concrete. Mechanical brushing is sometimes used where large numbers of parts are to be handled.

INSULATION FOR COLD WEATHER

Full recommendations for winter concreting practices established by ACI Committee 306 include the following statement regarding the value and method of form insulation.

"Arrangements for covering, insulating, or housing newly placed concrete should be made in advance of placement and should be adequate to maintain in all parts of the concrete the temperature and moisture conditions recommended.

"Since during the first 3 days requiring protection most of the heat of hydration of the hardening cement is developed, no heat from outside sources is required to maintain concrete at correct temperatures if heat generated in the concrete is suitably conserved. This heat may be conserved by use of insulating blankets on unformed surfaces and by insulated forms where repeated reuse of forms is possible. Temperature records will reveal the effectiveness of different amounts or kinds of insulation or of other methods of protection for various types of concrete work under different weather conditions. Appropriate modifications and selections can be made accordingly. Methods for estimating temperatures maintainable by various insulation arrangements under given weather conditions have been published. . . . The amount of insulation necessary for good protection can be determined from information shown [see Table 4-B] for various kinds of concrete work and for several degrees of expected severity in weather. For successful use, and efficient reuse, commercial blanket or bat insulation must be adequately protected by means of tough, moistureproof cover material from wind, and rain, snow, or other wetting. Moreover, it must be kept in close contact with concrete or form surfaces to be effective."

Several types of insulation material are suitable for or specially produced for formwork; among these are a spray-on type used largely for steel forms, a foamed polystyrene and polyurethane board that can be cut to fit between studs of vertical forms, and various kinds of wood and mineral bat or blanket insulation. These materials must be kept dry to maintain insulating values shown in Table 4-B. Prefabricated form panels

Table 4–B

INSULATION REQUIREMENTS FOR CONCRETE WALLS AND FLOOR SLABS ABOVE GROUND (Concrete placed at 50° F)

Wall thickness, ft	Minimum air temperature allowable for these thicknesses of commercial blanket or bat insulation, °F[a]			
	0.5 in.	1.0 in.	1.5 in.	2.0 in.
Cement Content (300 lb per cu yd)				
0.5	47	41	33	28
1.0	41	29	17	5
1.5	35	19	0	−17
2.0	34	14	− 9	−29
3.0	31	8	−15	−35
4.0	30	6	−18	−39
5.0	30	5	−21	−43

Table 4–B (continued)

Cement Content (400 lb per cu yd)

0.5	46	38	28	21
1.0	38	22	6	−11
1.5	31	8	−16	−39
2.0	28	2	−26	−53
3.0	25	− 6	−36	
4.0	23	− 8	−41	
5.0	23	−10	−45	

Cement Content (500 lb per cu yd)

0.5	45	35	22	14
1.0	35	15	− 5	−26
1.5	27	− 3	−33	−65
2.0	23	−10	−50	
3.0	18	−20		
4.0	17	−23		
5.0	16	−25		

Cement Content (600 lb per cu yd)

0.5	44	32	16	6
1.0	32	8	−16	−41
1.5	21	−14	−50	−89
2.0	18	−22		
3.0	12	−34		
4.0	11	−38		
5.0	10	−40		

Insulation Equivalents

Insulating Material	Equivalent Thickness of Commercial Blanket or Bat Insulation
1 in. cellular polyurethane foam	1.667
1 in. of commercial blanket or bat insulation	1.000
1 in. of loose fill insulation of fibrous type	1.000
1 in. cellular polystyrene foam	1.000
1 in. of insulating board	0.758
1 in. of sawdust	0.610
1 in. (nominal) of lumber	0.333
1 in. of dead-air space (vertical)	0.234
1 in. of damp sand	0.023

[a] *Notes to Table 4–B.* The tables are calculated for the stated thicknesses of blanket-type insulation with an assumed conductivity of 0.25 Btu per hr per sq ft for a thermal gradient of 1 °F per in. The values given are for still air conditions and will not be realized where air infiltration due to wind occurs. Close-packed straw under canvas may be considered a loose-fill type if wind is kept out of the straw. The insulating value of a dead-air space greater than about ½ in. thick does not change greatly with increasing thickness. Handbooks or manufacturers' test data should be consulted for more detailed data on insulations.

are now available with insulation sandwiched between two plywood faces or permanently attached to the outer face. Electric heating blankets have also been used to maintain desired concrete temperature during cold weather. (See Figs. 4-12 and 4-13.)

The several kinds of wood and mineral "wool" insulating bats for form-work come in 1- and 2-in. thicknesses with widths designed to fit between studs spaced 12, 16, or 24 in. apart. The insulation itself is about 1 in. less than these widths, and the outer casing material has reinforced flanges for nailing the bats to the studs. The outer covering or encase-ment may be made of polyethylene plastic, asphalt-impregnated paper, or a plastic-paper laminate, each meeting the general requirements of ACI Committee 306 for weather resistance.

Figure 4–12 Forms for a large bridge pier insulated with foamed polystyrene plastic.

Figure 4–13 Insulation blanket designed primarily for use on slab forms has been wrapped around column forms and tied in place.

This insulation may be stapled or attached with batten strips to sides of the form framing. The ends of the blanket should be sealed by removing a portion of the blanket and bringing the casing material together, then stapling or battening down the ends to form headers to exclude air and moisture. Corners and angles of forms should be well insulated with the material held in place by battens, staples, or tie wires (Fig. 4-14). For steel forms, the insulating blanket can be applied tight and held in place by wedging wood battens or by securely tying.

Where practical, the insulation or insulated form should overlay any previously placed cold concrete by at least 1 ft. Where tie rods extend through the insulated form, a 6×6-in. plywood washer can be placed on top of the insulation blanket and fastened securely.

Avoiding Thermal Shock

The need to limit excessive or rapid temperature changes before the strength of the concrete has developed sufficiently to resist temperature stresses is a factor that must be considered when planning form removal in cold weather. The following statement is by ACI Committee 306:

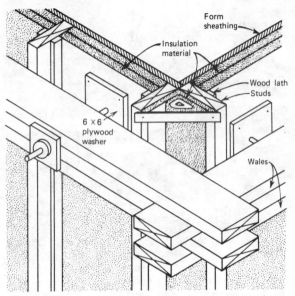

Figure 4–14 Typical method of attaching insulation bat to vertical form, showing corner detail and plywood washers used around tie rods.

"Winter concreting practices must be adequate to limit rapid temperature changes, particularly before strength has developed sufficiently to withstand temperature stresses. Sudden chilling of concrete surfaces or exterior members in relation to interior structure can promote cracking to the detriment of strength and durability. At the end of the required period, protection should be removed in such a manner that the drop in temperature of any portion of the concrete will be gradual and will not exceed, in 24 hours, the amounts shown below:

For very thin sections with ¾-in. maximum aggregate, 50° F.
For thin sections with 1½-in. maximum aggregate, 40° F.
Moderately massive sections, 30°F.
Massive sections, 20° F.

SECTION TWO
HANDLING CONCRETE

CHAPTER 5
TRANSPORTING

HISTORY OF READY MIX TRUCKS

Earliest Development

As early as 1909 in the progressive little town of Sheridan, Wyoming, concrete was being delivered by a horse-drawn mixer. This was probably an impressive sight and at that time very innovative, but it would be by today's standards, hardly an efficient way to transport concrete. The mixer was 2½ ft square and had a capacity of ¼ yd of concrete. All the ingredients—cement, water, aggregates—were put into the mixer while paddles, turned by the wheels of the vehicle, mixed the concrete en route to the job. A small water tank was attached.

Shortly after that time some mixers were mounted on Ford trucks. Materials had to be hauled along with the mixers.

In 1916, Stephen Stepanian of Columbus, Ohio, designed a self-discharging, motorized transit mixer that was basically similar to a modern ready mixed concrete truck. His patent application was denied, but his contribution to the ready mix industry was recognized in later years.

The 1920s

One serious handicap to developing transit mixers was the poor quality of motor trucks in the 1920s. A drum full of concrete overloaded the old trucks and the extra power needed to rotate the drum caused the trucks to overheat and stall.

In 1926, a Barrymore Mixer delivered concrete for the Children's Hospital in San Francisco. This mixer was developed from a dump body that had high sides and a horizontal shaft equipped with paddles to keep central-mixed concrete agitated. This was not a truck mixer as we know it today, but an agitator.

The Paris Mixer was developed in 1926 by the Paris Mixer Company in Seattle, Washington. It was the first horizontal, revolving drum, truck mixer to be regularly produced. In 1928, Transit Mixer Company purchased the manufacturing rights to the Paris Mixer and marketed the equipment in the eastern half of the United States. (See Figs. 5-1 to 5-4.)

The 1930s

In 1930, the Jaeger Machine Company entered the truck mixer field and in 1939, offered a high-discharge mixer truck. The Chain Belt Company of Milwaukee had been offering horizontal truck mixers since 1920, and in 1936, Blaw Knox entered the field. T. L. Smith Company and Rex Chainbelt, both of Milwaukee, introduced high-discharge truck mixers in 1939. (See Figs. 5-5 to 5-7.)

Figure 5–1   Mixer mounted on 1920 Ford truck.

Figure 5–2   Mixer with chute on 1920 Ford truck.

Figure 5–3   1928 model Barrymore mixer.

Figure 5–4   1928 Paris Mixer.

Figure 5–5   Chain belt model of the 1930's.

Figure 5–6   Chain belt model.

The 1940s

With heavier trucks and better engines, mixing drum capacities were increased. High-discharge drums were developed to a greater degree, and hydraulic jacks were employed to adjust horizontal drum heights. High-discharge mixers with 3, 4, and 5 cu yd capacity enabled producers

Figure 5–7   T. L. Smith high-discharge style—about 1939.

to expand at a spectacular rate in the early 1940s when demands for ready mixed concrete were astronomical because of defense construction. A decade earlier, it would have been impossible to supply the concrete.

With the end of World War II, returning servicemen created a demand for housing, and the resulting building boom further spurred the industry to develop new and better equipment.

The trend in recent years has been towards greater efficiency in materials handling and transporting equipment. Trailer-mounted mixers are in use where traffic conditions permit. Portable batch plants can be set up close to a building site and moved along with construction progress. (See Figs. 5-8 and 5-9.)

MODERN READY MIX TRUCKS

A ready mix producer now can choose from a variety of trucks depending on his budget and his needs. The sizes of present-day construction require much larger mixers. Fortunately, powerful engines have been developed that can handle modern, heavy equipment.

Figure 5–8 High-discharge type—about 1941.

Figure 5–9   Rex Roto Paver. Hydraulic jacks at rear provide adjustable heights.

Drums

Two types of truck mixers and agitators are:

horizontal axis revolving-drum type (Fig. 5-10).
inclined axis revolving-drum type (Fig. 5-11).

Maximum sizes for trucks used as mixers and as agitators are set forth in Table 5-A.

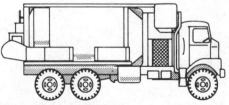

Figure 5–10   Horizontal axis, revolving-drum type ready mix truck.

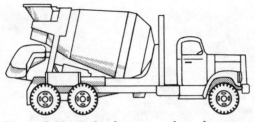

Figure 5–11 Inclined axis, revolving-drum type ready mix truck.

Mixer capacity must be stated in cubic yards of mixed concrete. Manufacturers must guarantee their trucks to mix at rated capacity. As agitators, the capacities can be higher than that specified for mixing, in which case the capacity must show on the manufacturer's data plate. The Truck Mixer and Agitator Standards of the Truck Mixer Manufacturers Bureau establish the maximum capacity of a truck drum, both as a mixer and as an agitator. Maximum mixer capacity is generally between 60 and 63% of the gross volume of the drum. The maximum capacity as an agitator is generally set at 80% of gross drum volume. However, most manufacturers recommend lower agitator ratings to prevent spillage, particularly when the truck is operated with high slump concrete in hilly terrain.

Table 5–A
TRUCK MIXER AND AGITATOR SIZES

Maximum Capacity in Cubic Yards of Mixed Concrete When Used as:	
Mixer	Agitator
6	7¾
6½	8½
7	9¼
7½	9¾
8	10½
8½	11¼
9	12
10	13¼
11	14¾
12	16
13	17½
14	19
15	20¼
16	21¾

Mixing speed for revolving-drum truck mixers is generally in the range from 4 to 18 rpm.

Agitating speed is from 2 to 6 rpm of the drum.

The Truck Mixer Manufacturers Bureau furnishes a rating plate for each ready mix truck manufactured. Before the truck can be rated, drawings of the drum with all the dimensions must be submitted to the bureau. Any changes must be submitted also. The plate must be attached to the mixing and agitating unit. Each manufacturer must attach a plate of his own design stating actual capacity as an agitator, and the minimum and maximum mixing and agitating speeds. He may want to add serial number, weight, and patent numbers.

The shape and blade arrangement of drum mixers cause the concrete to move from end to end parallel to the axis of rotation of the drum. Concrete should roll or fold over on itself. To do this the drum turns clockwise when observed from the back. To discharge the concrete at the end of the mixing cycle, the drum is reversed causing the concrete to climb back up and out of the drum. This is due to the spiral placement of the drum blades.

Drums are run by a chain-belt system with power usually taken from the truck engine or from the truck flywheel. They are either mechanically or hydraulically driven. Many of the trucks can be custom fitted with the power source the buyer prefers. Some units are equipped with separate gas or diesel engines to drive the mixer.

Water Tanks

Some truck mixers have no water tank or water system, but most have a flush water-tank system or a mix and flush system.

The waterless arrangement is usually used when the concrete is to be completely mixed at the central plant and just transported to the job site in the mixer. A few ready mix producers, prefer that there be no water system on the truck so the construction people will not be tempted to add more water to the mix.

A water tank holding flush water only is designed to allow the operator to clean the mixer after a load of concrete has been discharged.

The third optional arrangement has a two-compartment tank, one for mix water and another for flush water. The total capacity of the tank is usually not more than 50 gal per cu yd of maximum mixer capacity. The tank is equipped with a measuring device and a sight gauge on both compartments.

Water for mixing is pumped or delivered into the batch at the head section of the drum, or by dual injection into the head and discharge sections of the drum. The pump should be able to deliver not less than 45 gal of water per minute into the drum.

Charging

Most truck mixers are equipped with a charging hopper at the higher end of the drum. It is large enough to position easily under the batching plant. The hopper receives the materials as they drop and feeds them into the drum. The proper loading or charging of water, cement, aggregates, and admixtures is an important factor in maintaining batch consistency for each truck load.

Other mixers, including the horizontal axis type, have removable hatches that are positioned on the top when batching. The drum locks into place with the hatch at the top. In case the mix sets up too fast, the hatch can be positioned at the bottom of the drum so the concrete can be ejected easily. (Also, a man can enter the drum through this hatch to inspect and clean the interior.)

Discharge Chutes

Discharge is accomplished quickly without spilling by means of extendible chutes at the rear of the truck. The trend is toward wide, deep chutes which allow rapid discharge. Most trucks are equipped with chutes that can be rotated approximately 180°.

Revolution Counters

There are two basic types of revolution counters in current use. One is a counter that records only the total number of drum revolutions. The other can: (1) record only the number of revolutions at mixing, or (2) record the number of revolutions at mixing speed and also the total number of drum revolutions.

The first type is practically foolproof as well as being inexpensive, but some state highway departments feel it does not accurately tell true mixing time, since it records agitation speeds and mixing speeds without differentiation.

A timer is installed on some of the counters so that the mixing stops as soon as the revolution counter is turned off by the timer. It can only be restarted by resetting the entire unit. When the truck returns to the plant, the inspector can tell if the operation of the counter has been interrupted by checking the clock on the timer.

Special Mixers

Truck mixer manufacturers are constantly improving their products with adaptations that make possible increased capacity, quicker discharging, and more uniform mixing of concrete. Some of the innovations include a tilting mixer which has a drum that can be raised hydraulically to nearly horizontal for maximum discharge speed. This enables rapid discharge of low-slump concrete, while ordinary concrete discharges as fast as 5 sec per yd. This mixer can receive dry batch concrete quickly and mix it throughly in a very short time.

Another trend is to spread the load over a larger area by adding another axle. A truck that normally has a capacity of 7 to 8 yd can thus carry 2 extra yards without violating highway load laws. This is done by putting the charging and discharging hopper on wheels that are lifted and dropped hydraulically. When on the road, the wheels carry the load of the hopper about 6 ft behind the mixer. At the job site the hopper is raised, putting more weight on the drive axles and providing better traction and maneuverability. The operation of this mechanism is done from the cab of the truck. An inexpensive way to get additional payload is to add either tag axles or pusher axles to a standard tandem.

Trailer-Mounted Truck Mixers

To accommodate larger loads, mixers have been mounted on trailers. This is another way to increase the payload on a truck without breaking highway load laws. The laws base the load of any hauling unit on how the load is distributed over the vehicle's axles and wheels. Trailers spread the load over the longest possible distance.

An option on some trailers is a hydraulic-powered assist to the rear axle. When the trailer gets into muddy or sandy places, power in the rear axle sometimes makes the difference between getting out and being mired.

Cleaning Ready Mix Trucks

Ready mix trucks are huge investments, and conscientious maintenance will prolong their life considerably. Also, they are a means of advertising for the ready mix company. A clean, bright truck makes a much better impression than one that is poorly cared for. Preventing dirt, grime, and hardened concrete from sticking to ready mix trucks by daily cleaning is the easiest and least expensive type of maintenance. There are, at least, three ways of doing this.

Ready mix producers have different ways of cleaning with water, all of which are good if they do the job. Some hose down the back end of the truck, the chute, and the inside of the drum after discharging each load. From time to time, the trucks must be thoroughly cleaned, but this is not difficult when they are kept comparatively clean after each load.

Some ready mix companies do not feel that this is enough to prevent the buildup of concrete inside the drum, so they wash the truck down again every evening when it is returned to the yard. A mild solution of muriatic acid sprayed on the mixer and allowed to stand for 15 min before being rinsed off with water cleans most of the old concrete off. Then the drum is wiped with clean dry rags, and the truck again looks like new.

Commercial detergent powder will help to remove dirt and oil, but it has no effect on hardened paste or concrete. Some ready mix yards have jet sprays through which their trucks are driven each night. Concrete should not be allowed to build up to any degree; but if and when it does, it can be removed with a mason's hammer or a chisel. Drums should be inspected daily so that any concrete can be removed while it is still fairly soft.

Some companies wash their trucks and then apply a coating of wax or light oil mixed with kerosene. Application is either by hand or sprays wherever concrete tends to build up. But the compound can cause rubber parts to rot, so hand application is preferable. Daily application of oil to vulnerable parts of the truck makes cold-water washing at the end of each day effective.

Many products are on the market that claim to dissolve or remove concrete and to inhibit bond so concrete is easily cleaned off. When concrete has built up to any extent, many applications of any compound are needed to remove it, and the compounds are expensive. They seem to be most effective when used every day, followed by a water washing. Daily preventive maintenance consisting of washing and coating with light oil or a proprietary material is the real key to long service from this expensive equipment.

PORTABLE CENTRAL-MIXING PLANTS
FOR HIGHWAY PAVING

More and more concrete is being mixed in portable central-mixing plants and transported to the paving site in large capacity, quick-dump, non-agitating or agitating trucks. Portable plants are used mainly where distance makes transporting ready mixed concrete from stationary plants impractical. Development of highly portable, automated batching and mixing equipment capable of turning out high-quality concrete has made this possible.

When in travel position, the total height of a portable mixer, including wheels, does not exceed 14 ft, making passage under bridge viaducts possible. It is not unusual for a portable mixer to measure 70 ft or more in length when dismantled for transit.

There are even self-erecting mobile plants, complete with permanently mounted wheels for towing to a job. They can be quickly dismantled and re-erected with a minimum of down-time and erection cost. Many are self-erecting or require less than 8 hr to erect. They can be used for either dry batch or central mix.

Mixing time on projects involving central-mix paving has been reduced to 60 sec or less without any loss in quality of the concrete. Studies made recently by the Bureau of Public Roads reveal that mixing times as low as 30 sec yielded uniform quality concrete on well-adjusted plants. This means that timing of each ingredient during charging of materials into the mixer must be well controlled. Blending must be thorough. Poorly adjusted equipment requires additional mixing time to compensate for deficiencies in charging and blending (Fig. 5-12).

Figure 5-12   Portable central-mixing plant.

Automation and computer technology have helped make the central-mixing plant a reality. A single operator at a control panel (which can be installed in the comfort of an air-conditioned trailer) performs all functions of charging, mixing, and discharging. This automation increases capacity and usually ensures quality control of the concrete.

A slump meter is available with portable mixers. It measures the power required to turn the mixer and thereby indicates the stiffness of the mix. It is calibrated with slump tests and gives the operator an immediate reliable check on the slump of the concrete.

Push-button controls activate electronic and hydraulic weighing and recording equipment. An automatic moisture meter in the sand hopper maintains a continual record of the moisture content to maintain uniformity (Fig. 5-13.)

Figure 5–13 The control console for a central-mixing plant.

Capacities of portable mixers can be as high as 720 cu yd per hr. However, 250 to 350 cu yd per hr is more common for the road paving industry.

Haul Units

The use of central-mix concrete for paving has created a need for haul units to carry the mixed concrete from the plant to the construction site. With low-slump, air-entrained concrete, this is possible with no loss in the quality of the concrete.

Central-mix concrete is often hauled to the building site in trucks with special features that inhibit segregation and make discharging at the site easy. They are manufactured as both agitating and nonagitating types.

Federal specifications for ready mixed concrete (SS-C-618a) specify that:

> "Central-plant-mixed concrete may be transported in nonagitating equipment only when provision for such method of transportation is made in the contract. When the use of nonagitating equipment is so permitted, the following limitations shall apply:
>
> (a) Bodies of nonagitating equipment shall be smooth, watertight, metal containers equipped with gates that will permit control of the discharge of the concrete. Watertight covers shall be provided for protection against the weather when required.
>
> (b) The concrete shall be delivered to the site of the work in a thoroughly mixed and uniform mass and discharged with a satisfactory degree of uniformity as prescribed in (c). Discharge shall be completed within 30 min. after the introduction of the mixing water to the cement and aggregate.
>
> (c) Slump tests of individual samples taken at approximately the beginning, the midpoint and end of the load during discharge shall not differ by more than 1 in. for a specified slump of 3 in. or less and 2 in. for a specified slump greater than 3 in."

Side-dump bodies permit the dump truck to be brought up along the forms without any backing or maneuvering. The driver can see where his concrete is going and control it without getting out of the cab of the truck.

The body is made of smooth high-tensile steel. The rounded bottom allows all of the concrete to be discharged rapidly with minimum segregation. The smooth body is easily cleaned after each haul or at the end of the day.

The body can be raised and lowered quickly by hydraulic controls. A gate operated by hydraulic cylinder opens the full length of the body.

This equipment is mounted on a standard truck chassis and delivers 6 cu yd (12 tons). It can, of course, be used for hauling aggregate, sand, gravel, or other materials. Units mounted on trailers can deliver up to 12 cu yd of concrete in one haul (Fig. 5-14).

The high-tensile steel body of a rear-dump truck is shaped like a bathtub, with rounded corners to prevent buildup. The body can be tipped vertically for rapid dumping. All of the load can be discharged in 30 sec. A discharge chute that pivots 180° assists in discharging concrete into difficult-to-reach places.

Standard dump trucks have higher capacities than either of the above dump bodies. They may be used on slip-form paving jobs where the spreader can accommodate a large load of concrete. They are less expensive, do a very satisfactory job, and have the advantage of being multipurpose units. Dumping is from the rear of the truck. Special rounded corners can be placed in the truck body to keep concrete from lodging in corners.

Helicopters. In situations where trucks cannot be used, helicopters have been of value. Equipped with buckets, they can haul 5 tons of concrete per hour for short distances. This is an expensive way to transport concrete but, in special situations, it can save enough time and labor to make it less expensive than any other method.

Figure 5-14 Side-dump, nonagitating concrete truck.

Concrete-Mobil

A fairly recent development in concrete transporting equipment is a truck-mounted batch plant and mixer. This unit is capable of delivering separate materials as well as concrete mixed to any specifications. The unit is called the concrete-mobile. One man can turn out concrete continuously or intermittently at a rate of 1 cu yd every 3½ to 4 min.

The equipment is available in 5 cu yd and 8 cu yd sizes, mounted on a truck or trailer. When mounted on a truck, the operating power is supplied by a power takeoff. Trailer-mounted units carry their own power plant so the tractor can be removed to be used elsewhere while the concrete is being batched.

A large bin divided into two parts carries the sand and coarse aggregate. Separated from this large bin is a smaller bin holding the cement. Two tanks carry the water.

Batching is controlled by feeders attached to the cement and aggregate bins. The water is measured by a flow meter. When the equipment arrives at the job site, the operator sets his controls for the required mix design. The machine automatically proportions the materials. Fresh concrete is mixed and delivered by a screw in less than 20 sec. Concrete is deposited through a chute.

Since the proportions can be changed at any time to produce concrete of varying strengths, this equipment is especially useful for small jobs such as curb or backup work.

The equipment does not have to return to the plant between jobs unless recharging is necessary. Furthermore, it can be fully charged the night before and can be ready to go first thing in the morning. A very few minutes are needed to clean it out with the water pressure hose carried on the equipment.

CHAPTER 6
PLACING

EQUIPMENT

Concrete can be handled satisfactorily by several methods: wheelbarrows, chutes, buggies, hoists, cranes, buckets, conveyors, and pumps. A portion of the differences in cost for various jobs are due to differences in handling procedures. Undoubtedly, the most convenient way to handle concrete is to permit ready mix trucks to discharge directly into forms. It is worthwhile to spend a little time preparing the site when no road or driveway is adjacent to the forms so that the trucks can drive up to them.

However, the size or height of the project or the condition of the ground often makes it impossible for direct discharge of ready mixed concrete. When this is the case, another method of transporting the concrete from truck to forms must be determined. Several considerations will influence the choice of technique.

The two primary considerations in selecting concrete placing equipment are (1) job specifications, and (2) job conditions. Job specifications include such factors as type and size of aggregates, consistency of the mix, permissible time from mixing to final placing, and the allowable distance of free fall. The job conditions have to do with methods of batching, size of the mixers, types of forms, and hauling distance.

Good handling procedure is governed essentially by the need to prevent or minimize segregation and the need to deliver concrete of maximum uniformity to the formwork.

Segregation of the mix must be prevented at all stages of handling. Any type of transportation that causes part of the concrete to move faster than adjacent concrete will tend to cause segregation. Jolting of the mix causes the large, heavy particles of aggregate to sink to the bottom and water and fine aggregate to rise to the top, resulting in a high water-cement ratio at the top surface. In floors this usually causes dusting, crazing, and spalling. Segregation is aggravated when concrete is dropped from too high an elevation or at an angle. This problem is most acute with high-slump mixes and those containing low amounts of cement or high concentrations of large aggregates. (See Figs. 6-1 to 6-3.)

Handling procedures should be such that the mix quality and finishing characteristics will not be unfavorably affected. No equipment should be used that results in detrimental changes in mix proportions so that compaction by normal vibration is no longer practical. Loss of slump may lead to a need for retempering with additional water and a resultant reduction in concrete strength and other undesirable characteristics.

Loss of material, particularly cement paste, must be avoided. Equipment must be tight and must allow clean transfer of the mix. Paste loss can reduce a good workable mix to one that is harsh and difficult to place.

Capacity of any delivery system must be great enough to prevent the formation of cold joints (a layer of concrete that hardens before the next batch is placed against it, causing poor bonding). The mix should be deposited in horizontal layers not more than 2 ft deep. Each layer should be delivered while the layer below is still plastic so that the two can be vibrated to give integrity to the mix.

Hoppers

Concrete collection hoppers furnish an uninterrupted supply of concrete from the mixer to the handling equipment. Hoppers should provide a vertical drop at the discharge gate. If discharge is at an angle, the coarse aggregate will fall to the far side of the container being loaded and the fine material will fall to the near side, causing segregation. Center discharge is recommended. Discharge gates should be of the nonjam type with a positive-action, spring-loaded linkage. Hoppers can be mounted on wheels or skids to simplify movement.

Collection hoppers are available in three basic designs:
1. One side open to receive concrete from push buggies or power buggies.

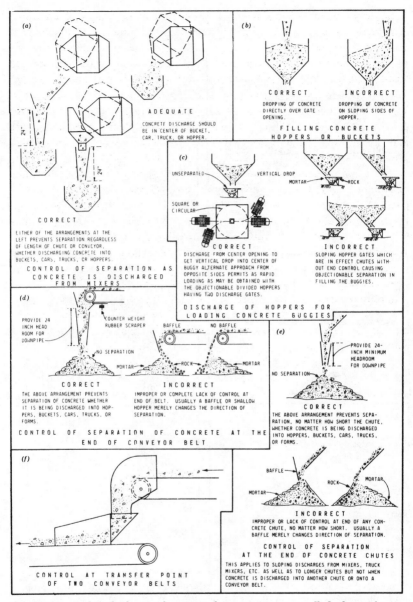

Figure 6–1 Unless discharge of concrete from mixers is controlled, the uniformity resulting from effective mixing will be destroyed by separation. (Illustrations courtesy of the *ACI Journal*, July 1972.)

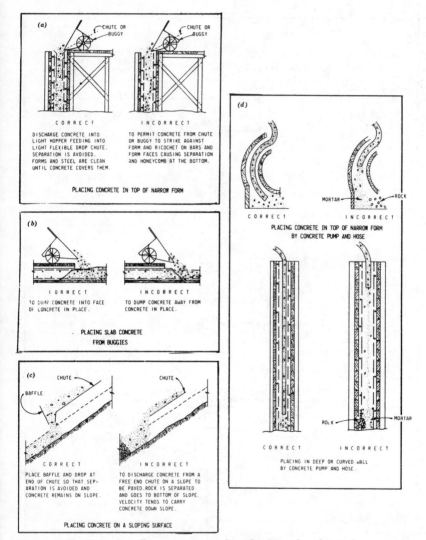

Figure 6-2 Concrete will separate seriously unless introduced into forms properly. (Illustrations courtesy of the *ACI Journal*, July 1972).

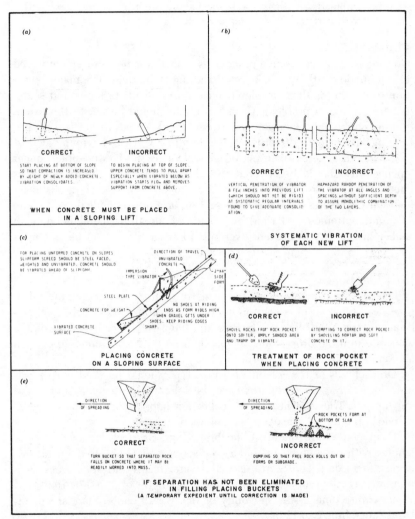

Figure 6–3 Correct and incorrect methods of consolidation. (Illustrations courtesy of the *ACI Journal*, July 1972).

2. Four sides for charging with normal-slump concrete by concrete bucket.
3. Extra-steep slopes, rectangular top, and large oval discharge opening, in combination with an oval-to-round transition drop chute section, to handle exrtemely low-slump concrete.

Chutes

Chuting is not as common a practice as it once was but is often employed in conjunction with other concrete handling techniques. The major objection to chuting is that low-slump mixes are difficult to move and segregation becomes a problem. Usually a chute should be flat to the extent that its horizontal dimension is between 2 and 3 times its vertical dimension. A baffle should direct the concrete into a vertical downpipe at least 2 ft long. In general, open chutes should be as short as possible; longer open chutes should be covered to reduce drying and slump loss. This method is now ordinarily reserved for moving concrete short distances and with concretes having slumps of 3 in. or more. The ideal open chute is round bottomed and rigid and made of metal, or metal lined, with sufficient capacity to prevent overflow. Flow should always be as a mass and not as a thin stream that can dry out.

Drop Chutes

When placing concrete in deep and narrow forms, it is more efficient to use a combination of collection hoppers and drop chutes. This combination permits the collection hopper to channel the vertical flow of the concrete into the drop chute, which confines it and prevents segregation as concrete moves into the form.

Drop chutes of the "elephant-trunk" type should be used where falls into formwork are greater than can be tolerated with open chutes and downpipes. They are fitted with collection hoppers or "elephant heads" of one of the three basic designs mentioned above.

Drop chutes are made of flexible rubber or steel sections. Rubber chutes with rectangular or round connection collars are used for narrow forms. Steel drop chutes are used for wider forms and open areas. Sections are usually round and are supplied in lengths of 2 to 3 ft. Hook and chain connections assure flexibility of the overall chute. Standard diameters range from 6 to 12 in.

Placing concrete by drop chute is normally restricted to distances of less than 25 ft. Segregation is usually a danger when drops exceed this limit. A special lightweight "elephant trunk" has been developed to overcome this limitation. The trunk is made of very thin reinforced rubber which, when suspended, does not round out like a conventional chute but collapses under its own weight and acts as a brake to slow the falling concrete. Such collapsible chutes have prevented segregation during 80-ft drops to hoppers at the bottom of a foundation.

Wheelbarrows

Wheelbarrows long served as the standard materials handling unit in construction. Although more efficient specialized equipment has come into wide use, the wheelbarrow still accounts for much of the movement of concrete on job sites. The conventional wheelbarrow is now used mainly for short hauls and relatively small amounts of materials. Depending on the strength of the workman, the usual maximum hauling distance of a wheelbarrow is 200 ft, and from 1½ to 3 cu yd can be placed per hour. Wheelbarrows should have pneumatic or cushion tires to minimize segregation by jolting. Runways should be as smooth as possible. The ends of planks forming a runway should butt and not overlap. Sharp turns and congestion should be avoided to assure productivity.

Buggies

Buggies or carts are, in effect, a variation on the wheelbarrow theme. Since they have two wheels and a lower center of gravity, buggies are more stable; they can also carry more material than wheelbarrows. A rocker attachment permits steeper dumping when handling stiff mixes. Rubber-tired buggies hold from 6 to 8 cu ft and are serviceable for hauls up to 200 ft. From 3 to 5 cu yd of concrete may be placed per hour with buggies.

Power Buggies

Power buggies are available in sizes from 9 to 12 cu ft and are practical for hauls up to 1000 ft. A rider-type diesel- or gasoline-powered buggy offers high hourly handling capacities. From 15 to 20 cu yd can be placed per hour on 600-ft flat hauls. With ramps they can be used to place concrete for second- and third-story work. Ramps should be at least 5 ft wide and solidly constructed to withstand the heavy loads and shock to which they are subjected (Fig. 6-4.)

SPECIAL EQUIPMENT FOR HIGHER STRUCTURES

The higher a structure, the more expensive concrete handling becomes. When concrete is moved above first-story levels, choice of equipment

Figure 6–4   Power buggies can carry from 9 to 12 cu. ft. of concrete.

depends on quick, easy erection and dismantling, and on good delivery rates of the equipment selected.

Booms

Hand guided booms or gantries are most suitable for use with a site mixer where relatively small quantities of concrete are handled over a scattered area. However, in recent years the use of ready mixed concrete, rapid buggy distribution, and greater availability of cranes has made boom handling uneconomical.

Another system uses a cantilevered boom centered on a turntable. The bucket can be fed from the outside of the delivery circle at any desired point. A big disadvantage of any boom system is the need for a relatively level site. Effectiveness is reduced when the mix must be placed at different levels. Power controlled hydraulic pumping booms are available for reaching about 100-ft heights. Concrete delivery lines are permanently incorporated into the boom.

Loaders

Elevator loaders are, in effect, power buggies that have the added ability of lifting the concrete hopper vertically. The use of elevator loaders is limited to no higher than third-story work. These loaders are speedy and eliminate the need for ramps. Models with tilting upper decks permit safe and level concrete handling on uneven ground. Swinging axles maintain a level load even if one wheel should be sunk in an excavation. Power steering is desirable to reduce operator fatigue.

Handling attachments for most loaders have a maximum capacity of 20 cu ft. Both hoppers and dumping buckets are available. They can be manually operated at dumping level or controlled from the operator's seat. Remote-controlled buckets are also available.

Hoists

Fixed hoists are most frequently used for transporting concrete for high-rise buildings and towers. The height of the jump tower is practically limitless, and the vetrical ascent of the hopper, when properly powered, is extremely rapid. A tower system usually consists of a central hopper at ground level, the tower itself, an elevator bucket, and a mobile hopper at the level where the concrete is to be placed. The concrete can then be transferred to hand or power buggies for distribution.

There are two basic types of elevator towers. With one, the concrete bucket is hoisted within the well of the structural frame. With the other, the bucket is external to the frame. Double-well towers are available for very large jobs.

Hoists offer lifting capacities of 3000 lb and greater. Hopper and bucket capacities are in the 1- to 2-cu yd range. Typical traveling speed is 150 ft per min when lifting and 200 ft per min when lowering.

Buckets and Cranes

Crane and bucket transport is an efficient means of handling concrete for most structural work. The mix can be delivered directly from supply point to forms in a single operation without intermediate handling. Small and large quantities can be handled with equal efficiency over almost any height. It is common practice to use three buckets per crane to avoid delay and to permit the greatest number of passes per crane in any given period of time. Choice of type of crane is largely dependent on the desired lifting capacity and on the required reach and degree of mobility. Buckets are available in various shapes and sizes (Fig. 6-5.)

The general-purpose round bucket is usually lightweight, with center discharge, and has a self-closing, controllable double-clamshell gate for handling normal-slump concrete. Sizes are usually from $\frac{1}{3}$ to 4 cu yd. Large, special buckets may handle as much as 12 cu yd. They are usually handled by truck or mobile cranes and are used on the construction of buildings, bridges, and other general structural concrete projects.

Figure 6–5 It is common practice to use three buckets per crane to avoid any delay in filling time. An empty bucket is left for filling while the full one is lifted to the forms.

Heavy-duty, low-slump buckets are designed for rough handling. They have steep side slopes and extra-large double-clamshell gates. They are used frequently on large dams and other projects that involve mass concrete in large volume. Sizes range from 1 to 12 cu yd. They easily handle dry or extremely low-slump mixes with aggregates up to 6-in. maximum size.

Laydown or rollover buckets permit convenient low-height loading from truck mixers or other sources. They are loaded in the horizontal position and shifted to the vertical position on lifting for transport to the discharge point. These types are available as both lightweight models for normal-slump concrete and heavy-duty models for low-slump concrete. The buckets usually have controllable center discharge and double-clamshell gates and come in sizes from ¾ to 5 cu yd.

Special concrete buckets of various designs have been developed to meet particular job requirements. Featherweight, adjustable-capacity buckets have been developed for helicopter and tower crane use. There are also underwater buckets with air or two-line gate control, with or without lids, buckets designed for handling with forklift trucks, and many other designs.

Attachments and accessories have been designed to meet particular job needs. For example, there is the accordion-type rubber collection hopper that is attached directly to the gate of the bucket, permitting discharge into walls or fixed hoppers. This accessory folds under the bucket when it is set set down to be recharged. There is also the suspended steel subhopper with long sections of rubber "elephant trunk," used for deeper, controlled placing.

Fork Lift and High Lift Loaders

Current models of fork lift and high lift loaders are capable of lifting concrete about 40 ft when a stable frame is supplied. Both storage hopper and buckets with capacities up to 20 cu ft are available. This equipment can be manually operated at the dumping level or can be controlled by the lift operator.

Tower Cranes

Self-climbing tower cranes are used in many complicated high-rise projects. They can operate about 200 ft without bracing and extend several hundred extra feet when braced. Where cranes rise up with the building, the height and capacity are limited by the height and structural strength of the building. The minimum economical height for tower crane operation is six stories. They can efficiently handle all materials on a site. Pickup and placing can be achieved with pinpoint accuracy within a full 360° arc and with a very long reach. The crane itself requires minimal space and is often mounted within the building. Electric power ensures nearly silent operation.

Labor requirements are reduced with tower cranes. One man with a remote-control box on his back can control all handling operations from any suitable location on the site.

The two basic types are the counterbalanced horizontal T-boom crane and the turntable crane with a telescopic boom. Both types can operate from either a static foundation or a track-mounted carriage. Both are safe to operate and are counterbalanced against overturning. The turntable crane is used mainly outside a building. The T-boom crane is most frequently located within a building (Fig. 6-6).

Cableways

Overhead cableways are used almost exclusively to handle concrete on large dam projects. Equipment investment is substantial and can be justi-

Figure 6–6 Tower cranes which rise with the building have the advantage of covering 360° and providing pinpoint accuracy for the whole construction site.

fied only for jobs involving vast quantities of concrete. The advantage of the method is that buckets with capacities as large as 8 cu yd can be transported at high speeds to the exact point of placement without further handling. Speeds range from 950 ft per min lowering a loaded bucket, to 1100 ft per min for a bucket coming up empty. Bottom-discharge type buckets with very wide openings allow quick, clean placing of low-slump concrete that may contain aggregates of 6- or 7-in. maximum size. Sometimes buckets are trucked into lifting position.

CONVEYORS

Concrete is automatically and quickly placed at many construction sites by a system that employs conveyor belts to transport the concrete from transit-mix trucks to the work areas. Conveyors help cut placing time because a steady flow of concrete can be maintained. Forms are filled just as fast as proper construction practices allow. A limited number of workmen are involved in placing and in vibrating the concrete. Conveyors have the important advantage that they can be used in conjunction with elevated hopper systems filling concrete buggies.

There are three basic types of conveyors. The first is the *feeder* conveyor used for transporting concrete over long distances. Maximum capacity for a 16-in.-wide belt is rated at 150 cu yd per hr. Jobs have been performed carrying concrete over distances of 400 to 500 ft. The second type of conveyor is the *portable unit*. It is used for short lifts such as carrying concrete from ground level to a second story or over an excavation. The third type of conveyor is the *side-discharge unit*. It is used to place concrete over a large area such as a bridge deck (Fig. 6-7).

Conveyors are supplied as standard units. The most common size portable conveyor has an effective working length of about 48 ft. Some have

Figure 6–7   Belt conveyors afford flexible access for unusual placement.

booms extending to 56 ft. Most feeder conveyors are about 30 ft long but can be made in lengths up to 85 ft. Side-discharge conveyors run between 22 and 103 ft in length. Belt widths are 16 to 18 in. Some may be 36 in. wide for heavy construction.

A series of conveyors can be arranged to feed each other consecutively. An average feeder-conveyor job will require five conveyors to provide a reach of 150 ft. Conveyors used in series often employ many belts and reach half a mile. Greater distances are technically and economically feasible.

With a feeder-conveyor system, concrete is transferred from one belt to the next by a baffle hood and shield arrangement that prevents spillage. Support for the units at transfer points is a simple swivel arrangement that makes movement easy. The swivel allows full 360° rotation.

With normal, smooth belts a cohesive concrete mix can be handled through an angle of 20 to 30° without difficulty. The wetter the mix, the lower the maximum slope. Portable conveyors have a corrugated top surface to resist flowback of the mix.

Conveyors are used as much for horizontal movement as they are for elevating concrete. The conveyor boom bridges over obstructions, excavations, or reinforcing steel which prevent the transit-mix truck from getting close enough to discharge directly to the point where the concrete is needed. Any type of mix can be handled, although maximum efficiency is achieved with 3-in. slump concrete. Belts must be kept clean during operation for maximum load-carrying ability and to avoid paste loss that could reduce workability. This is achieved by means of one or more scrapers that bear directly on the belt.

There is available a lightweight, truck-mounted, hydraulically powered conveyor that fits on the rear of a ready mix truck and delivers up to 75 cu yd per hr, 12 ft below grade to 12 ft above. When not in use, it rides over the truck fender.

PUMPING CONCRETE

Concrete can be pumped at rates up to 100 or more cu yd per hr. Distances range to 1500 ft horizontally, and more than 200 ft vertically, depending on the size of the transmission line, the mix design, and the size of the coarse aggregate. The delivery line can be laid over, under, or around obstructions. It can easily be run through windows or other small openings in the construction. This is a point-to-point delivery system— the concrete is delivered to its final destination in the forms. With this system there is a minimum of segregation.

Most of the successful pumps can handle properly designed mixes with slumps of 2 to 7 in. Concretes with slumps in excess of 7 in. tend to segregate, causing jamming of the transmission line. Difficulties are encountered in pumping some lightweight-aggregate concrete because pressures in the pumping systems tend to force mix water or paste into the more absorptive aggregates. A high-quality mix can be pumped at satisfactory rates. The adjustments in mix design necessary to ensure trouble-free pumping favorably affect properties of the concrete such as workability, strength, watertightness, abrasion resistance, and durability. Various placing defects are also prevented or reduced by pumping. The method ensures a steady, even flow of material (Fig. 6-8).

Large-Line Systems
The oldest methods of pumping concrete are the large-line systems. They

Figure 6–8 Pumping concrete can be one of the most versatile methods of delivering concrete from the ready mix truck to the form.

are best suited for large-volume placing over long distances where only limited flexibility at the discharge end is necessary. They are used mainly for mixes containing 2- and 3-in. maximum size aggregates, depending on line size.

The simplest type of pump used with large-line systems consists of a receiving hopper, two valves, a piston, and a cylinder. The hopper usually includes remixing blades to maintain the consistency of the concrete and to keep it moving. Pumping proceeds on a two-stroke action. The suction stroke of the piston is arranged to coincide with the opening of the inlet valve from the hopper and the closing of the outlet valve which controls the flow of the mix into the pipe. This suction stroke, assisted by gravity, draws the concrete into the cylinder. On the forward stroke, the valves reverse and a volume of concrete, approximately equal to the displacement of the piston, is forced along the pipeline. The cycle of the engine keeps the pipeline completely filled at all times so that there is continuous discharge at the form.

There have been several improvements on this basic pattern. Chief of these is the use of twin pistons, now quite popular.

Another procedure with large-line systems is the use of compressed air as the transport medium for the mix. Pneumatic systems operate with 4- to 8-in. pressure lines. They have few moving parts so that maintenance is reduced.

The mix is delivered into a 1- to 2-cu yd capacity pressure hopper which, when loaded, is sealed tightly by a circular clamped lid. Compressed air fed into the top of the hopper forces the concrete as a continuous slug through the transport line to a reblending discharge box. This box bleeds off the air and slows the concrete to prevent spraying and segregation on discharge. A new load can then be delivered to the pressure hopper. For very large jobs a series of hoppers can be used to ensure a more uniform supply of concrete.

Small-Line Systems
Small-line systems were developed for easier erection, dismantling, and handling than was found with the large-line systems. A small-line pump may have a 3- or 4-in. delivery hose and will pump concrete containing maximum size aggregate of 1 to 1½ in. up to 35 yd per hr, depending on hose size.

Some small-line pumps are used in conjunction with booms that carry the first 50 to 85 ft of delivery line. They can be remotely controlled from the point of discharge and simplify placement in difficult places. Most pumps use delivery lines consisting of a combination of metal tubing and rubber hose. Metal tubing has less friction resistance and is used whenever possible for straight runs. Rubber hose is used for the last 25 ft or so to provide greater flexibility in distributing the concrete at the discharge end.

Many small-line systems use piston pumps similar to those used for large-line pumping. A comparatively recent development is a pump that employs a squeeze principle to provide the pressure to transport the concrete. There are no moving parts in contact with the concrete in the squeeze-extrusion pump. The heart of this machine is a rubber pumping tube and rollers located in a vacuum chamber.

The rotating rollers squeeze the concrete through the pumping tube. The vacuum restores the shape of the pumping tube, creating a suction that draws additional concrete into the tube from the hopper. The standard machine, using a 3-in. line, can deliver concrete with 1-in. maximum size aggregate to distances of 300 ft horizontally and 100 ft vertically with a capacity of 30 to 35 yd per hr. A larger model, using a 4-in. line, places concrete at rates up to 60 yd per hr to distances of 400 ft horizontally and 125 ft vertically. This model can handle 1½-in. maximum size aggregate.

The most efficient setup for a pumping system is one that provides the straightest possible path between delivery hopper and discharge hose. Bends should be laid to give the smoothest possible flow. For example, 30 and 45° bends are better than 90° bends. Pumping has been used quite successfully to place concrete under water. Gas, diesel, or electric power units may be used to drive the pumps. Concrete may be pumped through rigid pipe of steel or plastic or through flexible hose—usually rubber, reinforced with steel wire or plastic. Aluminum pipelines are not recommended, since rough aggregate wears away the soft aluminum.

Grouted Aggregate

Grouted aggregate is a method of making concrete by placing coarse aggregate in the forms and pumping grout into the void spaces. The thick grout is pumped through pipes that extend to the bottom of the forms and are withdrawn as the grouting proceeds. Shrinkage is diminished with this method of placing because of the point-to-point contact of the aggregate particles. This method is used principally for restoration work and in the construction of tunnel linings, reactor shields, bridge piers, and underwater structures.

TREMIE CONCRETE

Concrete should be placed in air rather than under water whenever possible. When it must be placed under water, the work should be done under experienced supervision, and certain precautions should be taken to prevent pockets of water from being trapped in the concrete.

Several methods are used for placing concrete under water, the most common of which is with a tremie. A tremie is a straight pipe long enough to reach the lowest point within the form. Concrete is placed in the hopper at the top of the pipe; it flows down the pipe en masse, flushing the pipe of all water. As the concrete emerges from the bottom of the pipe, it begins displacing the water in the bottom of the form. As the concrete begins to fill the form, the tremie pipe is raised, but its end must never come above the surface of the concrete growing up in the form. In this way a uniform seal is kept between the concrete and water as the water is displaced by the concrete that is filling the form from the bottom up. Placing should be as continuous as possible, and the surface should be kept as level as possible to help maintain the seal (Fig. 6-9).

Tremie pipes are usually 10 to 12 in. in diameter and have a hopper attached to the top. A crane or hoist is provided to lower and raise the entire hopper and pipe during the operation.

Tremie concrete requires a high cement content: 660 lb of cement per cu yd is usually specified. Slump should be 7 to 8 in. The fine aggregate proportion should be somewhat higher than for normal conditions, often 45 to 50% of total aggregate. The maximum size of coarse aggregate should not exceed 1½ or 2 in. In most instances, the use of set-retarding, water-reducing admixtures with entrained air will give much better results in the flow qualities of the concrete. It will also diminish laitance and increase concrete uniformity.

Once concreting begins, it is important that the bottom of the pipe is not withdrawn above the concrete, since this causes loss of the tremie seal and can lead to voids, honeycombs, and excessive laitance. On very deep placements, the long tremie pipes must be raised as the concrete rises. For this reason most tremie assemblies are made in 10-ft sections that can be unbolted and removed one by one as the pipe is raised. When it is necessary to shift the position of the tremie, it should be lifted out of the concrete and moved to the new position.

Figure 6–9 Deposit of concrete through a tremie can result in void-free placements under water.

PNEUMATICALLY APPLIED CONCRETE

Pneumatically applied concrete, often called shotcrete, is a mixture of portland cement, aggregate, and water shot into place by compressed air. Aggregate graded up to ¾ in. can be used with certain types of equipment. The two basic processes for applying concrete pneumatically use dry mix and wet mix.

In the dry-mix process, the cement and aggregates are mixed in a relatively dry state and are conveyed through a hose to a nozzle where water is added. The dry materials are forced through the hose by, at least, 45-psi air pressure with a nozzle velocity of about 400 ft per sec. The water pressure at the nozzle should be, at least, 15 psi higher than the air pressure at the entrance to the hose. In the wet-mix process the concrete is premixed before pneumatic application.

Unit water content is kept to a minimum and high-strength, durable concrete can result. The quality of concrete is largely dependent on the man operating the nozzle. In the dry-mix process the nozzle man controls the amount of mixing water. In both processes he directs the nozzle, thereby controlling the thickness of the mortar layer and the angle of application.

The pneumatic application of concrete is used both for repair work and for new construction. The method is useful for conveying concrete into difficult locations and where relatively thin sections and large areas are involved.

Material that rebounds is seldom sufficiently well graded to be reused. Any sand that rebounds and collects on horizontal surfaces should be blown off to avoid leaving sand pockets.

CHAPTER 7
FINISHING, PART I

Concrete may be finished in many ways depending on the effect desired. Various colors and textures, such as exposed aggregates, may be called for. Some surfaces may require only screeding to proper contour and elevation, while in other cases a broomed, floated, or troweled finish may be specified. The tools that a concrete mason must know how to use are many, and his skill must be developed before he can work efficiently.

RULES

A 6-ft or longer folding rule is a must for every concrete mason. Its many uses include checking grade and laying out forms and sidewalk joints. There are two types of 6-ft folding rules, the carpenter's rule and the engineer's rule.

The carpenter's rule is marked to sixteenths of an inch on both sides.

The engineer's rule is marked to sixteenths of an inch on one side. The other side is marked to tenths and hundredths of an inch for ease in conversion from the decimal system. It is used by engineers in laying out plans in our "twelve-inches-to-the-foot" system.

Rules should be kept clean and the joints should be oiled for lasting use.

Straightedges—rigid pieces of wood or metal—are used to strike off excess concrete and level the surface to proper grade as soon as concrete has been placed. They should be approximately 2 ft longer than the distance between edge forms so that they can be handled at a slight diagonal. The straightedge is usually worked by two men who push it back and forth in a sawing motion as they move it across the surface of the concrete. A small amount of concrete is pushed ahead of the straightedge filling low spots and maintaining the plane of the surface.

Straightedges are made of a variety of materials. Each has advantages and disadvantages: wood will warp when moist, but can be made on the job and shaped to give the concrete surface a desired contour; aluminum and magnesium are light in weight, do not rust or warp, but require careful handling and maintenance because of the softness of the metal (Fig. 7-1).

Wood

The size of a wooden straightedge depends on the width of the surface to be worked. The 2×4-in. size should be made of well-seasoned timber up to 10 ft long. The 2×6-in. size can be as long as 16 ft, but it is hard to handle. A two-man, stand-up straightedge is made by attaching pieces of 2×4-in. lumber of equal size to both ends of a 2×6 in. straightedge. Handles are attached to the L-shaped ends forming triangles across the two pieces of lumber. The tool is used on large areas such as industrial floors. It is constructed on the site and discarded after the job is finished.

The wood straightedge must be cleaned after each use and checked for straightness for continued good finishing results. Irregular contours

Figure 7–1 The straightedge or strike-off rod.

can be corrected by inserting shims or small wedges between the 2×4-in. and 2×6-in. sections of the straightedge. Tightness of handles should also be checked and renailed if necessary.

Magnesium
The magnesium straightedge is called the magnesium screed by some manufacturers, although the term "screed" should more properly be applied to grade strips or side forms. It comes in a range of sizes. The 1×4-in. screed is either 8 or 12 ft long. Magnesium is a smooth material that will not stick to the surface when screeding air-entrained concrete.

Aluminum
When a 10- or 20-ft hollow aluminum straightedge is used to level and smooth plastic concrete, the ends are plugged to keep out concrete as it is moved across the surface. As the straightedge is pulled toward the concrete mason, it is tilted away to obtain a compacting effect rather than a tearing effect. Excess concrete may pile up in front of the tool; the excess should be moved ahead to the still unscreeded section.

Power
The power straightedge simplifies the work of striking off large areas. It also can be used on low-slump concrete. It will strike off and compact the concrete in one pass. One type of vibrating straightedge is made with a motor mounted on the center of a beam spanning the forms. Others have two straightedges instead of one, with the motor mounted between them (Fig. 7-2).

Wooden power straightedges should not be used on wooden forms because the friction caused by wood against wood makes pulling the straightedge difficult. Wood screeds are impractical for other reasons. The weight of a power straightedge makes it necessary to place braces about 1 ft apart to keep the screed from bowing and causing a dip in the floor. Also, the weight and vibration of the power straightedge can cause the wood between supports to crack and break. Metal edge forms are used when power straightedges are employed.

Roller
Roller-type straightedges are used frequently on large areas. The single roller variety is pulled manually by ropes or handles attached to inside, end sleeves. The double roller is power driven by a motor located between the rollers at one end.

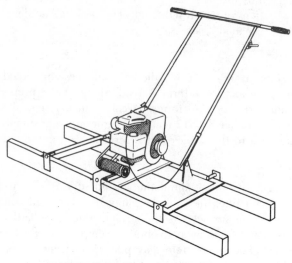

Figure 7–2   Power Straightedge.

TAMPERS

Tampers have been designed for use on low- or no-slump concrete. The solid hand tamper is made of 10- or 12-in.-square cast iron with a centered, upright handle. It is used to tamp small fill-holes on grade and the bottom course of two-course concrete where the first course is made of no-slump concrete and a grout mix is used for topping.

Jitterbug
The jitterbug, or hand tamper, is made of a metal grill, 6½ in. wide by 36 to 48 in. long. It forces large particles of coarse aggregate slightly below the surface, bringing up enough fines to permit the desired finish. It should be used sparingly and only with low-slump concrete. In most instances it is not necessary and is not recommended (Fig. 7-3).

Rollerbug
The rollerbug is used for the same purpose as the jitterbug. However, it has a long, attached handle that allows the concrete mason to work outside the forms. Additional handles can be attached depending on the width of the slab.

Figure 7–3   Hand tamper, commonly called a jitterbug.

Sometimes the rollerbug is used to roll in aggregate for exposed aggregate floors and walks. This use is restricted to narrow slabs and walks; if used on wide slabs, the weight of the rollerbug tends to push the aggregate into humps causing an uneven slab.

FLOATS

Bull Floats

Bull floats are used on fresh concrete before the water sheen appears to compact the surface of the concrete, to remove ridges and fill holes left by the straightedge, to push the aggregate beneath the surface, and to smooth slight humps and hollows. They are made of different materials in varying sizes (Fig. 7-4).

Ready-made bull floats are made of sturdy, laminated wood strips laid on edge. Poplar shows the best porosity and strength, and waterproof laminating prevents the tool from splintering and warping. They are 7½ in. wide and either 48 or 60 in. long and have lightweight aluminum extension handles, 4, 5, or 6 ft long, which permit large-size slabs to be bull floated.

Sometimes bull floats are made on the job using a 1×8-in. pine board that has its bottom edges planed off. To this, a handle of 2×2-in. wood is nailed. The corners of the handle are rounded off for a smoother, more comfortable grip.

Magnesium bull floats are the best type of float to use on air-entrained

Figure 7–4   Bull float.

concrete. They are light in weight, made of extruded magnesium, and are 8 in. wide and 42, 48, or 60 in. long. They have handles that can be connected to allow the mason to reach the concrete without walking on it. Handle lengths range from 4 to 6 ft. depending on the area to be covered and the man using the tool. The handle should not be so long that the function of the bull float is hindered.

One kind of magnesium bull float has a half-corrugated, half-flat face. The corrugated half is on the back or heel, and the corrugations are perpendicular to the float. When the float is moved out over the concrete, the corrugation depresses the coarse aggregate. When the float is pulled back, the flat half smooths the concrete surface.

The handle of a bull float protrudes some distance behind the man using it. He must be aware of the inherent danger and must make certain that he does not interfere with other workmen or hinder traffic. Insulated handles are available for working near power lines.

Darbies

Darbies serve nearly the same purpose as bull floats: compacting and smoothing the surface by cutting off humps and filling low places left by the straightedge. The concrete mason has a choice of size, shape, and material in deciding which darby to use for a specific job. Frequently, he will make his own darby.

A wooden darby is usually made of No. 1 pine, free of knots, and approximately 4 in. wide and 3 to 8 ft long. The higher end of the sloping handle is about 6 to 8 in. above the darby and is made of 2×2-in.

lumber with rounded edges to provide a good grip. It gives a sweeping pattern to the concrete finish as compared to the straight lines produced by a bull float. The wood darby does not work as well as magnesium does on lightweight concrete because of the pull of the wood against the concrete, and magnesium darbies are suggested for use on lightweight concrete (Fig. 7-5).

Small hand darbies are made of either wood or magnesium. They are used on sidewalks and small areas or when resurfacing or topping floors to level surfaces. When using the small darby on high-slump concrete, spiked knee boards are needed to move over the surface.

Hand Floats

Wood hand floats are used after darbys and bull floats to consolidate minor humps and fill in holes, leaving the concrete surface free of small depressions. They are constructed of many different types of wood, depending largely on the preference of the concrete mason. Mahogany and walnut floats are suitable for either air-entrained or non-air-entrained concrete because of their hardness. Redwood does a good job on non-air-entrained concrete, but it is a soft wood and is not durable. It should not be used on air-entrained or structural, lightweight concrete. Pine and cypress may also be used on non-air-entrained concrete (Fig. 7-6).

Manufacturers also make floats of laminated canvas-resin, teak, and other woods as well as specially shaped handles to suit the preferences of the concrete masons who will be using them. Many masons make floats and handles to suit their own requirements.

The magnesium hand float is best suited for use on air-entrained and structural, lightweight concrete. Other materials have been tried: wood produces too much drag, while steel is too heavy and seals the concrete

Figure 7–5   Darby.

Figure 7–6   Small darby or hand float.

surface. Lightweight, long-lasting, magnesium hand floats are made in different sizes with widths of 3½ to 3¾ in. and lengths of 12 to 20 in.

Cork floats, which can be purchased ready-made, are usually 12 in. long by 5 in. wide. A ¾- to 1-in. thickness of cork is attached either to a piece of wood or directly to the handle.

Although these floats can be purchased, many concrete masons prefer to make their own by buying the cork and attaching it to a handle. A handle can be bought or else can be made to fit the hand. Because cork floats are soft, they are used for patching large voids on concrete wall or ceiling surfaces. A cork float is used with a circular motion and will rub off excess grouting material used in the voids. Cork floats are often used to put a textured finish on exposed concrete.

A sponge rubber float is made by attaching a rectangular piece of rubber to a hard back fitted with a handle. It is used for the same purposes as a cork float. However, sponge rubber works better to fill air bubbles and other small imperfections in concrete walls, beams, and columns. It will also give a better "sand finish" to exposed concrete than the cork float.

Power Floats

A power float machine can only be used on no-slump topping of two-course concrete floors. Attached to the sides of the motor are hammers that pound the top of a disk which rotates on the concrete. This brings up the moisture in the topping for finishing. The power float is also used when metal aggregate shakes are applied to the floor. Latest models have a cam arrangement that is adjustable for vibration of a disk float. After the water sheen has disappeared, the float is used to hammer the metal aggregate into the topping and to bring up enough matrix to finish the slab. A valuable safety feature of the power float machine is the substitution of a gasoline-powered motor for an electric one; with the electric-powered motor there was danger to the operator who might not be wearing protective rubber gloves and boots (Fig. 7-7).

TROWELS

Hand finishing trowels come in various sizes for different uses. They are available in sizes 3, 3½, 4, 4½, and 4¾ in. wide, and from 10 to 20 in. long. A sill or step trowel is 3 in. wide and 6 to 7½ in. long. Margin trowels are usually 1½ to 2 in. wide and 6 in. long. The pointing trowel has a blade usually 4, 5, or 6 in. long with the back or heel no wider than 3 in. The large fresno trowel is 5 in. wide and 24 to 36 in. long. It has long handles and should be used only where a broom finish is required.

The metals used in the manufacture of hand finishing trowels are carbon tempered steel or top quality, tempered stainless steel. They are mounted on forged aluminum or steel shanks. A rounded wood handle is held on the shank with a washer and nut. Handles are either straight or are what is known as the camelback or California handle. A good steel will ring when plucked with the thumb. The handle should fit the hand and the trowel should have a slight concave curve upward (Fig. 7-8).

Uses

The many uses of margin or pointing trowels include: (1) cleaning other tools before washing them, (2) mixing mortar in buckets, (3) cleaning

Figure 7–7   Power float.

Figure 7–8 Trowel.

forms, (4) opening edges before running the edger, (5) cleaning out and patching honeycombs or voids, (6) finishing around pipes and other inconvenient places. The wider and longer trowels are used for the first passes. As the concrete hardens, the mason will use somewhat shorter trowels for second or third passes to allow increased unit pressures. The narrow and shorter trowels, sometimes known as fanning trowels, are used for hard, dense concrete finishes.

Troweling hard surfaces can wear down the edge of a trowel until it is razor-sharp. Such a trowel is hazardous and will not give a proper finish. Trowels should be filed when they become worn.

To file a trowel, a 16-in.-long steel file is laid flat on a wood surface. Gloves are worn, and the trowel is grasped by the handle and the back of the blade, then run back and forth on the file until the edge is straight and returned to its original thickness. The trowel should never be held on a flat surface with one hand and filed with the other, since one slip could cause the edge of the trowel to cut the hand. Also, a true edge cannot be produced with this technique.

Power Trowels

Power trowels are very useful when a floor is so hard that cement paste cannot be worked up with hand tools. They cannot be used on a soft concrete surface because they will tear it and make holes and other imperfections. There are several types of power trowels in sizes from 20 to 48 in. in diameter. The 20-in. trowel is used exclusively for home and light-duty industrial floors. It fits easily into the trunk of a car. Large power trowels are designed to float and trowel. Some manufacturers place the float blades above the troweling blades. The desired blade can then be flipped into place. Another type has separate float blades that slip over the troweling blades and are removed for troweling.

Since the machine generally rotates clockwise, pressure applied upward on the handle grip makes the trowel travel left while pressure downward makes it travel to the right. Once the basic movement is known, an operator can master proper operating techniques. A uniform lapping pattern should always be applied: keeping the blades perfectly flat the first time or two so ripples don't develop. Then the concrete should be gone over with a slight pitch on the blades. This increases the amount of pressure on the portion of the blade touching the concrete, further consolidating it.

If the blades penetrate a soft spot in the concrete, the trowel handle can throw the operator. Some machines have a safety feature which stops them automatically in such a case, thus avoiding a dangerous situation (Fig. 7-9).

The need to keep tools clean and in good working order cannot be

Figure 7-9   Power trowel.

overstressed. The pride that a concrete mason has in his job is increased by a clean, efficient set of tools. When tools are not cleaned before clinging concrete has time to harden, the job of getting them back into readiness becomes more difficult and time is wasted.

EDGERS

Finishing steps, curbs, gutters, sidewalks, highways, and the like, requires tools of special dimension and design to make uniform edges, rounded to constant radii. This compacts the concrete and prevents weakened planes and breakage. (It is easy to imagine what could happen if curbs or steps in public places were unevenly edged. Pedestrian safety is paramount here, but naturally appearance is important, too.) Edgers are usually made of stainless steel so that they will not rust. Heavy-duty bronze is sometimes used but, since steel is light, smooth, and gives a better finish, it is preferred. Some concrete masons like edgers with turned-up ends because they will not dig into soft concrete (Fig. 7-10).

Sizes, designated by the radius of the curved part, vary according to the job for which the edger is designed. Steps, sidewalks, patios, and

Figure 7–10   Edger.

pavements require an edger with a smaller radius. The large-radius tools are used on curbs to eliminate sharp edges and reduce damage to vehicle tires, and on gutters to facilitate drainage.

Edgers for sidewalks have a radius of ½ to 1 in. and are wide enough to remain flat when cutting the edge. Edgers with long handles, called walking edgers, eliminate bending down and kneeling, but can only be used when concrete is very soft—before bleeding takes place. A cleaner, better edge is achieved by a hand edger. To make certain that the handle will fit their hands comfortably, some concrete masons prefer to make their own edgers. Edgers for some special jobs cannot be purchased, in which case the mason will have to make a custom edger.

Step tools are made in matching sets called inside and outside tools. They will have the same radius to insure a uniform looking step. The radius should never be more than ½ in. or the step will become hazardous when wet. These tools are used after the riser board is removed and while the concrete is still soft enough to work but hard enough to hold its shape. Sometimes the outside step tool is called a corner tool. It is run over the top of the step to put a finish on the edge, part of the riser, and the tread. Some step tools have a right angle instead of a radius.

Pedestrian safety must be carefully considered in finishing curbs. Front curb edgers are the same radius as the gutter tool. The size of the edger is usually ¾- to 3-in. radius, 1 in. sidewall, 10 in. long and 6 in. wide. If the curb is 7 in. wide, many concrete masons select an edger that has the top flange wide enough to ride on top of the back curb form. This will help the concrete mason to keep the top of the curb flat. Gutter tools are available that match the dimensions of the curb tools.

The curb tool is used after the face board of the curb has been removed but while the concrete is still plastic. It is used to finish the outside radius of the curb to a smooth and even surface.

The gutter tool is used immediately after the curb tool in the gutter part of the curb and gutter. This will prepare the face and gutter of the curb for brushing.

A tool known as a "mule" is often used for simultaneously shaping curb and gutter.

GROOVERS (JOINTERS)

Groovers are made of stainless steel, bronze, or malleable iron. Most groovers come in 6-in. lengths and varying widths. The important part of the groover is the bit, which controls the depth of cut to be made in the concrete. The cut varies in width as well as depth. Control joints are generally cut one-fifth the depth of the concrete to prevent unintentional cracking. When it is used to make designs that are shallower than the control joint, a groover can be selected for the particular results desired. A thicker bit will give a wider joint mark. Care must be taken not to select too wide a groover for these designs, as this may cause accidents when small heels catch in the groove (Fig. 7-11).

Many different handles are attached to groovers. The straight handle attached at both ends of the groover is usually used for straight design and control joints. Another type of straight handle is attached at only one end.

Walking groovers have long handles to allow the concrete mason to stand up while cutting joints. These groovers must be used as soon as the concrete is bull floated. If a finished concrete sidewalk surface is to show a side-joint mark, the walking groover is used again after the final troweling. Walking groovers are not as easily controlled as hand groovers.

The multiple jointer and groover is often used to make a nonslip tread on a stairway or ramp. It is widely used on steel staircases where the metal pantreads are filled with concrete.

Figure 7–11   Groover or jointer.

BRUSHES

Small or large hand brushes, such as a 4- to 6-in. paintbrush, are a must for the concrete mason. They are used to wet holes to be patched, and later to wet the patch itself for floating and scrubbing with rubber or cork floats. They are also used to brush out light tool marks on risers of stairs and to brush the treads to make them skid-proof.

BROOMS

Nylon or fiber bristle brooms can be bought in a number of lengths. They are used to give the fresh concrete surface a rough texture, usually on pavements, driveways, curbs, and gutters.

A finer nylon or hair bristle broom is used mostly for small jobs such as sidewalk, stair treads, and wall finishes.

Various other broom sizes and degrees of coarseness are available for jobs ranging from texturing parking lots to exposing aggregate (Fig. 7-12).

KNEE PROTECTION

Pads

Knee pads are not tools, but they are a vital part of a concrete mason's equipment. They protect the knees against dampness and painful injuries caused by kneeling on small pebbles.

Pads should be made of good leather lined with a heavy felt pad. A buckled strap goes around the outside of the leg.

Figure 7–12   Broom.

Boards

Knee boards, which are wide, flat pieces of lumber, are necessary when hand floating and troweling large floor areas or wide sidewalks. They distribute the weight of the mason over a large area so that he will not mar or damage the plastic concrete (Fig. 7-13).

CONCRETE VIBRATORS

A vibrator is a mechanical oscillating device of high frequency that is used to consolidate fresh concrete. Vibrators are available in three types: internal, external, and surface.

The *internal vibrator* is usually called a "spud vibrator." It will consolidate concrete very well if used properly. A spud vibrator should be inserted in the concrete, pulled out, and reinserted every 12 to 18 in. It should never be left vibrating any longer than 5 to 10 sec because the concrete will tend to segregate leaving a matrix or paste around the vibrator and honeycombed concrete against the forms. Internal vibrators must never be used to move concrete (Fig. 7-14).

The *external vibrator* operates on the same principle as a hand or electric air hammer. Some have a flat piece of a 4×4-in. hard rubber attached to an iron plate on the end. This is held against the forms, and the vibration of the operating tool causes the plastic concrete to move into voids. Other vibrators may be attached at selected positions on the

Figure 7–13   Knee boards.

Figure 7–14 The man on the right is feeding an internal vibrator down between the forms to consolidate the concrete.

forms. These are particularly useful in the present-day precast industry. The external vibrator may also be used on each lift immediately following internal vibration to reduce the size and amount of surface air pockets.

The external vibrator should be used where internal vibrators would damage relief, rubber, or plastic molds that are attached to the forms for architectural concrete. The vibrator must not be held in one place too long, or it will cause structural defects.

The *surface vibrator* employs a portable horizontal platform on which a vibrating element is mounted. Some machines combine the surface vibrator with the power straightedge and accomplish both jobs at one time.

SANDBLASTER

Sandblasting is a method of cutting away or eroding the surface by a stream of sand ejected from a nozzle at high speeds by compressed air. Sandblasting concrete is usually done to expose aggregate. It is also used on old concrete floors before they receive another coating of concrete or other topping. This cleans the floor thoroughly to promote good bonding of the new concrete to the old.

Sandblasting is usually done with silica sand, but today there are also manufactured sands for blasting. The particular fineness of sand to be used depends on the surface hardness of the concrete. Proper clothing is important. The sandblaster must wear the following:

1. High-top shoes so that his trouser legs can be tied over the shoes to prevent sand from going into them.
2. A full protective leather apron to prevent sand from tearing clothing.
3. A full headgear covering head and shoulders.
4. Long leather gloves for protection of arms and hands.

GRINDERS

Hand rubbing stones are usually made of No. 20 grit silicon carbide and may be used for dressing down form and mold marks on exposed concrete. They are from 6 to 8 in. long, 2 to 3½ in. wide, and ¾ to 2 in. thick.

Both hand and power grinders are made with coarse and fine abrasives. The coarse abrasive is used to remove heavy seam ridges and protrusions left by tie rods and form indentations. The fine abrasive is used on architectural exposed concrete to rub off small protrusions left by indentations in the mold and to blend exposed concrete after filling holes. Also, honing and polishing will produce smooth exposed aggregate surfaces as in terrazzo work.

Operators of dry grinders should always wear goggles to protect their eyes and nose masks to protect their lungs. They should wear protective clothing over their arms in case a grinding stone should break while going at high speed. Hard hats are essential on all types of high-speed grinding jobs.

A No. 10 coarse grit abrasive should be used for high-speed dry grinding. For lower speeds, a No. 20 or finer grit should be used. The surface of the concrete and the imperfections to be removed will dictate the size of grit to be used.

The low-speed electric grinder, which has a motor with a flexible shaft, is used for grinding grout into walls. This grinder has a small water hose connected to it that supplies water to keep the wall wet during grinding.

Another type of grinder is an inexpensive one run by compressed air. It is used on jobs where many grinders are used at one time. A single air compressor can supply enough air to power as many as 20 grinders.

Power floor grinders are used to grind off undesirable humps or to smooth rough concrete floors that have been damaged or were not properly troweled. These grinders come with either electric motors or gasoline engines.

One type has a solid abrasive disk, while another has three abrasive stones attached to the bottom of a metal disk.

When grinding terrazzo floors, three different abrasive stones are used: first, a coarse No. 24 abrasive, second, a medium No. 18 abrasive, and third, a polishing stone, Nos. 8 to 12 abrasive.

Another type of grinder is the sanding disk used to remove trowel marks and rough spots on lightweight aggregate and later-fill floors so that when vinyl or asphalt tile or other manufactured floor coverings are installed, the trowel marks will not show through. Floors must be smooth to receive carpeting also. If they are not, the rug and padding will wear through over high spots.

Operators of floor grinders should always wear rubber footwear and rubber gloves. This is especially important when an electric grinder is used, since the operator will be standing in water. When dry grinding, the operator should wear goggles and a nose dust-guard.

CONCRETE POWER SAWS

The use of heavy-duty concrete saws in all branches of concrete construction and maintenance work has steadily increased. Progressive government agencies specify concrete sawing for controlling breakage when making service cuts and other openings in concrete.

Sawing before breakout will clearly define the areas to be removed and will prevent damage to adjacent concrete. It will produce a neat, smooth, straight edge to patch against, but jackhammer methods leave jagged edges that are subject to water seepage and patch failure. Concrete saws are also used for making contraction joints in streets, highways, airfields, and floors. Other uses include: taking test samples of concrete, and making openings for light poles, machinery bases and sewer openings (Fig. 7-15).

Concrete power saws are available in electric and gasoline models, ranging from 3 to 60 hp. They can be purchased singly, in tandem, or as an entire train pulled by a lead saw.

A concrete saw used for contraction joints will create a perfectly perpendicular joint. It will permit the concrete to crack beneath the joint instead of in a random, irregular manner. The irregular break around coarse aggregate particles interlocks and acts as a load transfer device between floor or pavement sections.

Concrete saws should be purchased with an eye to the future rather

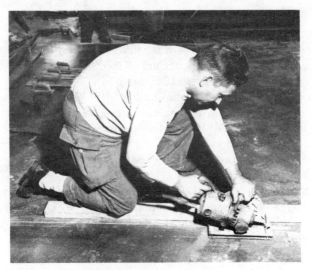

Figure 7–15   Hand operated concrete power saw.

than just for the immediate need. They are expensive investments so a versatile size and model should be chosen.

The blades of concrete saws are made of either carborundum or industrial diamonds. Diamond blades will cut concrete at any age, but they require water to cool them. Carborundum blades (called abrasive blades) should be used as soon as the concrete is hard enough to cut without raveling the edges of the partially cured concrete.

A concrete mason should study the manufacturer's manual to learn everything he can about the saw before using it. His own safety and the safety of those working with him can be assured if he takes the time to know his machine (Fig. 7-16).

Before using the saw, the operator should be sure that:

1. The water hose has enough pressure.
2. The guiding gauge will stay in alignment.
3. The motor has been serviced properly.
4. He wears rubber to keep his feet dry.

The electric hand concrete saw is used primarily on small jobs or if designs are to be cut in concrete slabs. When using abrasive wheels, the

Figure 7–16   Large concrete power saw.

operator should wear a face mask and goggles. The face mask will protect him from the dust created by the dry abrasive wheel, and the goggles will protect his eyes from small particles that fly from the wheel and the concrete.

BUSHHAMMER

Mechanical spalling and chipping of a hardened concrete surface is called bushhammering. Bushhammers are used primarily to give a rough texture to concrete surfaces and to expose the aggregate, usually about $3/16$ in. deep on the surface. This can be done with a single hand hammer that has a 2×2-in. head sawtooth both ways to points for cutting the concrete. Pneumatic tools are available with combs, chisels, and other multiple points. Tandem hammers have been used on the walls of multistory buildings.

A worker using any type of bushhammer should always protect himself with goggles, hard hat, and gloves.

SCARIFIER

A scarifier is a power-driven roller that has loose, hard-tool steel cutters to chip and dig into old concrete. There are two types: one that the operator walks behind for small jobs, and one on which the operator rides to prepare old highways and airport runways to receive new concrete.

CHAPTER 8
FINISHING, PART II

STAIRS AND STOOPS

Concrete steps make the approach to a house safer and more attractive. Concrete will not rot, and with adequate foundations, steps will not sag and become a hazard. Concrete steps are easy to keep clean by sweeping or hosing down. When properly built, they are slip-proof in wet weather and do not require painting or other maintenance.

Steps should be at least as wide as the sidewalk, and all risers should be exactly the same height. Maximum user comfort is achieved when risers are 7 in. or less and the tread is 10 in. or more. For aesthetic reasons, staircases are sometimes built with a rise as small as 5 in. and a correspondingly wider tread up to 14 in. A landing is desirable to divide flights more than 5 ft high (Fig. 8-1).

Preparation
Base. An economical way to keep steps from sinking is to dig two 6- or 8-in.-diameter postholes beneath the bottom tread. The holes should extend below the frost line and be filled with concrete. The top step or landing should be tied to the existing wall with two or more metal anchors. Well-tamped soil or granular fill may be used inside the forms to reduce the amount of concrete needed for the steps.

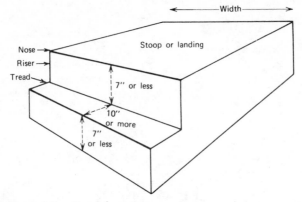

Figure 8–1   Parts of a staircase.

Forms. Staircases usually are built after the structural frame of the building is completed. They can be precast and delivered finished, or cast in place. Precast staircases have the advantage of not requiring forms or falsework on the job.

Staircase construction requires the same care and attention in placing, finishing, and curing as any other concrete construction. Forms must be strong, rigid, and tight. Since they will be removed before the concrete is completely set, they must be constructed so that they can be removed without damaging the concrete. Steel forms have been used successfully for staircase construction.

Forms must be checked to be sure they are level, plumb, and well braced. Braces should be no farther than 2 ft apart and should brace the bottom as well as the top of the form.

Where the top tread or stoop meets the building, an isolation joint should be formed and joint material should be placed against the wall.

Forms for staircases are usually removed after the concrete has set for 1 or 2 hr, providing the profile of the stair is such that it will not be damaged by early stripping. This leaves time for the nosing, riser, and ends to be finished before the concrete is too hard. Extra care is needed not to damage overhangs or nosings when removing forms.

Placement

Placing of concrete staircases should begin with the bottom step and proceed to the top, with each tread being floated off to the height of the riser. Placement will progress at a slower pace than when placing a flat slab, so each tread should be given a chance to stiffen a little before filling the one above it. This also gives the braces a chance to set tightly and to hold the riser board firm.

As stair forms are filled, the side wall forms should also be filled. These must be spaded or vibrated as progress is made up the stairs. Forms should be tapped lightly to release air bubbles.

After wall and stair forms have been filled with concrete, the stoop or entrance is filled and spaded or vibrated. A straightedge is used to strike off excess concrete. Finished surfaces will have to drain, so it is necessary to put a crown on the stoop or have it pitched ¼ in. for each foot from the building to the stairs (Fig. 8-2).

Figure 8–2 Steps with the top surfaces finished. When the concrete has set sufficiently to hold its shape, the forms will be removed, and the sides and front will be troweled.

Finishing

Finishing is begun at the top and progresses down. The stoop is first bull floated or darbied and then the edges are cut. Treads should be floated so that there is a pitch of about ¼ in. from back to front to insure that no water will lie at the back of the tread. The edger should then be run along the riser board, followed by the first troweling of the tread.

While using knee boards to protect the concrete, the next step is to determine whether it has set sufficiently to hold its shape when the forms are removed. If it has, the nails and braces are removed from the top riser first. Then the riser board should be tapped lightly and removed carefully. Each step should be finished before going on to remove the next board.

The riser must first be floated. It may require a little water to be sprinkled on it. With a pail of water and a brush handy for this purpose, the riser is floated to bring out the mortar in the concrete. A step tool having the same inside radius as the edger should be used on the stoop, running it back and forth until all holes are closed. The tread above the riser should then be troweled.

If not enough mortar can be worked out of the concrete to finish it, a little mortar should be put on the riser and tread. For this purpose, the concrete mason should have a pail of mortar consisting of:

1. 1 part cement.
2. 1½ parts fine sand.
3. Enough water to make a fairly stiff, but still workable, mortar.

Putting a little mortar where the riser and the next tread meet and running the step tool back and forth until all the holes are filled will produce a clean finish on the riser and tread. Tool marks left by the step tools should be removed with a margin or pointing trowel. Using a damp brush from which all excess water has been shaken, the riser and tread should then be brushed.

After the top riser and tread have been completed, the knee board is moved down to the next tread and the process repeated until all risers and treads are completed. Fast, careful work is essential, since too much time spent on one riser and tread may allow others to set too hard to be finished properly.

The following summarizes the sequence for finishing stairs and stoops:

1. Stoop
   a. Strike off
   b. Bull float or darby
   c. Cut edges
   d. Hand float
   e. Trowel

2. Tread
   a. Float off
   b. Edge
   c. Trowel
3. Riser
   a. Remove riser board
   b. Float
   c. Run step tool
   d. Trowel

Sides of Steps. When the side forms are removed the same day concrete is placed, a mortar mix made of 1 part cement, 2 parts sand, and enough water to make a workable mix is spread with a hawk and trowel on the walls and sides of the risers to a thickness of about ¼ in. This purge coat is then floated with a cork or rubber sponge float. If a smooth finish is desired, troweling follows. If a brush finish is called for, a damp brush is used after floating. Other designs can be made on the walls by using shallow groovers or the pointed head of a large (16d or 20d) nail. A pointing or margin trowel can be used to take off groover marks. Light brushing is all that is needed to finish the sides.

Tread Finishes. Finishes for treads may be dictated by architectural or practical considerations, but whatever finish is used, special care must be taken. When special aggregate is used, grinding can give an attractive finish. For staircases in basements, where decoration is not important, all that may be necessary is to rub them down with a carborundum stone and a mixture of portland cement and water, removing form marks and fill holes.

Wire-combing, brooming, sack-rubbing, swirl-troweling, or wood floating will give nonslip texture to the surface when appearance is not of utmost importance. Abrasive grits, such as silicon carbide or aluminum oxide, can be troweled into the treads during finishing operations. Also available are block or strip inserts made of the same material as the grit, which can be set into the concrete in patterns. Another way to prevent slipping is with special nosings and shallow grooves parallel to the nosings.

For decorative effects, treads can be painted with suitable concrete paints, or covered with tiles or carpeting. Paints wear away unevenly unless protected at points that get most of the wear and tear. Terrazzo or granolithic finishes give greatest durability and an attractive finish.

Curing procedures are very important in all concrete work. They will be discussed in detail in later chapters.

FLOORS AND SLABS ON GROUND

Introduction

A true craftsman understands the nature and properties of the material with which he works, and concrete has some interesting characteristics. Generally, all the dry materials used in making quality concrete are heavier than water. During and after placement, these materials will have a tendency to settle to the bottom and force any excess water to the surface. This reaction is commonly called "bleeding." More bleeding occurs with non-air-entrained concrete than with air-entrained. It is of utmost importance that the first operations of placing, strike off, and darbying (or floating) be performed before any bleeding takes place. The concrete should not be allowed to remain in wheelbarrows, buggies or buckets any longer than is absolutely necessary. It should be placed and spread as soon as possible, immediately struck off to proper grade, and then darbied (or bull floated). These last two operations should be performed before any free water has bled to the surface. The concrete should not be spread over a large area before it is struck off—nor should a large area be struck off and allowed to remain before darbying (or floating).

If any finishing operation is performed while the bleed water is present, serious crazing, dusting, or scaling can result. This point cannot be overemphasized and is the basic rule for successful finishing of concrete surfaces.

Striking Off

The surface is struck off by moving a straightedge back and forth with a sawlike motion across the top of the screeds or forms. A small amount of concrete should always be kept ahead of the strike-off or straightedge to fill in all the low spots and to maintain a plane surface (Fig. 8-3).

Tamping

On some jobs the next operation is to use the hand tamper, or jitterbug. This tool should be used sparingly or not at all. If used, it should be only on concrete having a low slump (1 in. or less) to compact the concrete into a dense mass. Jitterbugs are sometimes used on industrial floor construction because the concrete for this type of work usually has a very low slump.

The hand tamper, or jitterbug, is used to force large particles of coarse aggregates slightly below the surface so the concrete mason can pass his darby (or float) over the surface without dislodging any aggregate.

Figure 8–3   Striking off the surface of newly placed concrete.

Darbying

After the concrete has been struck off (and, in some cases, tamped), it is smoothed with a darby to level any high spots and fill depressions. This should fill in all surface voids and only slightly embed the coarse aggregate, preparing the surface for subsequent edging, jointing, floating, and troweling.

Bull Floating

The purpose of bull floating is exactly the same as that of darbying. On any job a choice must be made between darbying and bull floating, since the two operations are not necessary on the same surface. Because of its long handle, the bull float is generally easier to use on a large area.

Edging and Jointing

When all bleed water and water sheen has left the surface and the concrete has started to stiffen, it is time to begin the other finishing operations. No finishing operations should be started until the concrete will sustain foot pressure with only about ¼-in. (6-mm) indentation.

Edging. Where edging is necessary, this could be the next operation. On most floors edging is not necessary. It is most commonly used on sidewalks, driveways, and the like, and produces a rounded edge on the slab to prevent the chipping that would otherwise occur to sharp corners. The edger should be run back and forth until a smooth edge is produced. The concrete mason should be careful that all coarse aggregate particles are covered. He should keep the flat part of the edger on the same plane as the slab so that the edger will not leave too deep an impression in the top of the slab. Otherwise, the indentation may be difficult to remove with subsequent finishing operations (Fig. 8-4).

Jointing or Grooving. Immediately following edging, the slab should be jointed (grooved). The cutting edge or bit of the jointing tool cuts joints, called control or contraction joints, in the slab. These joints are used to control any cracking tendency in the concrete that may be the result of shrinkage stresses caused by drying or by temperature changes. The grooves create a weakened line that induces cracking to occur at that location. When the concrete shrinks, the cracks in these joints can open slightly, thus preventing irregular and unsightly cracks. In sidewalk and driveway construction, the tooled joints are usually spaced at intervals equal to the width of the slab provided that it does not exceed 15 ft. Double driveways require a longitudinal joint. This determines the spacing of the transverse joints, since they should be laid out as close as possible to square.

Figure 8-4   Edging.

For control joints, the jointer should have a ¾-in. bit. If the slab is to be grooved only for decorative purposes, jointers having shallower bits may be used (Fig. 8-5).

It is good practice to use a straight board as a guide when making the groove in the concrete slab. If the board is not straight, it should be planed true. The tooled joints should be perpendicular to the edge of the slab. The same care must be taken in running joints as in edging. A tooled joint can add to or detract from the appearance of the finished slab.

On large slabs it may be more convenient to cut joints with a power saw fitted with an abrasive or diamond blade. Joints made with a power blade should be cut within 4 to 12 hr after the slab has been placed and finished. Cutting of joints must be done as soon as the finished concrete surface is firm enough not to be torn or damaged by the blade, and before random shrinkage cracks can form in the concrete slab. Sawing should

Figure 8–5   Jointing.

produce slight raveling at the edges. If a clean cut results, sawing was probably done too late.

Floating

After edging and hand-jointing operations, the slab should be floated. Many variables—concrete temperature, air temperature, relative humidity, wind, and the like—make it difficult to set a definite time to begin floating; but, as a general rule, floating should be started when the water sheen has left the surface of the slab and the slab will support the weight of a concrete mason with a maximum indentation of ¼ in.

The purpose of floating is threefold:

1. To embed large aggregates just beneath the surface.
2. To remove slight imperfections, humps, and voids, and to produce a level or true surface.
3. To consolidate mortar at the surface in preparation for other finishing operations.

Aluminum or magnesium floats should be used, especially on air-entrained concrete. This type of metal float greatly reduces the amount of work required by the finisher because drag is reduced and the float slides more readily over the concrete surface while maintaining good floating action. A wood float tends to stick to the concrete surface and tear it; a light metal float forms a smoother surface texture than a wood float.

The marks left by the edger and the jointer should be removed by floating, unless such marks are desired for decorative purposes, in which case the edger or jointer should be rerun after the floating operation (Fig. 8-6).

Troweling

Immediately following floating, the surface should be steel-troweled. It is customary for the concrete mason to hand float and then steel-trowel an area before moving his knee boards. If necessary, tooled joints and edges should be rerun before and after troweling to maintain uniformity, thus producing neat, straight edges and control joints.

The purpose of troweling is to produce a smooth, hard surface. For the first troweling, whether by power or by hand, the trowel blade must be kept as flat against the surface as possible. If the trowel blade is tilted or pitched at too great an angle, an objectionable "washboard" or "chat-

Figure 8–6 This mason is doing final floating and initial troweling at the same time. The float is in his left hand and the trowel is in his right.

ter" surface will result. For first troweling, a new trowel is not recommended. An older trowel that has been broken in can be worked quite flat without the edges digging into the concrete. The smoothness of the surface can be further improved by additional trowelings. There should be a lapse of time between successive trowelings to permit concrete to increase its set. As the surface stiffens, each successive troweling should be made by smaller-sized trowels to enable the mason to use sufficient pressure for proper finishing. The steel-troweled surface leaves the concrete very smooth. However, since surfaces of this kind become quite slippery when wet, this is not always desirable.

Finishing Air-Entrained Concrete
Air entrainment gives concrete a somewhat altered consistency that requires a minor change in finishing operations from those used with non-air-entrained concrete.

Air-entrained concrete contains microscopic air bubbles that tend to hold all materials in the concrete, including water, in suspension. This type of concrete requires less mixing water than non-air-entrained concrete, and has improved workability with the same slump. Since there is less water and it is held in suspension, little or no bleeding occurs. This is the reason for slightly different finishing procedures. With little or no bleeding, there is less of a wait for the evaporation of free water from the surface before floating and troweling. This means that, in general, floating and troweling should be started sooner—before the surface becomes too dry or tacky. If floating is done by hand, the use of an aluminum or magnesium float is essential. A wood float drags and greatly increases the amount of work necessary to accomplish the same result. If floating is done by power, there is practically no difference between finishing procedures for air-entrained and non-air-entrained concrete, except that floating may start sooner on air-entrained concrete.

Practically all defects and failures on horizontal surfaces are caused by starting the finishing operations while bleed water or surface moisture is present. Better results are generally obtained, therefore, with air-entrained concrete. Air-entrained concrete is a must if the slab is to be subjected to freezing and thawing conditions.

Terrazzo

The word *terrazzo* is taken from the Italian word meaning "terrace," and has come to be defined as a "form of mosaic flooring made by embedding small pieces of marble in mortar, followed by polishing."

Today terrazzo is made by combining selected marble chips in a matrix of portland cement or in other matrices. Washed terrazzo or rustic terrazzo is a variation in which decorative quartz, quartzite, onyx, and granite chips are substituted for marble chips, and where, in lieu of grinding and polishing, the surface is washed with water or otherwise treated before it has completely hardened to expose the stone chips.

Terrazzo is an especially fine flooring material because it requires so little care. Two parts marble are mixed with one part cement, and additional marble is sprinkled on the surface during installation so that a minimum of 70% marble shows. Since the surface is at least 70% marble, a very low-porosity material, it absorbs very little of anything. Waxes are not necessary and are not recommended because a sealer is used that soaks into the mixture, sealing off the pores. Maintenance of the surface is very simple, making terrazzo a popular flooring material.

Marble for terrazzo is carefully selected for color and size. The quarried stone is crushed by a process that largely eliminates flat or slivery chips. In this case marble refers to all calcareous rocks (including onyx, marble, and travertine) and serpentine rocks (hydrous magnesium silicate) that can be ground and polished.

Either gray or white portland cement is used in terrazzo. White portland cement is more expensive, but offers a clearer background for the marble chips and also is more successfully combined with pigments.

Industrial Floors

Concrete floors for industrial buildings are subjected to many types and rates of traffic and exposure. To give good service, floors generally must have these qualities:

1. *Durability.* Depending on conditions, the floor must be able to withstand the effects of weathering, freezing and thawing, or attack by aggressive agents such as mild acids.
2. *Adequate Load-Carrying Capacity.* If the floor is not strong enough to support the loads to which it is subjected, it will become badly cracked.
3. *Wear-Resistant Surface.* If the surface erodes away under traffic, there will be a continual problem of dusting and maintenance, and areas with heavy traffic may become rutted.
4. *Freedom from Random Cracks.* If cracking is not controlled, many jagged cracks will form. These are unsightly and conducive to future problems. Sealing or protecting the edges of these cracks may be difficult.
5. *Suitable Texture and Nonabsorptive Surface.* Surfaces should be constructed so that they are true within $\frac{1}{8}$ in. in 10 ft. They should require little or no maintenance.

Classes of Floors on Ground

There are two types of floors on ground: one-course and two-course floors. The most common type is the one-course floor (Classes 1 to 5, Table 8-A). In one-course floors, the single thickness of slab provides both the strength and the wearing surface. In most one-course floors the same concrete mixture is uniformly distributed from top to bottom. The surface may be textured or it may be smooth. In some instances, a wearing surface of extra-hard aggregate is provided by applying what is generally referred to as a "dry shake." Certain extra-hard natural aggregates such as quartz, traprock, or granite, graded from $\frac{3}{8}$ in. down, are applied to the surface of the concrete just before power floating. Sometimes graded iron filings are used. Other times, artificially manufactured aluminum oxide aggregates are specified.

A two-course floor (Classes 6 and 7, Table 8-A) is made up of a base slab and a wearing surface, generally called a topping. The topping may be bonded so that it acts integrally with the base slab, or it may be unbonded. It should never be partially bonded. Bonded toppings should be ¾ to 1 in. thick, and precautions must be taken to insure a complete bond. For heavy-duty floors, a very high quality nonplastic mix, not more than 1 in. thick, is recommended. This mix should be compacted with specially designed equipment. Poor consolidation results from thicker layers. Nonbonded toppings should be at least 2½ in. thick, and precautions must be taken to insure that no bond occurs. Nonbonded toppings are recommended only in areas where it is not possible to protect the floor from strong acids or other aggressive agents. Floors in slaughter houses and chemical plants are sometimes made of unbonded toppings so that when the surface is eaten away, topping can be readily removed and replaced.

One-Course Floors

Subbase. Floors on ground depend on the subbase for support. Concrete floor slabs distribute loads to the subbase over relatively large areas; to do this, the subbase must be uniform in character, and its load-carrying capacity must not vary with changes in moisture. To accomplish this, the subbase is often constructed of granular material such as sand, gravel, crushed stone, or slag. Cinders or other friable materials that break up easily are not acceptable. Clay and loam should never be used as a base, since they shrink and swell with changes in moisture. Granular fill should be put down in layers not more than 6 in. thick. These layers should be parallel to the final surface of the floor. To compact the fill material, plate-type vibratory compactors or very heavy steel-wheel rollers should be used. The vibration created by vibratory compactors literally shakes the granular particles into their most compact positions. Equipment such as trucks and bulldozers should not be used for this operation because they do not compact the fill uniformly.

Joints. For concrete floors on ground to function properly without random and excessive cracking, it is necessary to provide adequate joints. These joints must accommodate horizontal movement and, in some cases, both horizontal and vertical movement. One type of movement associated with the hardening of the concrete is drying shrinkage. In addition, since all parts of a structure will not settle uniformly, it is necessary to provide for some differential in the vertical movement. For example, columns

Table 8–A
FLOOR CLASSIFICATIONS

Class	Usual Traffic	Use	Special Consideration		Concrete Finishing Technique
1	Light foot	Residential or tile covered	Grade for drainage		Medium steel trowel
2	Foot	Offices, hospitals, schools, churches	Nonslip aggregate mix in surface		Steel trowel; special finish for nonslip
		Ornamental residential	Color shake, special		Steel trowel, color, exposed aggregate; wash if aggregate is to be exposed
3	Light foot and pneumatic wheels	Drives, garage floors and sidewalks for residences	Crown; pitch joints; air-entrainment		Float, trowel and broom
4	Foot and pneumatic wheels	Light industrial commercial	Careful curing		Hard steel trowel and brush for nonslip
5	Foot and wheels—abrasive wear	Single-course industrial, integral topping	Careful curing		Special hard aggregate, float and trowel
6	Foot and steel-tire vehicles—severe abrasion	Bonded two-course heavy industrial	Base: Textured surface and bond		Surface leveled by screeding
			Topping: Special aggregate, and/or. mineral or metallic surface		Special power floats with repeated steel troweling
7	Unbonded toppings	Classes 3, 4, 5, 6	Mesh reinforcing; bond breaker on old concrete surface; minimum thickness 2½ in. (65 mm)		—

and walls that rest on footings may settle either more or less than a floor slab resting on the subgrade. If the floor slab is tied into the walls or columns by reinforcement, bond, or keyways, cracks will occur because of this differential settlement. It is, therefore, necessary to isolate the floor slab from other parts of the structure.

Control joints can be formed by tooling, but for industrial floors this is generally unsatisfactory, since they will cause small-wheeled vehicles to jolt as they pass over the joint. Control joints also can be formed by inserting metal or fiber strips immediately after the concrete is placed, but this often complicates the finishing operation because it is difficult to keep them vertical and in place. One of the most satisfactory methods of forming control joints is to saw them into the concrete after it has stiffened. The resulting void is then filled with molten lead to prevent chipping of the edges. The joints should divide the floor into approximately 15-ft. squares. It is important that control joints be spaced close enough so that aggregate interlock is maintained. The horizontal movement or joint opening is a function of the joint spacing.

Isolation joints are joints that separate the floor slab from walls or columns or other parts of the building. They permit both horizontal and vertical movement. These joints do not employ interlocking aggregate, keyways, or reinforcing steel.

Construction joints are stopping places in the concreting operation. They can be made rigid, but this is usually difficult. The lack of aggregate interlock can be remedied by forming a keyway to provide for load transfer across a construction joint.

Mix. A proper concrete mixture must be used for industrial floors. Of prime importance is the water-cement ratio which, for industrial floors on ground, should not exceed 0.50. The flexural strength of the concrete measured in the standard manner should be at least 600 lb per sq in. To obtain this, the compressive strength generally must be, at least, 4500 lb per sq in. on standard 6×12-in. cylinders. The cement content should be at least $5\frac{1}{2}$ sacks per cubic yard for $1\frac{1}{2}$-in. maximum size aggregate or 6 sacks per cubic yard for $\frac{3}{4}$-in. maximum size aggregate. To minimize bleeding and segregation, it is recommended that low-slump, air-entrained concrete be used. The air content need not exceed about 5%.

The concrete should be as stiff as possible. Rarely should the slump exceed about 3 in. for industrial floors on ground. Most of the problems that occur in floors can be traced to overwet concrete mixtures. The aggregates should be the hardest and toughest available. If an extra hard surface is needed, it may be obtained by constructing a two-course floor or by use of a dry shake. In either case, selected hard, tough aggregates should be used. The cost of finishing a floor is generally about the same regardless of the type of aggregate. The additional cost for using the best aggregate available is generally a small percentage of the total cost.

Reinforcement. To be effective, reinforcement should be located about 1½ in. from the top of the slab. Enough steel should be provided to eliminate random cracking. Some authorities feel that to do this, the cross-sectional area of steel should equal ½ to ¾ of 1% (0.005 to 0.0075) of the cross-sectional area of the slab. This means that for a 5-in.-thick floor, ⅜-in. bars on 4-in. centers, running in each direction, would be the minimum required. Random cracking can be prevented easily and often less expensively by providing adequate control joints. The common practice of laying wire mesh on the subbase and then pulling it up through the concrete often results in the steel's being near the bottom, or under the slab, where it is ineffective for resisting cracking. Steel should not be used across a control joint since it will not allow the joint to function.

Thickness of Floor Slab. The thickness of a floor slab is dependent on the maximum load that must be carried on a relatively small area. Generally, this is the heaviest wheel load anticipated. The greater the wheel load or the smaller the contact area, the thicker the slab must be. The load that a slab of given thickness can support is a function of the flexural strength of the concrete. The stronger the concrete, the greater load the slab can carry. In other words, for a given loading condition, the required thickness of a slab can be reduced when stronger concrete is used. The poorer the subbase, the thicker the slab must be to support a given load. It is recommended that industrial floors on ground never be less than 5 in. thick.

Surface. It is equally important that the wearing surface be hard and smooth, especially if small-wheeled vehicles will use the floor. If floors are not hard and smooth, even light traffic on hard wheels will erode the surface.

In order that the concrete has the desired hard, dense surface, it is important that it is adequately finished. The four most important factors to observe are:

1. To use low-slump, air-entrained concrete, and be sure to get complete consolidation. This will minimize bleeding and segregation.

2. To delay floating until the concrete is stiff enough to withstand the weight of a man.
3. To trowel the surface repeatedly with time lapses between passes until the trowel rings as it passes over the concrete.
4. To make neat control joints. If they are tooled, use a small-radius tool. If they are sawed, cut them as soon as the concrete has hardened sufficiently so as to ravel only slightly during the sawing operation.

Curing. Concrete floors must be cured a minimum of five days under ideal conditions. A longer curing time is preferable. Adequate curing depends upon moisture, time, and temperature. It is necessary that adequate moisture be provided and that the concrete not be allowed to dry out until the curing period is completed. The lower the temperature, the longer curing must be continued. In cold weather, a minimum of seven days is recommended. Waterproofed curing paper is an excellent means of curing concrete floors on ground. Sheets of curing paper should be overlapped and taped. Other methods include the use of damp burlap, polyethylene sheeting, or a curing compound. After the curing period, floors should be allowed to air-dry for a few days before traffic is permitted.

Dry-Shake Method. Dry shake is applied and finished the same way for industrial floors as when applying color by this method, except that the materials may be job mixed. Mineral or metallic aggregates are mixed with dry portland cement and, if job mixed, must be proportioned according to the aggregate manufacturer's directions (usually one part cement to two parts aggregate by weight). An iron aggregate surface mix should not be applied to floors containing calcium chloride because of the increased tendency toward rust staining. The materials should be thoroughly dry mixed, applied, and finished in the following manner.

1. Float the surface (by hand or power) after all water and excess moisture have evaporated or have been removed.
2. Evenly distribute about two-thirds the specified amount of material.
3. Float again to incorporate the dry-shake aggregates into the concrete.
4. Apply the remaining one-third of the specified amount at right angles to the first application.
5. Float a third time.
6. Trowel (hand or power). No troweling or use of power trowel blades should be permitted before this time.
7. Trowel as many additional times as necessary. (See Fig. 8-7).

Figure 8-7 Application of iron dry-shake to an industrial floor to produce a very wear-resistant surface.

Two-Course Floors—Class 6 (Table 8-A)

Two-course floors of Class 6 are classified as heavy-duty industrial floors. The top course must be at least ¾ in. (20 mm) thick. The concrete used should have a maximum slump of 1 in. (25 mm). Because of the relatively small amounts of concrete in the top course and the low slump required, the concrete for the top course should be job mixed. The base course should be screeded and darbied (or floated). Close surface tolerance of the base course is important. Two kinds of floors, integral and bonded, may be made.

Integral. For these floors, the top course is placed before the base course has completely set. Any excess moisture or laitance should be removed from the surface before the top course is placed. At the time that the top course is being placed, the concrete should be sufficiently hard so footprints of the workmen are barely perceptible. The use of a disc-type power float is suggested, followed by a minimum of two power trowelings on the top course. Finally, hard troweling should be done by hand to remove any trowel marks left by the machine. The concrete is then moist-cured for maximum hardness.

Bonded. To prepare the base course for a bonded topping, locations of joints must be marked so that joints will be placed in the top course directly over those in the bottom course. After the base course has partially set, the surface should be brushed clean with a broom. This removes laitance and helps to assure bond of the top course. The base course should be wet-cured a minimum of three days. If the topping is to be applied without further delay, the curing cover or water should be removed from the slab and any collected dirt and debris should be washed off. After all free water has evaporated or has been removed from the surface, a grout should be applied. The grout should be a 1:1 mixture of portland cement and sand passing the No. 8 sieve (2.38 mm), mixed with enough water to produce a creamy consistency. The grout should be applied to the floor in segments, keeping only a short distance ahead of the concrete-placing operations that follow it.

While the grout is still damp, the top course should be spread and screeded. The use of a disc-type power float is suggested, followed by a minimum of two power trowelings. Trowel marks left by the machine should be removed by the final hard-steel troweling done by hand.

If a short time is to elapse between the end of curing of the base and the placing of the top course, the surface of the base course should be protected from dirt, grease, plaster, paint, or other substances that would interfere with the bond. Prior to placing the top course, the base course should be kept wet overnight. Immediately before placing the topping, the base course should be thoroughly cleaned by scrubbing with a brush and clean water. All excess water should be removed and a thin layer of grout should be scrubbed in. While this is still damp, the top course should be placed and screeded.

If a long time (from a week to several months) elapses between the end of curing and the placing of the top course, with the floor being subjected to heavy construction traffic, a more thorough cleaning is needed. The surface is scrubbed with water containing detergent. If oil or grease has been spilled on the floor, a mixture of sodium metasilicate and resin soap is useful. The floor should then be rinsed until the wash water is neutral or only slightly alkaline to pH paper. Commercial hydrochloric acid, diluted 1:10 with water, should then be applied at about 1 gal. per 100 sq ft. This can be applied through a sprinkler made by drilling $\frac{1}{16}$-in. (1.5-mm) holes in a plastic hose at 2-in. (50-mm) intervals. The acid should be broomed over the surface by workmen wearing rubber boots, gloves, and goggles. Respirators and rubber suits may be necessary. Exposed areas of the skin should be protected from the acid with grease.

After the acid stops foaming, the floor should be flushed with water and scrubbed with stiff brooms to remove reaction products and any sand that has been loosened. Washing should continue until the surface liquid is neutral to pH paper and all loose material has been removed. The surface should then be inspected, and any remaining laitance or unsound concrete removed by repeated treatment. Finally, the surface is allowed to dry until no free water remains. A grout is scrubbed in and the topping applied as described above.

In some circumstances it may be convenient or desirable to bond the topping with an adhesive. Certain epoxy and epoxy-polysulfide resins have been used successfully for bonding fresh to hardened concrete. Since floor hardness is important, moist curing is essential.

Two-Course Floors—Class 7 (Table 8–A)

The two-course floors of Class 7 are unbonded, and are used when floors will be subjected to strong acids or other destructive matter. Chemical plants and slaughter houses are examples of where floors of this kind might be useful. When the topping has been eaten away in one area, it can be removed and replaced easily. Care must be taken to insure that the topping does not bond at any point.

The topping must be, at least, 2½ in. (65 mm) thick, but thicker toppings are recommended. The base course, whether old or new, should be covered with plastic sheet, or with felt, spread as wrinkle free as possible. Sufficient wire fabric, supported on "chairs," is placed to control shrinkage cracking in the finished slab. Relatively dry concrete is used for the top course. Power floats and power trowels are usually required, but final, hard troweling should be done by hand. In all other respects, the unbonded toppings are constructed in the same manner as those already discussed.

Exposed Aggregate on Grade

Exposed aggregate is one of the most popular decorative finishes for flat concrete slabs. It offers an almost unlimited color selection and a wide range of texture variation. Properly prepared exposed aggregate finishes have a rugged, nonskid surface that is highly resistant to wear and weather. A project in Meridian Hill Park, Washington, D.C., originating in 1918, still shows no signs of weathering (Fig. 8-8).

Selection of Aggregate. Things to consider in selecting decorative aggregate include:

1. Size and color, to produce the desired effect.

Figure 8–8 An attractive exposed aggregate drive.

2. Shape and hardness, to produce a good-looking, long-lasting finish.
3. Soundness. Aggregate containing soft particles or particles that oxidize and cause stains should be avoided.

Size of the aggregate can vary from pea gravel to fist-size rocks, depending on the desired texture. Recommended sizes range from ⅜ to ¾ in. Aggregate less than ¼ in. in diameter should be avoided. Either natural or white portland cements, with or without color oxides, can be used to augment the aggregate color. For many jobs, natural portland cement with locally available aggregate is acceptable. Native gravels are usually found in a range of browns. Some of the more attractive surfaces are obtained with aggregates containing a large percentage of black particles.

Aggregate chosen should be hard and sound. Flat, silver-shaped particles, or aggregate that is too smoothly polished or ball shaped, should be avoided. Such aggregates do not bond properly and are easily dislodged.

Preparation of Aggregate. Aggregate should be cleaned prior to use. It should have no fines or coatings that might affect bonding. A test panel should be made using the selected aggregate before the job is started. This will insure that the material chosen will produce the desired finish.

Aggregate should be damp with all free water drained off. Degree of dampness depends on weather conditions (hot, windy weather requires more surface moisture than cool, humid weather). Specifications may call for aggregate to be saturated or surface dry.

Placing Methods. There are two methods of placing aggregate and concrete for an exposed aggregate finish. One is an all-in-one method where aggregate and concrete are mixed and placed together. The other involves placing the slab and laying the aggregate as a topping course.

All-in-one method. The following mix is suggested:

Portland cement (Type 1A), 1 part by weight
Concrete sand, 2 parts by weight
3/8-in. and smaller pea gravel, 3 parts by weight

The above mix can be varied somewhat as to aggregate content, but experimentation should be conducted on small slabs until a satisfactory appearance is obtained. Mix water should be carefully controlled to hold the slump as low as practicable for workability. A stiff mix prevents too much paste from rising to the top, but the mix does not have to be too stiff for reasonable workability. A 2- to 3-in. slump is recommended.

Before placing starts, individual slabs should be laid out in areas between 100 and 200 sq ft in size. If too large an area reaches the proper degree of set at one time, the washing procedure cannot be completed before some of the slab sets too hard.

A crew of two finishers and two helpers for an area of about 800 to 1000 sq ft per day has proved satisfactory. The individual slabs can be separated nicely with 1×4-in. redwood boards arranged in suitable patterns.

One ready mix truck is adequate for a crew of four because the delay in returning to the ready mix plant for additional loads allows enough time for the washing operation as the respective areas reach the proper degree of set.

While placing the concrete, the usual procedures should be followed with respect to spreading and striking off. Metal strike-offs should be used since they produce a minimum of mortar on the surface. Following the straightedge, the slab should be floated with magnesium darbies or bull floats only until the surface is level and properly sloped to drains. Too much working at this stage will depress the aggregate too deeply into the cement paste.

Exposing the aggregate (all-in-one method). From the floating operation to completion, timing is critical. Consideration must be given to weather conditions in judging whether a slab is ready for washing. In hot, humid weather the slab will set uniformly from bottom to top without excess drying of the surface. For that reason, hot, humid weather is best for finishing exposed aggregate slabs of this type. When the relative humidity is low, the slab will tend to dry out on the surface prior to setting. This tendency is aggravated if the wind is strong. A fine mist may be sprayed continuously on the surface, starting as soon as the water sheen disappears, or the slab should be covered with a waterproof sheeting to prevent premature drying. During cool, humid weather, the concrete sets properly, but slowly.

The slab is ready for washing when the initial set has occurred, the water sheen has completely disappeared, and the concrete will support the weight of a man without deflection. This condition is generally reached between 45 min and 5 hr after the concrete is struck off. The surface should then be brushed without water. If the surface is too hard, water may be gently applied to remove surface mortar uniformly until aggregate is visible. Two to three washings should follow with a time lapse between each. When clear water flows off the slab, the job is almost completed. After curing, the exposed surface should be lightly washed with a hydrochloric (muriatic) acid solution (20%). The acid wash will remove any light residue of cement and brighten the colors of the aggregate. Vigorous scrubbing and flooding with clear water will remove any traces of acid.

The type of brush used is important. The bristles should be nylon because they retain their stiffness after becoming wet, whereas natural fibers soften when exposed to water. The washing operation requires two finishers—one to man the broom, the other to handle the water hose. The washing operation should begin at a corner of the slab on the high side. Good exposure should be attained in one pass so that coarse aggregate will not be dislodged. Areas that have too little exposure can be reworked later.

While washing, care must be exercised to avoid cutting too deep. Many finishers feel that a large percentage of the stones must be exposed, but attempts to wash too deeply usually result in displacement of large balls of concrete. A good compromise is to remove concrete to a depth of one-third the average diameter of the aggregate. If the individual stones are not left adequately embedded in the cement paste, they soon will be dislodged under traffic. For northern latitudes it is especially important to broom and wash only enough to expose the very tops of the aggregate, to minimize the loosening effect of freezing.

Topping-course method. A concrete mix containing about 550 lb of cement, with a maximum slump of 3 in., should be used for this method. Aggregate should be cleaned and soaked thoroughly before use to prevent mixing water being drawn from the concrete. Soaking also serves to lubricate the entry of the aggregate into the surface of the slab. After soaking, however, the aggregate must be drained to prevent free water being carried into the concrete.

Immediately after the slab has been screeded and darbied, the selected aggregate should be scattered by hand. It should be evenly distributed so that the entire surface is completely covered with a single layer. Special care should be taken in covering the edge to insure a uniform appearance.

The initial embedding of the aggregate is usually done by patting with a darby or bull float or with the flat side of a strike-off board. After the aggregate is embedded and as soon as the concrete will support the weight of a mason on kneeboards, the surface should be hand floated with a magnesium float or darby. This operation should be performed thoroughly, so that all aggregate is completely embedded just beneath the surface. The grout should completely surround and slightly cover all aggregate, with no holes or openings left in the surface.

Following this floating, a reliable retarder may be sprayed or brushed on the surface according to the manufacturer's recommendations. On small jobs a retarder will probably not be necessary. Retarders are generally used on large jobs for better control of exposing conditions. When a retarder has been used, exposure of the aggregate can be delayed. Care must be taken, however, to avoid heavy accumulations of retarder in low areas (Figs. 8-9 to 8-12).

Figure 8–9 The selected, decorative aggregate is distributed to cover the entire surface evenly.

Figure 8–10 The aggregate is initially embedded by patting it with a darby.

Figure 8–11 The surface is then darbied to completely cover the aggregate with grout.

Figure 8–12 Exposing the aggregate is done by simultaneously brushing and hosing with water.

Exposing the aggregate (topping-course method). Proper timing is quite critical in this method of exposing aggregate. A test panel should be made and tested to determine the best time to begin. As soon as the grout can be removed, exposing should be done by simultaneously brushing and flushing with water without overexposing or dislodging the aggregate. With either method, if it is necessary for cement masons to move about on the surface during exposing operations, kneeboards must be used with special care. This should be avoided if possible because of the risk of breaking aggregate bond.

Curing. Curing of the slab after washing presents no special problems and normal curing practices should be employed. However, membrane curing compounds should not be used, in case the surface of the slab needs an acid wash. Nothing that might discolor the slab should be used to retain the moisture. Straw, earth, and some kinds of building paper are examples of things that could discolor the slab.

Colored, Nonslip and Decorative Flatwork

Variations in texture and pattern of concrete surfaces are limited only by the imagination of the designer and the skill of the craftsman. Many interesting and practical finishes can, in fact, be produced with less effort and expense than a smooth-troweled surface. They can provide a nonslip finish, as well as being attractive.

Integral Color. There are several methods of using integral color in concrete flatwork. The choice of the method to use will depend on how much wear and tear the area will be subjected to, the size of the area, and the cost of the project. Only mineral oxides should be used since other pigments may fade.

One-course method. In the one-course method, the pigment and concrete are placed in the mixer, and the concrete is laid in one course. Mixing time should be longer than normal to assure uniformity of color. The dry pigment and cement should be thoroughly blended before being added to the mix.

Two-course method. A regular concrete base is placed and allowed to stiffen for the two-course method. Then a 1- or ½-in.-thick colored topping is put on and finished with a wood float or, if a smoother finish is desired, a steel trowel. Overtroweling should be avoided. This method requires more labor, but the economy of using pigment in only the top layer can more than offset the cost of the extra labor.

Dry-shake method. A mixture of dry cement, colored pigment and well-graded sand are applied to a floated surface. Eighty percent of the sand should pass a No. 8 sieve, and not more than 3% should pass a No. 30 sieve. The proportions will vary, but generally 1 part of pigment, 10 parts cement and 10 to 15 parts sand are used. The maximum amount of pigment that should be used is 10% of the cement by weight. The mixture is blended and two-thirds of it is uniformly sprinkled over the concrete surface after the surface water has disappeared. It is floated into the surface immediately, and then the rest of the dry shake is applied in a direction perpendicular to the first and floated in. The edges and joints should be run both before and after application. Air-entrained concrete may or may not be recommended for this method, depending on weather and job conditions.

Nonslip Surface Treatment

Dry shake. Slip-resistant aggregates should be hard and nonpolishing. Fine aggregates are usually emery, corundum, or a manufactured abrasive. The slip resistance of concrete may be improved by replacing the fines with those of a more slip-resistant aggregate.

Hard aggregates such as quartz and trap rock, as well as metallic aggregate, are frequently used in making heavy-duty floors. These will be discussed later.

Before being applied to the surface, the slip-resistant material should be mixed with dry portland cement. Proportions usually range from 1:1 to 1:2. However, the manufacturer's directions should be followed if given. The procedure for nonslip monolithic surface treatment is exactly the same as that outlined for colored surface treatment.

The slab must be well cured after treating and finishing.

Acid etch. Another method of making a surface nonslip is the acid etch, which is done after the concrete has hardened. The surface must be cleaned of all oil, grease, and dirt, and a solution of diluted hydrochloric (muriatic) acid sprinkled on the floor. There are disadvantages to this method.

1. The results are extremely varied and uneven.
2. Liquid acid drains to lower spots and crevices.

3. There is a possibility of injury to workers and adjacent surfaces from careless sprinkling.

A safer method of applying the acid has been developed. Sawdust is saturated with the dilute acid and put on the concrete surface in a ¼-in. layer. It is allowed to remain for one hour. If the sawdust is raked several times during the hour, it releases gasses that will give better results.

The solution used in this method is 1 part hydrochloric acid to 10 parts water by volume. One gallon of the solution will saturate about 1½ lb of dry sawdust. Approximately 120 sq ft can be treated with 25 lb of saturated sawdust.

Textured coatings. Textured coatings are compounds of polyurethane, epoxy resins, or other materials that are applied after the concrete has hardened. These are applied, allowed to harden, and subsequently textured. Epoxy resins applied in a ¼-in. topping are the most popular coating compounds. They are easy to apply, will take several antiskid textures, have greater ability to adhere to concrete than other compounds, and will retain the applied surface texture under extreme weather and traffic conditions.

Polyurethan coatings provide a long-lasting thin coating that has the advantage of permitting evaporation of water vapor through the topping. They are not recommended for use on concrete where there will be much standing water.

There are a variety of materials available for use as nonslip toppings. Some are composed of two materials, while some have only one component. They vary greatly in quality and durability. Manufacturer's directions should be followed when using them.

Nonslip textured surface—swirl or broom design. A nonslip surface texture that can be decorative as well as functional is the swirl design. This swirl texture can be produced on a slab by using a magnesium or aluminum float or a steel finishing trowel. When a float is used, the finish is called a swirl float finish; when a trowel is used, the finish is called a swirl or swirl trowel finish.

After the concrete surface has been struck off, darbied, or floated and steel troweled, the surface is ready to be given either the swirl float or swirl trowel finish. The float should be worked flat on the surface in a semicircular or fanlike motion. Pressure applied on the float with this motion will give a rough-textured swirl design (Fig. 8-13).

Figure 8–13   Swirl trowel finish.

With the same motion, but using a steel finishing trowel held flat, the cement mason can obtain a finer-texture swirl design on the concrete surface. Moist curing of the slab is the final operation.

The wavy broom design may be used to add interest to a nonskid textured surface. The design is given to the surface after all the conventional operations of striking off, darbying, floating, and troweling have been done. The broom is drawn across the slab in a wavy motion. The coarse texture is obtained by using a stiff-bristled broom. A finer texture may be achieved by using a soft-bristled broom (Fig. 8-14).

After the concrete has set sufficiently so that this surface texture will not be marred, the slab must be moist-cured.

Nonslip mechanical treatment. In recent years sawing has been used to produce a slip-resistant surface in hardened concrete. This technique is usually employed when conditions prevent application of a nonslip surface at the time of casting or when discovery of the need for slip resistance is made after the concrete has hardened. For example, a contractor installing a roadway encounters heavy rain that removes the textured surface and the concrete hardens before a new texture can be applied, or application of the nonslip surface is forgotten during the casting operations, or accidents reveal the existence of a slippery surface after the concrete has been in service for some time.

A mechanical treatment is usually more economical than retopping a slab. Various methods are available. The choice will be governed by the condition of the concrete and the size of the area to be treated.

Figure 8–14 Wavy broom finish.

Diamond-drum roughening is the most effective method. The concrete should be scored transversely on a single pass. The cutting drum should preferably have not less than 50 circular, segmented diamond saw blades per 12 in. of width. It should be set for a depth of $\frac{1}{8}$ in. To insure uniform scoring over the entire surface, it is best for the drum to be mounted on a multiwheeled articulated frame with outrider wheels. Mechanical sawing also provides a satisfactory surface texture, but it is frequently considered less satisfactory because a large quantity of water is needed during the operation. Saws also tend to ride over low areas, thus giving a less uniform treatment.

Flail grooving can provide concrete with adequate nonslip resistance. With this method, a number of hard steel disc flails cut grooves 1 in. apart, $\frac{1}{8}$ in. wide, and between $\frac{1}{16}$ and $\frac{1}{8}$ in. deep. Roughening of a surface is also possible with high-speed percussion rotary hammers and by thermal shock treatment with an oxyacetylene flame. Both of these methods are only suitable for use on comparatively small areas since they are rather slow and expensive.

Decorative Finishes. Random flagstone patterns make an attractive finish for patio, sidewalk, or poolside slabs. However, producing them is very time-consuming. Such patterns are made by embedding prepared wooden "joints," which are removed after the slab has hardened. Wood lattice stock ($\frac{1}{4} \times 1\frac{1}{2}$ in.) is cut into varying lengths of from 4 to 32 in. Individual pieces are finished by carving the sides into irregular jagged shapes. Edges must be undercut to ease removal and prevent locking of the wood into the slab surface. Curling of the wood strips may be prevented by thoroughly soaking them.

After the slab has been screeded and darbied (or floated), it is allowed to set until ready for finishing—in this case somewhat sooner than normal. The wooden joints are laid out on the slab in the desired pattern. Very small or complex shapes should be avoided.

As soon as the pattern is laid out, the wooden strips are pressed into the concrete and the surface is floated. The top of the wood should be just flush with the surface of the concrete. The concrete may then be troweled for a normal gray color, or if desired, a different color can be added by the dry-shake method. (Figs. 8-15 to 8-19).

After floating and before second troweling, the tops of the wood strips are cleaned of any paste that may have been carried over them. A patching trowel or putty knife can be used to scrape the wood clean. A final steel troweling will then bring the wood joints and concrete to a uniform surface, which can be left smooth or can be lightly brushed.

The following day the curing cover can be taken off, and the wooden strips removed by lifting with a patching trowel. The slab may be left

Figure 8–15   Wood "joint" forms for flagstone finish.

Figure 8–16 Laying out flagstone pattern with wooden strips.

Figure 8–17 Floating in wooden strips.

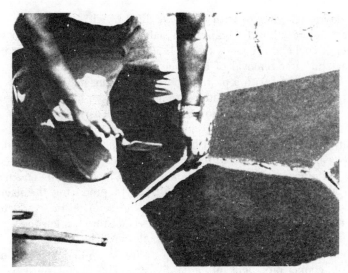

Figure 8–18 Removing joint forms.

Figure 8–19 Filling joints with contrasting mortar.

with depressed joints or with the joints filled in with a different colored mortar. Before filling the joints, the slab should be flooded with water to keep it cool and the joints damp. Immediately before troweling in the mortar and after removal of any free water from the joints, the joints should be brushed with a paste of portland cement and water, mixed to the consistency of heavy cream. Care must be taken to prevent smearing paste or mortar outside the joint area. A large coarse synthetic sponge and bucket of water will be useful in cleaning the joint edges. In this operation, it is best for two finishers to work together, with one painting in the cement grout ahead of the mortar and cleaning up the joint edges, while the other concentrates on packing in the mortar firmly and neatly.

An alternate method (Fig. 8-20 to 8-22) of producing the flagstone pattern involves tooling the joints into the surface. After the concrete has been screeded and darbied, and after excess moisture has left the surface, the slab is scored in a random flagstone design. This can be done with a jointer or groover. It is best, however, to use an 18-in.-long piece of ½- or ¾-in. copper pipe bent into a flat S-shape. This tooling must be done while the concrete is still very plastic, since coarse aggregate must be pushed aside by the tool and embedded into the slab. The first jointing will leave blurred edges. After the water sheen has disappeared, the

Figure 8–20 Tooling joints in slab.

Figure 8–21 Troweling after intial jointing.

Figure 8–22 Touching up joints with soft brush.

entire area should be floated and the jointing tool run again to smooth the joints. A careful troweling follows. The final operation is a light brushing of the troweled surface along with a careful touching-up of the joints with a soft-bristled paintbrush.

Applied designs. The leaf impression is an interesting decorative treatment that may be used as a border around a patio or along the edges of a garden walk.

Fresh leaves are taken from local trees. After the concrete has been floated and troweled, the leaves should be pressed carefully, stem side down, into the freshly troweled concrete. This is most easily done by using a finishing trowel. The leaves should be so completely embedded that they may be troweled over without dislodging them, but no mortar should be deposited over the leaves. After the concrete has set sufficiently, the leaves are removed.

The circle design is an interesting surface that can be used in many ways, such as a border around a patio slab, over the entire slab, or in alternate squares. Circles of different sizes and overlapping circles add interest.

After the concrete has been placed, struck off, darbied, floated, and steel troweled, the surface is ready to be given the circle design. If cans of various sizes are used, the largest should be applied first. The open end is pressed into the freshly troweled surface and twisted slightly to insure a good impression. After making a number of large circular impressions, the next size can is used and the operation is repeated. This is continued until the desired number and sizes of circle impressions have been made. If the slab is exposed to weather, the surface should be given a lightly brushed nonskid finish and the slab must be moist-cured (Fig. 8-23).

The keystone finish is a special one that has a travertinelike texture. It can be used for a patio, garden walk, driveway, a perimeter around a swimming pool, or in any location where an unusually decorative flat concrete surface is desired.

After the concrete slab has been struck off, darbied or floated, and edged in the usual manner, the slab is broomed with a stiff-bristled broom to insure bond when the finish (mortar coat) is applied.

The finish coat is made by mixing 1 bag of white portland cement and 2 cu ft of sand with about 1/4 lb of color pigment (usually yellow is used to tint the mortar coat, but any mineral oxide colors may be used). Care must be taken to keep the proportions exactly the same for all batches. Enough water is added to make a mixture the consistency of thick paint.

Figure 8–23 Designs and patterns can be created by many means.

This mortar is placed in pails and thrown vigorously on the slab with a dash brush to make an uneven surface with ridges and depressions. The ridges should be about ¼ to ½ in. high. The surface is allowed to harden enough to permit a cement mason on it with kneeboards.

The slab is then troweled with a steel trowel to flatten the ridges, leaving the slab surface smooth in some places and rough or coarse-grained in the low spots. Depending on the amount of mortar coat, many interesting textures can be produced. The slab should then be scored into random geometric designs before curing (Fig. 8-24).

A texture similar to the keystone finish is produced by scattering rock salt over the troweled surface and then pressing it into the concrete with a roller. After the concrete has completely hardened, the salt is washed away by thorough flooding with water. This dissolved salt will leave pits or holes in the surface.

The keystone and rock salt finishes should not be used in areas subject

Figure 8–24 This random-scored keystone finish makes an attractive patio area.

to freezing weather. Water freezing in the recesses of these finishes tends to break up the surface.

Sidewalks, gardens, and patios can be enhanced by special tools made of lightweight aluminum which are used on freshly finished concrete. They come in a variety of designs resembling brick or tile.

The tools are actually platform walking units. One tool is placed in front of or beside another and the concrete mason walks over the surface of the concrete. As he does this, the design is stamped into the concrete. For this type of tool, the concrete should not contain aggregate larger than 1/4 in., and the cement content should be at least 550 lb per cu yd.

The slab should cure for at least three days before any more work is done on it. At the end of this time, the joints may be partially filled with mortar. This will make an easier surface to walk on, especially with high heels.

These stamping tools are reuseable and do not require a great deal of skill to use, making this an economical way of decorating a concrete slab.

CHAPTER 9
FINISHING, PART III

FINISHING PAVEMENTS

The finishing of the surface is perhaps the most important step in pavement construction. The public demands a smooth-riding, skid-resistant surface. Properly constructed joints, uniform texture of final finish, and freedom from irregularities are all important in obtaining good riding qualities.

Machine Finishing

After the concrete has been deposited on the subgrade and spread to the desired thickness, it must be consolidated and finished. This is accomplished by mechanical finishing machines on most large projects. Large finishing machines have two screeds that not only give the concrete its proper contour but also consolidate it by pressure. The front screed is about $3/16$ in. high to allow for consolidation by the rear screed. Other machine adjustments include those for variation in crown, forward speed of the machine, frequency of transverse screed movement, and front-to-back screed tilt. The finishing characteristics of a wide range of concrete mixtures can be handled by careful adjustment of these features (Fig. 9-1).

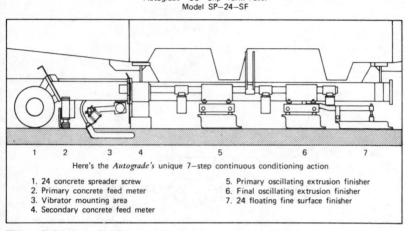

Here's the *Autograde's* unique 7—step continuous conditioning action

1. 24 concrete spreader screw
2. Primary concrete feed meter
3. Vibrator mounting area
4. Secondary concrete feed meter

5. Primary oscillating extrusion finisher
6. Final oscillating extrusion finisher
7. 24 floating fine surface finisher

Figure 9–1 A profile diagram of a slipform paving machine.

The added workability of air-entrained concrete usually requires that the frequency of transverse screed movement be increased. To prevent tearing of the concrete, the leading edge of screeds should be raised ⅛ to ³⁄₁₆ in. above the rear edge. In all cases, the amount of concrete carried ahead of the screeds should be limited to 4 to 6 in. in diameter for the front screed and 1 to 2 in. for the rear. Excess concrete tends to flow under and lift the screed which may cause irregularities. If there is not enough material "rolling" ahead of the screed there are liable to be low spots and inadequate consolidation.

At intersections and other places where highways are widened, finishing is often done by hand or with a pipe screed becaused the machines are not able to work on pavements that widen gradually.

Hand Finishing

Where the size of the job is not large enough to justify the use of a finishing machine, or where special conditions prohibit its use, hand strike-off may be used. The strike-off screed is made of metal, or of wood with a piece of steel along the bottom. It is shaped to the crown of the pavement, and plow handles are attached at each end. It usually weighs at least 15 lb per lin ft and must be rigid enough to maintain its shape. The screed is cut to leave the concrete a little above the grade—usually ¼ in.—so that it will have the proper elevation after it is compacted.

The screed should rest on the forms and be drawn forward with a sawing motion. A depth of, at least, 2 in. of concrete should be carried in front of the screed for the full width of the pavement.

Following striking off, the concrete should be compacted by tamping. This consists of resting one end of the screed on the form and lifting and dropping the other end on the concrete. The opposite end is then lifted and dropped. The screed is moved forward in such a manner that the whole surface is struck at least once, and neither end is advanced more than 1 ft ahead of the other.

Vibration

Vibration may be either the surface or the internal type. Surface vibration is usually done by a pan-type vibrator mounted on the concrete spreader, or by vibrating the front screed of the finishing machine. Internal vibration may be done by a transverse tube supported ahead of the forward screen of the finishing machine, or by a series of vertical tubes mounted in a straight line across the pavement.

For a given cement content, a 10% increase in flexural and compressive strength can be obtained through the use of proper vibration. These results, however, can only be obtained if the mix is designed for use with vibratory equipment. Best results come from a mixture having a slump of about 1 to 2 in. Proportions should be adjusted to use a somewhat higher ratio of coarse aggregate to fine aggregate than would be satisfactory with nonvibrated concrete. Concrete with excessive mortar or a high slump is apt to result in segregation when it is vibrated. Coarse aggregate size is usually 1½ to 2 in. in paving concrete.

Scraping with a Straightedge

When required, the surface is scraped to remove small ridges left by the transverse finisher or the screeding and consolidation operation to produce a smooth-riding pavement. The scraping is done with a 10-ft-long straightedge mounted on a long handle.

The straightedge should have enough weight and rigidity to act as a cutting and smoothing tool and to retain its shape. For proper operation in constructing one lane at a time, the straightedge should be placed on the pavement at the near edge and, with the handle at knee height, it should be pushed across the surface to the far edge. Then the handle should be lifted to shoulder height and pulled in the same path back to the near edge. In two-lane construction the straightedge is pushed to the center of the pavement, then lifted over any excess material, gently replaced on the slab and pulled back. After this operation is completed, the straightedge should be moved forward half its length and the process should be repeated.

Checking the Surface

Following the scraping-straightedge operation, an inspector should check the surface. This is done at intervals of 3 or 4 ft transversely, with a light 10-ft. straightedge parallel to the center line (Fig. 9-2). Any high spots discovered are removed and low spots are filled. Disturbed areas are smoothed with a long-handled float and the surface is again straight-edged to see that it is uniform. Slip-forming has eliminated much of the need for hand straightedging.

Final Finishes

There are several methods of producing nonskid textured surfaces on concrete. Extensive field and laboratory investigations have shown that long-lasting skid resistance of concrete pavement is determined by the type of aggregate, the amount of mortar, and the surface texture result-

Figure 9–2   Even the early slip-form pavers required hand finishing.

ing from the combination of these materials. Initially, the skid resistance is related to the type of surface finish as well as the cement paste and fine aggregate. However, as the original finish wears away, the coarse aggregate becomes exposed and takes on more importance. The surface finish and the type of aggregate, therefore, are both of direct concern to the highway engineer in providing a high degree of skid resistance throughout the life of the pavement.

Tests have shown that the more abrasive and coarse grained the texture of the surface, the higher the friction values are likely to be for such surfaces when wet. Under dry conditions, airplanes can land on concrete runways safely. When it is wet they are susceptible to skidding and hydroplaning. When an airplane touches a wet airstrip, there is a film of water between the tires and runway that has a lubricating effect. As the depth of the water and the speed of the aircraft increase, the water builds up a resistance—in effect, a wedge—under the tires. This wedge tends to lift the tires away from the runway so that there is little area of contact. If the water becomes deeper than the tread on the tire, the aircraft is floating. This high-speed floating process is known as hydroplaning—a dangerous situation because friction and subsequent control are virtually eliminated.

Road and highway pavements require nonskid finishes, as do floors and sidewalks. The texture will vary according to the use for which the pavement is planned. Regardless of the method used to produce texture, the timing is of major importance. If started too soon, while mortar is still too soft or plastic, a dense, slick surface will result. On the other hand, if finishing is delayed too long, it will be ineffective in producing the desired texture.

Burlap Drag. A burlap drag is the most commonly used final finishing method for pavement. Generally it is in four layers, each weighing 10 oz per sq yd, at least 3 ft of which are in contact with the surface. The burlap must be kept reasonably clean and free of hardened mortar. This method produces a longitudinally grooved texture. The first pass is made before the water sheen has disappeared from the surface. Additional passes are made if required.

Transverse Brooming. Combing with a natural- or synthetic-bristle broom with reasonably coarse and stiff bristles can produce excellent skid resistance. A long-handled broom is pulled across the width of the pavement, giving a grooved texture perpendicular to the direction of the traffic.

Other Methods. As mentioned previously in the discussion of roughening concrete flatwork, sawing or roughening with a diamond drum, flail grooving and acid etching are all useful methods of roughening old concrete that has had its texture worn away.

SLIP-FORM PAVING

The slip-form paver spreads the concrete over the subgrade, vibrates and tamps it, strikes it off, and shapes it to the desired crown and thickness. Also attached to the machine is a mechanical transverse reciprocating rubber belt and, sometimes, a burlap drag for final finish. If necessary, minimum hand finishing similar to that used in the conventional paving method can be done ahead of the burlap drag. The result: a smooth-riding, economical, concrete pavement built in one pass of a single machine without the use of side forms.

History

On November 22, 1947, engineers of the Iowa Highway Commission formed a finished slab of concrete 18 in. wide and 3 in. thick without the use of side forms. In February 1948, by using a slightly improved and enlarged model, the same engineers succeeded in placing a slab 36 in. wide and 6 in. thick. In 1949, a much larger slip-form paver placed a ½-mile section of county road consisting of two 10-ft-wide, 6-in.-thick lanes. By 1969, slip-form pavers had been used in 43 states and several foreign countries (Figs. 9-3 and 9-4).

Characteristics

Sliding Side Forms. Sliding side forms eliminate the need for pre-erected side forms and all the equipment necessary to move them from one place to another. This reduces overhead and labor costs.

Subgrade Accuracy. The smoothness of the pavement is strongly dependent on the preparation of the subgrade. This requires special techniques in constructing the subgrade to accurate grade and cross section. The paver runs on tracks that require an area just outside each pavement edge to be shaped and compacted to comparable grade tolerances. The paver's tracks must be on carefully graded paths to make a smooth pavement. Final grade regulation is determined by accurately placed, taut wires attached to stakes alongside the path of the paver. A two-finger sensing

Figure 9–3   Conventional paving train requiring forms.

Figure 9–4   Slip-form paving train consisting of a placer spreader, mesh-placer and depressor unit, slip-form paver, and tube finisher. Notice that this train is set up to also place and depress continuous reinforcing steel. Photo courtesy of CMI Corporation.

element attached to the paver automatically adjusts the final screed to conform to the guide wires (Fig. 9-5).

Concrete Uniformity. There is a great need for batch-to-batch uniformity in the concrete. It should have the lowest slump that can be effectively compacted and finished. Minor variations can adversely affect concrete uniformity. Greater control is necessary to insure proper mix temperature, handling, absorbed moisture in the aggregate, and the like. The batching trucks must be kept moving regularly so that delivery will be prompt and evenly spaced.

Reinforcing Steel Installation. Some paving jobs require reinforcing steel to be laid before the concrete, and this can be done by special machines. The steel, however, can be easily dislodged by surplus concrete ahead of the main screed unless it is adequately tied and firmly anchored.

Mix Compaction. Mix compaction can be a problem in slip-form paving because of the nature of the paver. It moves 6 to 12 ft per min and has to thoroughly compact a 12-, 24-, 36-, or 48-ft-wide slab which is 6 to 10 in. thick and is shaped to crown. All this is done by a single machine.

Figure 9–5 The final smoothness of the pavement is regulated by the carefully placed stringline against which delicate sensing devices ride and modulate the machine's elevation and direction. Photo courtesy of CMI Corporation.

Advantages

Economy. Probably the greatest advantage is the economy possible with slip-form paving. It saves approximately 50 to 75 cents per square yard. This means more highway for less taxes. The initial capital investment for conventional paving equipment, including paving machines, forms, and the equipment necessary to handle the forms, is approximately $250,000. Slip-form equipment costs about $125,000, which is an initial saving of 50%. Fewer men are needed with slip-form equipment, so that there is also a savings in wages. The job is done more quickly, saving time and labor costs, and enabling the contractor to move the paver on to other jobs sooner.

Product Quality. Pavement produced with slip-form pavers gives improved riding quality. The better mix control required means increased strengths and better uniformity from one area of the pavement to another.

Operation Efficiency. One machine replaces all of the following:

1. Spreaders.
2. Finishers.
3. Float finishers.
4. Belt and burlap drag machines.
5. 10,000 to 12,000 line ft of steel paving forms.
6. Equipment necessary for handling and rehandling forms.

Less machinery means less manpower, which makes job management easier. All the working elements are kept in a smaller working area, providing for better control and supervision.

Versatility. Slip-form pavers can be used:

On large projects or small projects.

On thick (12-in.) pavement or thin (6-in.) pavement.

On new construction or repaving.

On mainline or ramps.

On level terrain or mountainous terrain.

On rural areas or urban areas.

On wide (48-ft.) pavement or narrow (10-ft.) pavement.

With reinforced concrete or unreinforced concrete.

For plain surfaces or crowned surfaces.

With an on-site concrete mixer or a central plant mix or combination.

Problems

The advantages of slip-form paving have been described. Naturally, there are also problems connected with it, and some of them will be discussed here.

Personnel. More training is required for field supervisors, operators and mechanics. Field supervisors should have technical training to know the nature of the equipment with which they are working. Operators need several months of training with other experienced operators. This requires careful planning ahead, so that new men are available as they are needed. Manufacturers often have men trained to give technical assistance and to help train new operators.

Care and Maintenance of Paver. Each time the paver is moved from one location to another, it must be inspected for misalignment, warping, breakage, and inadvertent changes in settings of various control elements. Special mechanics must be trained who are well versed in slip-form pavers, and who can cope with any mechanical problems that arise on the job. It is important that the paver keep moving to insure uniformity of pavement.

Preparation of Pavement Subgrade. Rough pavement is produced where there is inadequate preparation of the subgrade. Untreated base materials are easier to prepare than cement treated materials because they can be reworked until correct, but treated bases give a more stable, all-weather working platform.

Uniform Concrete Mix. A uniform delivery rate of concrete is important. Improved batching equipment has made this possible. High-frequency vibrators operating at 3000 to 5000 vibrations per minute have helped ensure adequate density of the concrete.

Edge Slump. Problems with the edge slump are usually due to a lack of uniformity in the mix. A slump of more than 3 in., long, trailing forms, poor alignment and control of the paver, damp weather, rain before the concrete hardens, or a mix with "bleeding" characteristics are other problems. Plastic sheeting should be used to cover the pavement if rain is likely. When rain starts, paving should be discontinued and the slab should be quickly finished. Steel forms or lumber should be kept handy to use if pavement begins to slump.

Preparation of the Subgrade

The type of subgrade on which pavement is placed should be determined by local experience and anticipated traffic. A granular subbase may be required. On an existing gravel road, there may be sufficient granular material in place to provide a good subbase. If the pavement is to be placed over an existing roadbed, the roadbed should be scarified to a depth of not less than 6 in.; it should then be recompacted and shaped to the correct line and grade to provide uniform support. Construction line and grade stakes are set as they are with conventional paving.

Special care should be taken to insure that the subgrade near the edge of the slab is compacted to a degree equal to that of the central portion of the roadbed. The nature of the existing material will determine the type of compacting equipment to be used.

After the subgrade has been graded and compacted, stringlines are set from the grade stakes on each side of the road. A form grader is used to cut a pathway at the proper grade for the tracks of the fine grader and the slip-form paver. This is the preferred method, although a motor blade is sometimes used to cut this path on sandy subbases. After the pathway has been cut, a second stringline is usually set on the inside of the path. It is set to a constant height above the path on steel pins and then is adjusted to a smooth grade line. Measurements are taken continuously to make sure that the pathway is parallel to the smooth vertical alignment of the string. Minor hand grading is done to correct any deficiencies. This double check insures an accurate pathway elevation for the paver track, the key to a smooth-riding pavement.

Fine Grading. Fine grading may be done by either of two methods. On soils that are easily worked, such as sandy soils, fairly good results may be achieved with a subgrade planer with tracks similar to those on the slip-form paver. This machine is usually pulled by a crawler tractor. However, a self-propelled mechanical fine-grading machine mounted on tracks, similar to the slip-form paver, is preferred for fine grading and is more adaptable to all types of subgrade soils. Excess material is dumped outside the track line without the need for any hand shoveling. After the fine grading has been completed, a tandem or steel-wheel roller may be used to recompact material loosener during fine grading.

If the shoulders are not wide enough to handle a concrete mixing operation, it may be necessary to operate on the prepared subgrade. In such instances, the fine grader may be equipped with a bridge so that the batch trucks can pass over it without interruption. Operation of trucks on the prepared subgrade should be avoided. If it is necessary, truck operations should be spread over the entire width of the roadbed and should not be concentrated in a single lane, since this will destroy uniform subgrade compaction.

If the paving mixer and batch trucks are operated on the prepared subgrade, a subgrade planer must be attached to the rear of the mixer to plane off any irregularities formed by the mixer trucks and truck wheels. While this operation cuts off high spots caused by subgrade rutting, it does not provide compaction of material in low areas.

The final subgrade preparation consists of wetting the subgrade ahead of paving to prevent absorption of the mixing water by the subgrade. In some areas, waterproof subgrade paper or plastic sheets are used to eliminate the need for wetting the subgrade.

Concrete for Slip-form Paving

Production and Transportation. Concrete mixes for slip-form paving can be produced in several ways, depending on the size of the job and the proximity to ready mix plants.

Portable high-production, central-mixing plants are widely used for many reasons. They can accommodate a large job easily. They can fill large capacity, quick-dump hauling units by employing either single- or dual-drum mixers. They are portable, so they can be moved in a short time and relocated even during one particular job. Approved mixing drums can complete the mixing in 50 to 60 sec, so hauling units can be filled and moved on rapidly.

Another method of mixing and transporting concrete to slip-form pavers is in tilt-drum, truck-mounted mixers filled at low-profile batching plants. On-site mixers supplied by batch trucks are also used very successfully on small jobs.

Delivery of Concrete at Paver. Mixing time and the amount of water used must be carefully controlled to have a uniform mix. One of the most important considerations in slip-form paving is rate of delivery of the mix to the paver. Dump men are usually employed to coordinate the rate of delivery and to insure a uniform speed of operation. The need for smooth pavement, as well as for economy in operation, makes this necessary.

No matter what means is used, delivery must be uniform and dumping must provide a constant level of concrete in front of the main screed. The operator can vary the speed of the paver to control the height of the concrete in front of the screed, but unless there is enough concrete, there will be low spots that will result in rough pavement.

Spreader boxes have been developed with devices to hold the wheels while concrete is being dumped into them from hauling units. These work very well—so well, in fact, that some contractors using pavers equipped with receiving hoppers use spreader boxes instead.

It is relatively much better practice to slow the paver to creeping speed rather than stopping it completely during any interruption of the truck cycle. Operators must anticipate the rate at which concrete mix is arriving at the paving site as far in advance as possible and must adjust paver speed accordingly, rather than paving rapidly while trucks are on hand and having to wait between trucks.

Final Finishing

Usually a burlap drag is used to give texture to the surface after slip-forming. This is done by having two burlap drags attached to wheel-mounted, hand-pushed, foot bridges. A similar bridge is often used to carry curing liquid to be sprayed on the surface, although the curing bridge is self-propelled (Figs. 9-6 and 9-7).

Slip-form paving does not always leave a smooth-riding surface, but concrete planers can be employed to cut away high spots. One type of planer has 2-ft-wide cutting heads with about 100 diamond saw blades. Another has a heavy cylinder with diamonds set in the surface in a spiral pattern. Bumpcutting leaves a disagreeable looking surface and is very expensive. It is much more desirable to construct a smooth surface from the beginning.

Sometimes the surface is not skid resistant enough for the intended use because of weather conditions, equipment failure, and the like, that delay the finishing work. In these cases, scoring or scarifying can be done. Grooves about 1/8 in. wide and 1/8 in. deep are cut in the pavement. Texture provided by this operation has shown great promise in preventing the condition of hydroplaning, which occurs with aircraft taking off and landing on wet pavement, as well as with cars on highways.

Figure 9-6 Final finishing is often done by a machine like this tube finisher which is adding a texture to the concrete with a burlap drag. Photo courtesy of CMI Corporation.

DISCOLORATION

Concrete can be used as a decorative as well as a utilitarian material. As a result, our standards for its surface appearance have risen considerably. We expect concrete to be attractive, but it can become discolored as a result of conditions present in the concrete itself, or as a result of external factors.

Causes of Discoloration—Mix Design and Concreting Practices

Discoloration caused by mix design and ingredients, or by poor concreting practices, generally takes one of these forms:

1. Large areas of contrasting color.
2. Black spots or dark "leopard" spots.
3. White patches caused by efflorescence.

Areas or sections of slabs that contrast with other areas often result because of a change in the cement being used. Cement from different

Figure 9–7 This machine is scarifying the surface and then spraying a curing compound on newly placed concrete. Photo courtesy of CMI Corporation.

manufacturers can contrast markedly in color before and after hydration. If more than one ready mix supplier is being used for a job, the resulting color of the concrete could be distinctly different if they use different brands of cement in addition to different mix proportions. Aggregates also can vary in color from one source to another, but this is usually a minor factor.

There is no way (except by painting) of reducing this contrast in color once the concrete has been placed, since it extends through the concrete section. If uniform color is important for a job, care should be taken to insure that all exposed concrete will be made from similar materials.

Even with the same ingredients, a distinct change in color can result from a change in the water-cement ratio. A cement paste with a low water-cement ratio is characteristically darker than one with a high ratio. Construction practices and mix designs which could result in localized variations in the water-cement ratio—such as conditions conducive to bleeding, and indentations in the surface which could collect water—must be avoided.

Nonuniform curing and curing procedures also have a bearing on variation in concrete color. Concrete that is not moist-cured at all exhibits the greatest amount of discoloration. Moist-cured concrete usually exhibits the least variation.

Uneven curing will often result in uneven coloring. Therefore, if even coloration is important, care must be exercised to insure that the curing membrane is applied over the entire surface of the slab in a reasonably uniform amount. Moisture barriers such as polyethylene film and water-proof paper must be thoroughly sealed at joints, must be anchored carefully along edges, and must be as wrinkle-free as possible.

When these materials are allowed to become rippled over concrete containing calcium chloride, a pronounced mottled appearance, called the "greenhouse effect," often results. On warm days each ripple becomes, in effect, a miniature greenhouse wherein a water evaporation-condensation cycle is repeated many times. The results are white efflorescence, plus variations in color caused by buildup of condensate water at the perimeter of the ripple and at low spots in the concrete surface. It is extremely difficult and time-consuming, on a project of anything but the smallest size, to smooth out the wrinkles in a polyethylene film. Other means of curing are recommended where uniform surface color is important.

Calcium chloride in concrete can contribute to discoloration. The effect of calcium chloride depends on the alkali content of the cement. If there is a relatively small amount of cement alkalies, compared to the calcium chloride, light spots may occur on a dark background when the concrete is not moist-cured. If the converse is true—a high ratio of cement alkalies to calcium chloride—dark spots on a light background often develop in the absence of moist curing.

Spotting is more obvious on hard-troweled slabs, but infrequently noticed on rough-finished surfaces (burlap drag, rough broomed, or floated) because the rough texture camouflages the discoloration to some extent, and spotting is not frequently encountered with this type of surface finish.

Blackening of the surface is especially common when the flatwork is "burned" by late and vigorous troweling. This blackening apparently occurs as a result of a combination of hand rubbing of the metal from the trowel on the hardened concrete and a pronounced reduction of the water-cement ratio at the surface. Curing and subsequent corrective treatments ameliorate this problem somewhat.

Other factors that have an effect on discoloration include subgrade variations, applying cement as a dry shake, and applying mortar or cement paste to the surface. Concrete placed on a subgrade with varying absorptive capacities will vary the water-cement ratio according to the amount of water absorbed by the subgrade. This, in turn, will produce dark and light areas. Similarly, concrete that is not protected from drying winds will experience spotty curing and deposition of salts on the surface. Dusting overly wet concrete surfaces with cement to speed finishing time and smearing of plastic mortar or cement paste on concrete surfaces that have hardened too much to complete finishing will produce discoloration.

Avoiding Discoloration

Careful preparations should be made. A uniform, nonabsorptive subgrade should be developed; it should be moistened or covered with a moisture barrier if necessary. Concrete deliveries and work crews should be scheduled to avoid variations in placing and finishing procedures. Adequate protection of the concrete must be arranged.

Ordering concrete from more than one source should be avoided, if possible. If cement or aggregates or admixtures must be ordered from different suppliers, it is important that they be of the same type and quality. The use of calcium chloride should be avoided, especially in concrete containing low-alkali cement.

Concreting and finishing practices must be standardized and the concrete must be protected from drying between finishing operations.

Troweling must be done early enough to prevent burning the surface.

The concrete should be thoroughly and uniformly cured, using a wet-curing technique if possible.

Corrective Measures

If, in spite of all these precautions, concrete becomes discolored in areas where it is important that it have uniform color, the following measures can be taken.

1. *Large areas of contrasting color caused by different concrete-making materials*: The only corrective treatment for this type of discoloration is the application of some kind of opaque coating. This coating could be paint; or, in the case of colored concrete, one of the floor waxes offered by some manufacturers to match their dry-shake coloring materials.

2. *Dark spots*: In concrete not containing calcium chloride, dark spots can often be removed with a single washdown of water. In concrete containing calcium chloride, it is often necessary to wash the area several times. It appears that best results are obtained when such discolored concrete is washed soon after the spots appear. Also, in concrete containing calcium chloride, tests have indicated that di-ammonium citrate can be a safe, highly effective means of removing discoloration; unlike acids, it is not dangerous to handle. When applied to a dry, discolored slab, it penetrates the surface and promotes hydration of the cement compounds that contain iron, the process responsi-

ble for dark discoloration in many instances. After the treatment, the surface should be scrubbed and wetted down thoroughly for several days. This water can then penetrate more easily to promote further hydration, resulting in lightening of the concrete.

3. *Light spots:* Light spots are usually more difficult to remove. Repeated washings or weathering cause the dark background to lighten, approaching the shade of the spots. A chemical corrective treatment often used to eradicate light spots is the application of a strong lye wash. With this type of treatment, a 10% solution of sodium hydroxide (caustic soda) is distributed over the dry slab. This is allowed to remain in place for 1 or 2 days, and then is removed by thoroughly washing the surface. This type of corrective treatment is most effective when applied soon after the concrete has been cured. Caustic soda is harmful to skin and eyes; protective clothing (rubber boots, gloves, goggles, etc.) must be worn when using it.

Strong acid treatments generally have proved costly and dangerous to apply, difficult to control, and not particularly effective in reducing this type of discoloration.

4. *Efflorescence:* The most effective treatment for efflorescence is generous, prolonged flushing of the surface soon after the efflorescence appears. Scrubbing the set concrete with a stiff nonmetallic brush also will sometimes help. If these two techniques fail, a very dilute (1 to 2%) phosphoric acid solution will remove calcium carbonate efflorescence from hard-troweled slabs. Or a 3% solution of acetic acid or phosphoric acid can be employed.

Discoloration Caused by Outside Factors

Concrete can be discolored by a number of materials if they come in contact with it. Listed here are some stains that are commonly encountered, together with means of identifying and eradicating or minimizing them (Table 9-A).

Table 9–A

Type of Stain	Identification	Corrective Technique
Iron stains	Rust coloring or proximity to iron or steel objects.	*For light-colored stains:* mop the surface with a solution of 1 lb oxalic acid powder per gallon water; after 2 to 3 hr rinse surface with water and scrub with stiff brushes or brooms. *More pronounced stains:* trowel onto stained surface a paste consisting of 1 part sodium citrate and 6 parts water that has been mixed thoroughly with

Table 9–A (cont.)

Type of Stain	Identification	Corrective Technique
		an equal volume glycerine and tempered with whiting. After paste has dried, it should be replaced in a few days with a new application of the moist paste.
		Another method for exceptionally dark stains: Cover surface with cloth or cotton batting soaked with solution consisting of 1 part sodium citrate crystals in 6 parts water. Allow to stay in place 10 to 15 min. On horizontal surfaces sprinkle surface with thin layer hydrosulfite crystals which have been moistened with water and covered with stiff paste of whiting and water. On vertical surfaces, a whiting paste that has been sprinkled with hydrosulfite and moistened slightly should be applied. Remove these pastes in 1 hr (if left on too long they will cause a black stain). The surface should be thoroughly flushed with clear water.
Aluminum stains	White deposit	Scrub surface with muriatic acid solution. On ordinary gray concrete—10 to 20% solution. On colored concrete it should be weaker.
Copper and bronze stains	Green or brown	Apply paste dry, by mixing 1 part ammonium chloride (sal ammoniac) and 4 parts powder talc with ammonia water added afterward. Several applications of this paste may be needed.
Ink stains	Ordinary ink colors, red, green, violet, etc.	Mix a strong solution of sodium perborate (available at most druggists) in hot water and mix with whiting to form thick paste. Apply in $\frac{1}{4}$-in. layer and let dry. If blue stain remains, repeat again. If brown stain remains, treat with sodium citrate paste described in the section on iron stains.
Indelible ink containing silver salts	Black stain	Apply ammonia water. Other materials effective in removing certain ink stains are: Javelle water, chlorinated lime, strong soap. Javelle water may be purchased at drugstores or prepared as follows: Dissolve 3 lb washing soda in 1 gal water. In separate container, slowly add water to 12 oz chlorinated lime to make

Table 9-A (cont.)

Type of Stain	Identification	Corrective Technique
		smooth paste. Mash lumps. Add paste to soda solution and enough water to result in total of 2 gal. Place in stoneware jar and let settle. Clear liquid on top is Javelle water.
Tobacco and urine stains		*Light stains:* gritty scrubbing powder mixed with enough hot water to produce mortar consistency. Apply in a ½-in. thick layer and let dry. Several applications may be needed.
		Deeper, more stubborn stains: dissolve 2 lb trisodium phosphate crystals in 1 gal hot water. In separate shallow enameled pan, mix 12 oz chlorinated lime to pastelike consistency by adding water slowly and mashing lumps. Pour trisodium phosphate solution and paste into stoneware jar and add enough water to produce 2 gal solution. Stir well, cover jar and allow lime to settle. Apply layer ¼-in. thick of clear top liquid converted to thick paste by addition of powdered talc. Let dry and scrape off with wooden tool. Do not spill liquid on either metal (which it corrodes), or on colored fabrics (which it bleaches).
Fire or rotten wood stains	Dark chocolate-colored stain	Scouring surface with powdered pumice or gritty scrubbing powder will remove loose surface deposits. Then scrub surface thoroughly with solution glycerine diluted in 4 times its volume of water. Afterward, the solution of trisodium phosphate and chlorinated lime mentioned under tobacco and urine stains is applied to surface by soaking flannel cloth and pressing firmly against concrete by means of slab of concrete or glass. Should be repeated as often as necessary to remove stain.
Oil stains	Black deposit usually on garage floors	Area should be mopped soon after oil is spilled and area covered with dry powdered material such as dry portland cement, Fuller's earth, hydrated lime, or whitning, so no stain will form. If stain is allowed to set, cover it with piece of flannel soaked in equal parts acetone and amyl acetate, which in

Table 9–A (cont.)

Type of Stain	Identification	Corrective Technique
		turn is covered with concrete slab or pane of glass. A thorough scrubdown with benzine or gasoline will sometimes remove all stains.
Coffee stains		Apply cloth soaked with solution of 1 part glycerine to 4 parts water (by volume). Javelle water and the solution described for use on fire stains are also effective.
Iodine stains		Iodine stains will slowly disappear of their own accord. However, they can be removed by wetting them with alcohol, and afterward coating them with talcum powder or whiting. On vertical surfaces, a paste of talcum and alcohol may be applied to the stain.

CONCRETE REPAIR

No matter how much care is taken in placing and curing concrete, minor defects caused by air holes, form wire ties or bolts, fins formed in the joints of the formwork, and the like, will occur. Damage to concrete surfaces from traffic or weathering is almost inevitable. These defects can be corrected so that they are virtually invisible, and the concrete will be as strong, or stronger, than before. Several methods of repairing concrete are discussed in this section.

Correcting Surface Defects

Small surface fins or bumps can be removed by rubbing with a fine or medium carborundum stone. When concrete is more than one or two days old, any air holes can be filled with a mortar consisting of 1 part portland cement and 1½ parts sand. About 25% of the cement may be white portland cement, since patches have a lower water-cement ratio and dry darker than surrounding areas. Where the surface is exposed to view, the mortar color should blend into the existing surface. This mix should be rubbed in with a cork or rubber float, since steel will cause the area to darken.

Holes left by form bolts or ties require that the concrete be chipped away, at least, ¾ in. and the hole cleaned out, making sure that no form oil is left.

Because concrete shrinks as it dries, the mix to be used for patching should be earth-moist (often called "dry pack"), having just enough moisture to be workable. After the hole has been cleaned, it should be moistened. The mortar should be rammed into the hole and hammered in place with a rod and mallet to pack it tightly.

The outer surface should then be wiped with a damp cloth or rubbed with a wood float. When the surface is even with the old concrete, it should be cured by fog spraying or covering with a wet cloth. Some smoothing with a carborundum stone may be necessary after curing.

Honeycombing is corrected in much the same way. All weakened concrete should be cut out as far as the reinforcement or even behind it. The surface should be cleaned and dampened. Neat cement grout is then brushed into the surface, and the hole is filled with a patch of earth-dry mortar that is tamped into the hole. Finishing should be done with a wood float. Curing is done by a water spray or a wet cloth.

Epoxy Resins. Epoxy resins have proved to be very successful adhesives for repairing structural concrete. In cases where dismantling whole structures was the only alternative method of repair, expoxies have been used without interrupting use of the structures, thus effecting substantial savings in time and money. Strength tests on structural members repaired by epoxy have shown the area of repair to be stronger than the original concrete.

There are many epoxy resins on the market. The decision as to the proper one to use for a specific job should be made only after much study and testing. Preparation of the surface for repairing with epoxies is nearly the same as for repairing with mortar. The surface must be clean and free of all moisture. A wash of dilute muriatic acid, followed by scrubbing and flushing with clear water, is desirable where feasible. Cleanup may be done with sandblasting followed by air-water jet washing and thorough drying. The area should be dried for, at least, 24 hr, and should be at a temperature between 65 and 80° F.

Since there are many epoxies on the market, each with special characteristics, the manufacturer's recommendations must be followed carefully. Dangers inherent in using chemicals require that safety precautions be taken by workmen when using epoxies. Efficient and effective repairs can be made on both old and new concrete when the directions of the manufacturer are followed.

Resurfacing Bonded Concrete Resurfacing. Resurfacing is largely the same, whether it be a pavement or a floor, a large area or just a patch. There are some processes special to each, but there are more similarities than differences. Table 9-B gives a general picture of steps in resurfacing bonded and unbonded floors, bonded pavement, and patches. We discuss bonded resurfacing in detail so that this chart can be used as a comparison with better understanding.

Resurfacing of pavement is done either to add strength to inadequate pavements or to level old, rough pavement surfaces. When the purpose is only to level rough surfaces, a thin resurfacing layer will perform as well as, and frequently better than, a thick slab *if* the bond between the old and new concrete is good enough.

Preparation of surface. Proper preparation of the surface of the old concrete is the most important step in securing a satisfactory bond. On all jobs, any loose or poorly bonded material should be removed from the old surface. Jackhammers or chipping tools may be used on small areas. At locations where the old concrete is sound, or where there is a thin layer of laitance but no scale, the detergent and acid treatment will provide a satisfactory surface for bonding.

All bituminous material should be removed from the joints and from the surface of the old concrete. Paint markings should also be removed. These operations can be performed with a scarifying machine or by scarifying in conjunction with sandblasting and using hand tools. Solvents should not be used.

Sometimes sound concrete must be removed to provide the proper thickness of resurfacing.

Unless the edge of the proposed resurfacing is at a joint, a saw cut at least ½ in. deep should be made to provide a straight line where the old and new surfaces will meet.

After the area to be resurfaced has been scarified and bituminous material removed, the surface should be dry-swept and blown to clear it of debris and dust. The surface should be washed with water, thoroughly brushed and, finally, inspected for loose areas and oil spots that may have been missed. These should be removed by additional scarifying and should be followed by another washing and brushing of the surface if necessary.

In areas where it is sound, the old surface may be cleaned with detergent and acid. The surface is first washed with water. Oil drippings are then removed by spreading a detergent and scrubbing vigorously with power or push brooms. After the detergent has been scrubbed, the residue

Table 9–B
RESURFACING PROCEDURES

	Causes of Need for Resurfacing	Considerations	Preparation of Old Surface	Tools	Topping
Unbonded floors	Poor original construction.	Can floor take added load?	Roughen surface.	Pick or bushhammer or grinding machine.	2 in. thick reinforced with wire mesh 30 lb per 100 sq ft placed at mid-depth of slab.
	Damaged surface.	Can floor level be raised?	Remove loose particles. Remove grease and oil. Remove paint and dirt. Remove food, etc.	Air hose. Rub with detergent. Chip or sandblast. 10 to 20% solution muriatic acid or strong washing soda.	
Bonded floors	Same	Same	Wet surface overnight. Remove old floor to depth of 1 in.	Hose. Jackhammer.	1 in. new concrete without reinforcement.

	Slush Coat or Bonding Course	Method of Application	Concrete Mix Proportions	Finishing	Curing
Unbonded floors	Portland cement and water = consistency of thick paint.	Brushing well to avoid thick layer.	Aggregate: pea gravel, crushed stone or other. Graded from ⅛ to ⅜ in. Well-graded sand.	Power float. Heavy rollers. Screeding. Floating. Troweling.	Careful attention is necessary. Evaporation from floors is rapid and may cause crazing and cracking.
	Thin layer dry portland cement on wet concrete.	Brooming into surface and spreading uniformly.	1 part portland cement. 1 part sand. 1½–2 parts coarse material. Water—little as possible. Not more than 0.45 water/cement ratio, including free water in aggregate.	No dust coat of dry cement to take up excess water.	If floor exposed to hot sun or warm winds, place wood falsework and covering of tarpaulins over floor immediately after finishing each section. Sprinkle and cover with wet sand, burlap or paper as soon as possible without damaging surface.
Bonded floors	¹/₁₆ to ⅛ in. layer mortar or grout: 1 part concrete sand (plus No. 8 material removed). ½ part water thick creamy consistency.	Brooming into fogged surface just ahead of resurfacing concrete. Should never dry to white appearance before concrete is placed.	Aggregate: maximum size about half thickness of resurfacing concrete. Workable mix. Water-cement ratio not more than 0.45 water/cement ratio. Air-entrained.	Similar to conventional paving operations. Either vibrating or tamping is essential for intimate contact between old and new concrete.	More important than in ordinary concrete work. Start as soon as possible after finishing is completed. Early and adequate coverage with white membrane curing compound. Temperature higher than 90° F or low humidity and high wind, can fog with light spray until finish hard enough cover with wet burlap. Keep wet 8 to 24 hr.

Table 9-B (cont.)
RESURFACING PROCEDURES

	Causes of Need for Resurfacing	Considerations	Preparation of Old Surface	Tools	Topping
Bonded pavement	For added strengths. To even up rough surfaces.	Should pavement be widened at same time? Can pavement level be raised?	Saw at boundaries at least ½ in. deep. Remove poor concrete. Remove bituminous material from joints. Remove paint. Dry sweep or blow debris and dust. If surface sound; no scarifying needed. Wash with water. Remove oil: spread detergent on surface and scrub. Clean and etch with acid. Wash and brush until water neutral to pH paper.	Concrete saw. Jackhammer. Chipping tools. Scarifier or sandblaster. Same Broom or air hose. Water and power or push brooms. 1-2 lb. detergent per 100 sq.ft. Commercial hydrochloric acid (20° Baume scale) 1 gal per 100 sq.ft. Brooms, goggles, rubbers, gloves.	As specified.
Patching	Improper jointing De-icers.	Extent of area to be repaired.	Same Remove rust from exposed reinforcing steel. Force it down so concrete will cover 1½ in.	Same	Same

	Slush Coat or Bonding Course	Method of Application	Concrete Mix Proportions	Finishing	Curing
	Same as for bonded floors	Same	Same	Same	Same
Bonded pavement Patching	Water-cement paste or 1 part cement 1 part sand from which plus ⅛ in. size removed. Water to make thick	Broom vigorously into exposed concrete to displace air films and cover surfaces to uniform thickness of $\frac{1}{16}$ to ⅛ in.	See Table 6–1	Vibrate to give good bond. Hand tamping ok for smaller patches. After concrete is spread evenly slightly above surface, draw vibrator evenly without stopping. Strike off. Consolidate. Float. Finish as adjacent concrete.	Same

should be removed by washing and brushing until the wash liquid is neutral or only slightly alkaline to pH paper. The entire surface should then be cleaned and etched with acid. Commercial hydrochloric acid (20° Baumé scale) should be applied at a rate of 1 gal. per 100 sq ft. To allow some dilution and assist in distribution, the pavement surface should be moist, but without standing water, before the acid is applied.

The acid should be broomed over the surface and into joints and cracks with pushbrooms. Workmen need to be protected with goggles, rubber boots, and gloves. Under unusually windy conditions, respirators and rubber suits may be needed. Exposed areas of the skin should be protected with an application of grease.

After the acid stops foaming, the pavement should be flushed with water under pressure and scrubbed vigorously with stiff power brooms to remove acid residue and sand that has been loosened by the acid treatment. The washing treatment should continue until all loose material is removed and the surface liquid reacts neutrally to pH paper. A second treatment may be necessary if any laitance or dirty concrete remains. The surface should be clean and gritty enough to secure an adequate bond.

Before the bonding course and resurfacing are placed, the dirt or dust that may have accumulated should be blown or washed from the surface.

Forms. Some solid support will be required along the sides of the area resurfaced in each operation. This is necessary for the support of mechanical or manual equipment used in the finishing operations. If the resurfacing is part of a large area such as an airport runway, taxiway, or apron, the concrete adjacent to the lane to be resurfaced may be used as a form.

Widening. Pavements that require resurfacing are generally old pavements, too narrow for present-day traffic. These should be widened at the same time that the surface is restored. Two methods may be used:

1. The widening and resurfacing may be done in one operation, using the concrete mix specified for the resurfacing.
2. Widening can be placed up to the level of the old slab, using the standard highway concrete mix. After the concrete for widening has hardened, the resurfacing is placed over both the old slab and the widening in one operation. The concrete for the widening should *not* be finished smooth, but should be left rough to provide a proper surface for the bonded resurfacing.

Bonding course. Just before the resurfacing concrete is placed, a bonding course should be applied to the cleaned surface. This course should consist of a $1/16$- to $1/8$-in. layer of mortar or grout containing 1 part portland cement, 1 part concrete sand from which the plus 8 material has been removed, and about $1/2$ part water to give a thick, creamy consistency. Prior to placement of the grout, the surface (if dry) should be fogged very lightly with a water spray so that it is moist but without pools of water. Excess water should be removed with compressed air. Although on some projects the grout has been applied pneumatically, brooming the grout into and over the surface is recommended because it is more effective. It removes entrapped air films at the pavement surface and aids in incorporating into the grout any loose material still on the surface. The grout should be applied just ahead of the resurfacing concrete. In no instance should the grout be allowed to dry to a whitish appearance before the concrete is placed.

Mix design. Two prime considerations govern the concrete mix design:

1. The maximum size of the coarse aggregate should be about half the thickness of the resurfacing.
2. The mix should be workable—at a water-cement ratio of not more than 0.45.

All concrete should be air entrained for resistance to freezing and thawing, particularly where it will be subjected to salt applications.

Recommended trial mixes for various thicknesses of resurfacing are shown in Table 9-C. The slump should be varied within the limits given in the table, depending on weather conditions. On hot, windy days, a slump of 2 to 4 in. may be desirable. On cloudy, cool days, a slump of 1 to 2 in. may be preferable. The weather conditions, the absorptive properties of the old concrete, and the placement equipment will govern concrete consistency on each project.

Concrete can either be mixed at the site or delivered by ready mix trucks. When delivery of dry batches of ready mixed concrete is being arranged, it should be remembered that, depending on the relative thickness of resurfacing, each cubic yard will cover from about three to nine times as much area as during conventional paving. For this reason the progress of thin resurfacing is usually dependent on the ability to complete the finishing operations, rather than on the capacity of the concrete mixing equipment.

Table 9-C

RECOMMENDED TRIAL CONCRETE MIXES FOR THIN RESURFACING

	Thickness of Resurfacing, Inches			
	1/2	1	2	3
Cement content	100 lb	100 lb	100 lb	100 lb
Total water[a]	42 lb	42 lb	42 lb	42 lb
Fine aggregate[b]	190 lb	170 lb	190 lb	180 lb
Coarse aggregate				
Maximum size	3/8 in.	1/2 in.	1 in.	1 1/2 in
Amount°°	115 lb	170 lb	230 lb	305 lb
Air content	9 to 11%	6 to 8%	5 to 7%	4 to 6%
Approximate cement	850 lb	750 lb	700 lb	650 lb
factor	per cu yd	per cu yd	per cu yd	per cu yd
Slump	1 to 4 in.	1 to 4 in.	1 to 4 in.	1 to 4 in.

[a] Including free moisture in the aggregates.
[b] Based on saturated surface-dry aggregate, S.G. 2.65.

Placing and finishing. The placing and finishing operations are generally similar to those used in conventional paving. To secure a dense surface and intimate contact between the old and new concrete, either vibration or tamping—mechanically or by hand—is essential. The final finish should be such that the surface texture of the resurfacing will be similar to that of the adjoining pavement.

Jointing. The jointing of thin concrete resurfacing must be given careful consideration if the pavement is to function properly. It is important that all joints in the original pavement be reproduced in the new surface. Joints in the new surface must be directly over the joints in the old pavement and must be of equal or slightly greater width. If any joint in the resurfacing is narrower than the one in the pavement below, and if the joint closes, extremely high pressure will develop in the new surface. This may break the bond or may result in an unsightly spall.

The formed type of dummy groove joint or a sawed groove may be used to form contraction joints in the resurfacing. Steel strips can be inserted in the old joints to form the contraction joints. If placed before the concrete resurfacing, they should be inserted so that the top of the strip is below the surface of the new concrete. If they move under mechanical finishing equipment, they can be adjusted afterward without difficulty. After the concrete has stiffened, the steel strips should be removed and the joint edged to a radius of about 1/8 in.

If the old slab was built with a longitudinal center joint, a similar joint should be constructed in the new slab directly over the original joint. The most feasible method of forming these joints is to saw them. They should be cut the full depth of the resurfacing. The old slabs must be marked with stakes, with string stretched between them, at the exact location of the original joint.

Curing. Proper and adequate curing is more important in bonded resurfacing than in ordinary concrete work. It is essential that curing be started as soon as possible after finishing.

Under most conditions, early and adequate coverage with a white-pigmented membrane curing compound will result in satisfactory curing. If the temperature is above 90° F., or if the humidity is low and the wind velocity is relatively high, it may be advisable to dampen the surface with a light fog spray until the concrete can be covered with wet burlap without marring it. This should be kept wet for 8 to 24 hr.

Slab-jacking

Slab-jacking consists of maintaining or correcting the crown and profile of a concrete pavement by injecting grout under the slab. The grout fills the void under the slab, restoring uniform support. The purpose of slab-jacking is to restore or maintain a smooth-riding pavement. Jacking does not strengthen the pavement slab itself; it only reestablishes support or brings the pavement into alignment (Fig. 9-8).

The Principle of Slab-jacking. The principle of slab-jacking is simply that as grout is pumped under pressure through a hole in the pavement, it creates an upward pressure on the bottom of the slab in the area around the hole (Fig. 9-9).

Stiff grout will not flow as easily as a more fluid grout. Therefore, the stiffer the grout, the smaller the area affected; the more fluid the grout, the larger the affected area. In addition, the depth of grout in the cavity under the slab will affect the pressure over a given area. The thinner the cavity or layer into which the grout is forced, the greater the upward pressure in the area of the hole. An overly stiff mix may form an undesirable cone or pyramid under the slab, leaving unfilled cavities (Fig. 9-10).

Concrete pavements do not require exceptionally strong subbase support, but they do require reasonably uniform support. Any condition that causes nonuniform slab support must be corrected as soon as possible if the pavement is to have the long service life for which it was designed. Slab-jacking can correct most of these conditions, including the following:

Figure 9–8 Slab-jacking is a method of correcting the level of an area of concrete pavement by forcing grout under the slab to lift it to its desired height.

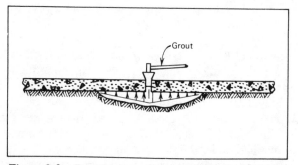

Figure 9–9 Pumping grout into the hole creates an upward pressure on the bottom of the slab in the area around the hole.

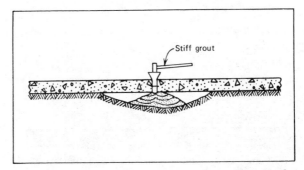

Figure 9–10 A stiff grout may form a pyramid under the slab, leaving unfilled cavities.

1. *Pumping.* Frequent heavy loads passing over the pavement cause the slabs to deflect at joints. This repeated flexing action can force a mixture of free water and subgrade soil up through pavement joints and cracks and along pavement edges. Continued pumping will eventually remove enough soil from under the slab to cause a void. Pumping does not occur if a sufficient depth of well-drained granular subbase is included in the pavement structure (Fig. 9-11).
2. *Advanced pumping and faulting.* Failure to correct pumping in its early stages will lead to settlement of the unsupported slab, resulting in a faulted joint or crack (Fig. 9-12).
3. *Faulting caused by densification.* The action of heavy traffic over undoweled pavement joints may cause the subgrade or subbase to consolidate in the vicinity of the joint, usually under the slab ahead of the joint. This densification will eventually lead to a faulted joint (Fig. 9-13).
4. *Embankment settlement.* Subsidence of embankments may cause pave-

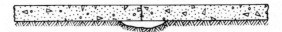

Figure 9–11 Mud-pumping causes a void under the slab.

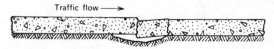

Figure 9–12   Settlement of unsupported forward slab.

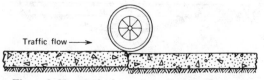

Figure 9–13   Faulted joint caused by heavy traffic.

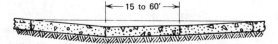

Figure 9–14   Subsidence of embankment may cause pavement settlement over an extended section.

ment settlements over extended sections (Fig. 9–14). If the settlement is uniform and does not affect the comfort and safety of traffic, and if the pavement is well supported, there is no need for correction. Differential settlement will generally warp the pavement, and some crown and profile correction will be required.

5. *Settlement over culverts.* Settlement over culverts generally occurs on short sections of pavement (Fig. 9–15). The cause of the failure should be determined and corrected before the pavement is repaired.

Figure 9–15   Settlement over culvert.

6. *Settlement of approach slabs.* One of the most common conditions requiring slab-jacking is the settlement of approach slabs to structures. This is often caused by sub-grade consolidation adjacent to bridge abutments where adequate compaction was not obtained during construction. A dip usually occurs at the end of the slab and a bump occurs at the bridge. Approach slabs must be kept at proper elevation to avoid impact on the bridge structure and to eliminate a hazard to high-speed traffic (Fig. 9-16).

Location of Holes. Slab-jacking is more of an art than a science. This is particularly true concerning the location of holes for injecting the grout.

Generally, holes should be spaced not less than 12 in. nor more than 18 in. from a transverse joint. The holes should be spaced not more than 6 ft center to center so that not more than about 25 to 30 sq ft of slab is raised by pumping any one hole.

The proper location of holes varies according to the defect to be corrected. For a pumping joint where faulting has not yet occurred, a minimum of two holes can be used. For a pumping joint with one corner of the slab faulted, the hole at the low corner should be set back to avoid raising the adjacent slab. The holes at the inside corners are used for inspection and for filling the void. Additional holes may be required to insure filling all the voids under the slab. If both corners of the slab are down, the inside hole should be relocated accordingly. Where the pavement has settled and slabs are in contact with the subbase, a single hole located in the middle of the panel, about 3 ft from the faulted joint, is sufficient.

Holes 2 to 2½ in. in diameter are drilled by either a core drill or a pneumatic drill. In some instances, to prevent spalling, a core drill is used to start the hole; the rest of the hole is completed with a pneumatic drill.

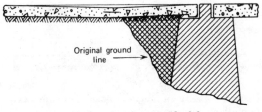

Original ground line →

Figure 9–16   Settlement of approach slab to structure.

Where the slabs are tight against the subbase, the use of an air line or blow pipe may be necessary to form a cavity exposing enough slab area for the grout pressure to take effect (Fig. 9-17).

Where the concrete slab is placed on and bonded to a cement-treated or other stabilized subbase, grout holes should be drilled all the way through the subbase. The grout should be injected beneath the subbase rather than between the concrete slab and the subbase.

No pumping or grouting should be done when the ground is frozen or when the air temperature is below 40° F.

Grouting. The following list of equipment is the minimum required for an efficient slab-jacking operation.

 1 concrete or pugmill-type mortar mixer.
 1 hydraulic jacking unit of the positive-displacement type, capable of instantaneous control of grout pressure.
 1 concrete buggy to transport grout from the mixer to the jacking unit.
 1 water tank truck with a minimum capacity of 250 gal.
 1 dump truck for hauling grouting materials and for towing the portable mixer.
 1 portable air compressor.
 1 pneumatic hammer with a 6 point, 2½-in.-diameter, pneumatic drill bit.
 10 tapered wooden plugs.

Generally, a crew of from 6 to 10 men is used for slab-jacking. The crew should have a foreman and, at least, one flagman.

A variety of grout mixes have been used successfully for slab-jacking. They generally consist of 4 or 5 parts of fine sand or finely ground limestone and 1 part of portland cement, with water added to produce the

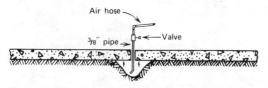

Figure 9–17   An air hose may be used to form a cavity for the grout pressure.

desired consistency. In areas where ground limestone is not readily available, hydrated lime has been used.

One of the most important characteristics of grout is its "flowability." Generally, the finer the material and the more uniformly graded the material, the greater will be its capacity to flow.

Wetting agents, additives that increase flowing ability, may also be used in the mix. The use of a wetting agent lubricates the grout and permits runs up to 6 ft; it also tends to eliminate pyramiding and increase subbase permeation. A mix of stiff consistency is used for raising pavement slabs, and a more fluid mix is used for filling voids.

Slab-jacking Procedure. Before work is started, some method of controlling the amount a slab is to be raised and the finished elevation of the pavement should be determined. A straightedge may be used for correcting faulted slabs. For short dips up to about 50 ft in length, a tight chalk line is adequate, providing the points used are in a true plane with the adjacent pavement in each direction. For dips in excess of 50 ft, a precise level and rod should be used to check the profile well beyond the dip. This will avoid building a bump into the pavement (Fig. 9-18).

For correcting a dip or sag in the pavement, jacking should begin at the low point of the sag and progress longitudinally. The holes should be staggered transversely and pumping should be continued until the slab has been raised. Then all holes should be pumped again to make certain no voids remain under the slab.

Slabs should not be raised more than 1/4 in. while pumping in any one hole at any one time. No part of a slab should lead any other part of the slab or any adjacent slab more than 1/4 in. at any time. When using two jacks, it is desirable not to work adjacent holes simultaneously. This may cause a line of stress that could crack the slab.

In all slab-jacking operations, the slab should be raised slowly with a uniform pressure. Pyramiding should be avoided by making sure that the grout "shows" in adjacent holes, that is, the movement of the grout can be observed.

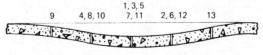

Figure 9–18 Jacking sequence.

A common problem in raising slabs is leakage or blowouts along the edge of the pavement or out on the side slope. This can usually be controlled by varying the consistency of the grout. Stiffening the grout stops many leaks.

Leakage may be avoided by pumping a few holes along the outer edge of the slab and allowing the material to set before the slab-raising is started. A very severe condition may require that additional holes be drilled and a stiffer grout used to stop the leak. In some instances, because of a weak berm (soil shoulder), it may be necessary to do most of the raising from holes along the center of the slab.

When the nozzle is removed from a hole, the grout should be cleaned from that hole. Additional pumping may be necessary to bring the slab to the desired grade, and hardened grout would then have to be drilled out. After jacking operations are completed, all holes should be filled with a stiff 1:3 cement grout, which is then tamped into place and floated to a smooth finish.

When it is properly used, slab-jacking provides a convenient, efficient and economical means of raising concrete pavement slabs. Cost will depend on the amount of material needed to fill voids in the subgrade before lifting starts, the distance the pavement is to be raised, the area of pavement involved, and other conditions. Generally, slab-jacking costs less than building up the surface with bituminous mixtures and gives better riding qualities and a better looking surface. It can save a great deal of expense if done soon enough to avoid pavement deterioration.

CHAPTER 10
JOINTING

PAVEMENTS

Joints are placed in concrete pavements to control transverse and longitudinal cracking by relieving stresses resulting from shrinkage of the concrete and differential temperature and moisture conditions between the top and bottom of the slab. Also, stresses caused by vehicle loads must be considered in joint design.

When the pavement is properly designed, the jointing pattern will control cracking by relieving these stresses and providing adequate load transfer. Joints will also divide the pavement into suitable increments for construction and will accommodate slab movements at intersections with structures, fixed objects, or other pavements.

The objective is to design joints for individual projects so that the load-carrying capacity and riding quality of the pavement are achieved at the lowest possible annual cost.

Stresses and Cracking
Before 1925, many pavements were built without joints of any kind except transverse construction joints, used at the end of each day's work or when paving was interrupted by rainfall. Unjointed pavements have also

been constructed in a few states in the past two decades. Crack behavior on these unjointed pavements shows how joint designs are used to control cracking.

To get the workability needed to place and finish concrete pavements, up to 50% more water is used than is needed for hydration. As the concrete hardens, this excess water evaporates from the concrete. As a result, the pavement occupies less volume after drying than while it was hardening. Heat of hydration during the hardening process, followed by lower temperatures after hardening takes place, further reduces the concrete volume.

Contraction from these sources is resisted by subgrade friction, causing tensile stresses in the concrete. (For average conditions, the tensile stress in a 9-in. slab, 100 ft long, is about 75 to 80 psi.) A typical crack pattern due to restrained contraction is shown in Fig. 10-1. The crack spacing will vary from about 60 to 150 ft, depending on the tensile strength of the concrete and subgrade friction (Fig. 10-1).

After hardening, temperature and moisture gradients between the top and bottom of concrete pavements cause stresses due to restrained warping. At and near the slab bottom, daily changes in temperature and moisture content are small. However, the exposed slab top has fairly large daily variations in temperature and moisture content. At night, the top

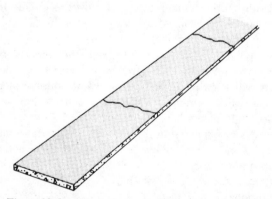

Figure 10–1 A typical uncontrolled crack pattern from contraction shrinkage.

is often cooler than the bottom. As a result, the top of the slab tends to contract and warp the slab edges upward, causing tension in the top of the pavement.

During the day, the top of the slab is often warmer than the bottom. The top then tends to expand and warp slab edges downward. This tendency is restrained by the subgrade and the stress is reversed, with compression in the top of the slab.

Differences in moisture content between the top and bottom of the slab cause similar but less severe stresses. Decreases in moisture content cause shortening or contraction, and increases cause expansion. The influence of restrained warping stresses on pavement design is complicated because temperature and moisture differences often have opposite stress-producing effects. For example, when the top of the slab is warmer than the bottom, causing expansion in the top of the slab, the bottom of the slab will usually have a higher moisture content than the top, causing expansion in the bottom of the slab. In such a case, the amount of stress due to warping will depend on the influences of the opposing factors, and will be substantially less than stress due to temperature or moisture differences alone.

Because restrained warping stresses are complex, and because repetitive traffic loads compound the problem, the only reliable basis for crack control is the performance of unjointed pavements. This performance shows that restrained warping stresses, acting in combination with loads, will cause additional transverse cracks between the initial shrinkage cracks and a longitudinal crack along the approximate centerline of pavements with more than one lane of traffic (Fig. 10-2).

One might wonder, why not build pavements without joints and allow them to crack randomly? This would certainly be cheaper initially, but maintenance and serviceability problems are increased. The irregular longitudinal crack makes it almost impossible to properly delineate the traffic lanes, especially when the longitudinal crack is sealed. The main problem, however, is progressive opening and infiltration of incompressible materials into these cracks, with a resultant loss in serviceability.

Infiltration is also a serious problem at transverse cracks, especially at the initial contraction cracks. At spacings of 60 to 150 ft, these cracks open up so much that sealing materials fail either in adhesion to the concrete or by ruptures. The results are infiltration, with resulting spalling, poor riding quality, and costly maintenance.

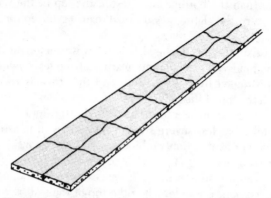

Figure 10–2 The crack patterns in a typical unjointed highway pavement subject to the combined effects of initial shrinkage, restrained warping, and traffic stress.

Transverse Joints

The simplest method for crack control in concrete pavements is a plain slab design with aggregate interlock joints, spaced to control transverse and longitudinal cracking. Joints to control cracking are made with a saw cut or groove in the slab, just deep enough so that cracks form below the saw cut or groove.

For effective control of transverse cracking, the joint depth should be at least, one-fifth the slab thickness. If transverse joints are sawed, it will also be good practice to make the saw cut no deeper than the diameter of the largest size coarse aggregate. Otherwise, cracks may go around pieces of large aggregate near the pavement surface instead of following the saw cut. The result may be unsightly spalls during warm weather when the concrete is expanding and in compression.

The irregular slab edges, on either side of the cracks below the joints, provide the needed interlock between slabs. To insure a long service life at the lowest cost, the designer must select a joint spacing that will both control cracking and have adequate interlock. Truck traffic volume and climate have a great influence on joint spacing. Where the environment is not severe, joint spacings between 15 and 20 ft have controlled cracking effectively. The design objective is the maximum spacing that will control cracking and preserve serviceability. To do this, and thus get the lowest possible annual cost, it is essential to carefully evaluate joint performance on projects where traffic, climate and joint spacing are relevant.

An important advantage of the plain slab design is that the short joint spacings keep seasonal movements at the joints very small. As a result the low-cost poured sealants keep joints effectively sealed against infiltration of incompressible materials for many years. Even though more joints are required, the joint seal lasts longer and prevents damage from spalls and other causes. The overall results are lower maintenance costs and longer service.

Skewed Joints

One method now widely used to improve joint performance, and thereby extend the life of plain pavements, is to prescribe skewed transverse joints (Fig. 10-3).

The skew of 4 or 5 ft in a 24-ft pavement width is just enough to move wheel loads of each axle across the joints one at a time. Skewed joints have these advantages:

1. Reduced deflection and stress at joints, thereby increasing the load-carrying capacity of the slab and extending life
2. Less impact reaction in vehicles as they cross the joints and, hence, a smoother ride

The present-day skewed joint design is an improvement of transverse joint design, showing consistently superior performance. It merits further use as an effective method of getting the maximum in joint performance and the lowest possible annual cost.

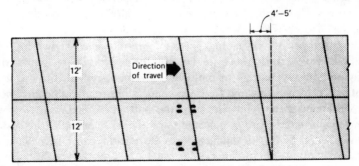

Figure 10-3 Skewed joints allow the wheel load of each axle to move across the joints one at a time.

Randomized Joints. Another method for improving performance and extending the service life of plain pavements is to use skewed joints at randomized or irregular spacings. A spacing pattern of 13-19-18-12 ft prevents rhythmic or resonant responses in cars or trucks moving at various speeds. Other spacing patterns will be effective if multiples of 7.5 ft are avoided. It is essential to keep the maximum joint spacing small enough to control transverse cracking.

Cement-Treated Subbases
Cement-treated subbases can materially increase the life and serviceability of plain concrete pavements. The most important functions with respect to joint design are (1) increasing slab support, (2) preventing consolidation, and (3) providing strong, stable joint support.

Longitudinal Jointing
Figure 10-2 shows an irregular longitudinal crack along the central part of an unjointed pavement. Cracks of this kind develop after pavements are subjected to traffic and are the result of the combined effects of loads and warping. Use of longitudinal joints to control these cracks began in 1918, and was widespread by 1930. In current design practice, these criteria are useful guides for spacing of longitudinal joints:

1. On both two-lane and multilane highway pavements, a spacing of 12 to 13 ft serves the dual purpose of crack control and lane delineation.
2. Longitudinal joints are not required for crack control on one-way ramps where the slab width is not more than 18 ft. Normal ramp width is about 16 ft, with increases to about 18 ft on short radius curves. These criteria are guides rather than hard-and-fast rules.

The two types of longitudinal joints in current use are shown in Figure 10-4. The tongue-and-groove joint at the top of this figure is prescribed for lane-at-a-time paving, for multilane pavements where the full width is not paved in one pass, and for ramp connections to mainline pavements. The weakened-plane joint shown at the bottom of Figure 10-4 is used where two or more lanes of pavement are placed at once. With slip-form paving, 3- and 4-lane pavements, 36 to 48 ft wide, can be paved in one pass.

Both types of longitudinal joints that depend on a keyway or aggregate interlock to maintain structural capacity and serviceability. To insure full

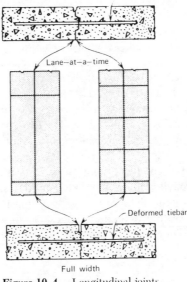

Figure 10–4 Longitudinal joints.

joint interlock, tiebars or tiebolts are used across longitudinal joints in most highway pavements. However, tiebars or tiebolts are not required where there is adequate restraint and confinement without them. This is true for urban pavements where materials placed back of curbs keep longitudinal joints closed without using the ties.

It is important to edge longitudinal construction joints as is shown at the top of Figure 10-4. If edging of the second lane is carelessly done, so that concrete rests on and over the groove formed when the first lane was edged, unsightly spalls occur. To prevent this, the lip of the edger used for the second lane should extend at least ¼ in. below the lip of the edger used for the first lane. To avoid these edging problems, many agencies now specify sawing a shallow groove (about ¾ in.) in longitudinal construction joints. This improves riding quality between lanes and makes for more durable concrete along the joint edges.

It is essential to make the depth of weakened-plane longitudinal joints at least equal to ¼ the slab depth. If this is not done, longitudinal cracking will not be controlled. Workmen will sometimes neglect to make allowances for blade wear when sawing long stretches of longitudinal joints. Hence, it is important to make frequent checks of joint depth while sawing is underway. It is equally important to check the depths of formed grooves and inserts in longitudinal joints. As further insurance against random longitudinal cracking, many agencies now specify a longitudinal joint depth of one-fourth the slab thickness plus ⅜ in. This takes account of variations in slab thickness that are certain to occur.

A recent development in weakened-plane longitudinal joints is the use of 4 to 10 mil polyethylene strips to form the joints. These strips are installed from a machine attached to the rear of a slip-form paver. Rolls of the polyethylene strips are placed on freely moving spindles. From these spindles the strips are fed into the plastic concrete through a guide that keeps the strips vertical and flush with the pavement surface. Strips are placed for one, two, or three longitudinal joints, depending on the width of slip-form paving underway. Placing the strips completes all work on longitudinal joints. There is no expectation that these joints will require any further sealing or maintenance. The result is a substantial reduction in overall joint costs.

Expansion Joints

Until about 1918, all transverse joints in concrete pavements were expansion joints. Spacings varied from 25 to 100 ft with joint widths of ¼ to 1 in. Since 1918, expansion joints have been gradually replaced with weakened-plane contraction joints, and today expansion joints are rarely used except at fixed objects and at unsymmetrical intersections.

Several factors have been responsible for this change from expansion to contraction joints. These factors include:

1. Recognition by highway engineers that cracking of unjointed pavements was the result of initial shrinkage and restrained warping, rather than expansion of the concrete.
2. Unsatisfactory performance of expansion joints where they are not constructed with meticulous care.

Engineers had long understood the principles involved in a weakened-plane contraction joint, and had recognized that this joint design could improve pavement performance and reduce costs. It was evident that cracks formed at weakened-plane joints, spaced to relieve tensile and warping stresses, would provide expansion space between the short slabs formed by the controlled cracks. It was also evident that this expansion space would reduce compression stresses that develop when high temperature causes the pavement to expand. It was reasoned that the reduction in compressive stresses would be enough to permit a longer spacing between expansion joints or to eliminate the need for expansion joints, except for special conditions. This analysis led to a decrease in the use of expansion joints and, by 1931, several states installed them only at bridge abutments.

However, some engineers still believed that expansion joints were needed to relieve compression stresses during hot weather. Experimental projects, along with data from surveys of pavements in service, are the basis for current practice regarding expansion joints. This practice prescribes that expansion joints are required only at fixed objects, unsymmetrical intersections, and other similar points of stress concentration, provided that:

1. The pavement is constructed of materials that have normal expansion characteristics.
2. The pavement is constructed during periods when temperatures are well above freezing.
3. The pavement is divided into relatively short panels by contraction joints spaced to control transverse cracking.
4. The contraction joints are properly maintained to prevent infiltration of materials.

Figure 10-5 shows a typical expansion joint installation. Expansion joints are usually placed at least one slab length away from bridge abutments and are generally ¾ in. wide.

Cypress and redwood board joint fillers have given the most consistently satisfactory performance. These boards resist decay, are resistant to compression during pavement expansion, and recover satisfactorily during periods of pavement contraction. The boards are placed 1 to 2 in. below the slab surface to allow space for sealing material. Plain round steel dowels spaced every 12 in. are the most widely used type of load transfer. One end of each dowel is equipped with a cap into which the dowel can move when the concrete is expanding. The half of the dowel with the capped end must be lubricated to break bond and permit horizontal movement.

Some highway departments install more than one expansion joint at the regular joint interval on each side of bridges and other structures. This aspect of joint design merits further study. Anchors designed to control slab movement and cast integrally with the pavement could eliminate

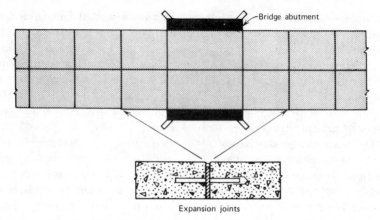

Figure 10–5 Expansion joints used on both sides of a bridge.

the need for expansion joints at structures. Such anchors have been constructed both in this country and in Europe to control slab movement.

Doweled Contraction Joints

For projects where it is evident that a plain slab with undoweled contraction joints will not control joint faulting, doweled contraction joints are used. Good planning requires a careful economic analysis of two alternative designs. The first design is shown in Figure 10-6.

Figure 10-6 shows a plain pavement with doweled joints spaced to control transverse cracking. With the additional load transfer provided

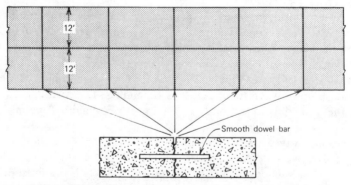

Figure 10–6 Plain pavement—doweled contraction joint.

by the dowels, the designer selects the maximum joint spacing that will control transverse cracking. The selection of joint spacing is usually based on prior experience with conditions similar to the project under design. Dowel dimensions are the same as those given for expansion joints. Caps are not required, but half of each dowel must be lubricated to permit slab movement. If dowels are used in conjunction with skewed joints, the dowels must be placed parallel to the longitudinal axis of the pavement, rather than perpendicular to the joint.

The other design employs doweled contraction joints and distributed steel. A typical design is shown in Figure 10-7. With this method, the distributed steel is designed to prevent slab faces from separating as widely after cracking occurs, and joint spacing is based on analysis of the relative cost of distributed steel and doweled joints for various joint spacings.

The amount and cost of distributed steel increase as slab length is increased, and the cost of joints increases as slab length is decreased.

The costs of distributed steel and doweled joints will vary considerably, depending on local material and labor costs for the various elements involved. Detailed cost studies made for several midwestern states show that the most economical joint spacing is usually about 35 to 40 ft. For comparison, the cost of joints per four-lane mile for a plain pavement, with doweled contraction joints spaced at 20 ft, was $18,300 on one job. The cost of joints and distributed steel per four-lane mile would have been $22,200. The coarse aggregate used on this job made a joint spacing of 20 ft more than adequate to control cracking. The plain pavement design saved almost $4000 per four-lane mile.

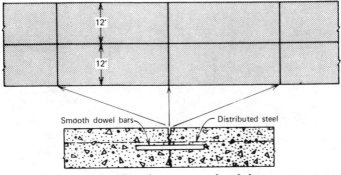

Figure 10-7 Distributed steel pavement—doweled contraction joint.

Construction Joints

Longitudinal construction joint details are shown in Figure 10-4 and were discussed in the section on longitudinal joints.

Transverse construction joints are designed for planned interruptions of paving operations, such as occur at the end of each day's paving. Emergency interruptions due to rain or to equipment failures must also be anticipated.

Figure 10-8 shows typical details for planned and emergency construction joints where pavements are constructed lane-at-a-time, or where there are more than two abutting lanes. Planned construction joints are installed at normal joint locations and consist of a butt-type joint with dowels for load transfer. Dowel size and spacing are the same as at other doweled joints. To ensure joint movement, the dowel ends extending through the butt joint must be lubricated before paving is resumed.

If an emergency makes it necessary to install a construction joint between normally spaced joints, this emergency joint should be a tied tongue-and-groove joint as is shown in Figure 10-8. If the emergency joint is not tied to prevent joint movement, an unsightly crack will develop across the adjacent panel. Either deformed bars or hook bolts may be used to tie the emergency joint together.

Figure 10-9 shows typical transverse construction joint details for full-width paving. Since both planned and emergency construction joints extend across both lanes being paved, it is not necessary to install tied tongue-and-groove joints for emergency work stoppages between normally spaced transverse joints. However, tie tongue-and-groove joints are generally used because doweled butt joints are more costly.

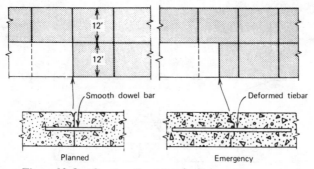

Figure 10–8 Construction joints for lane-at-a-time paving.

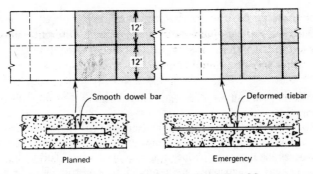

Figure 10–9 Construction joints for full-width paving.

FLOORS AND SLABS ON GROUND

The successful performance of slabs and floors depends primarily on how closely the basic construction steps—preparing the subgrade, concreting, finishing, curing and jointing—have been followed. The choice of locations and types of joints is decided according to the particular variables of each job. The purpose of joints is (1) to limit the unit size of concrete placement, and (2) to relieve stresses other than applied loads.

Fortunately, analyzing variables and deciding joint locations and types can be accomplished easily by merely keeping in mind four characteristics of concrete slab-on-ground construction:

1. All plastic concrete contains more water than is needed for hydration of the cement. When this extra water starts to evaporate, drying-shrinkage of the slab begins creating tensile stresses in the concrete. These tensile stresses must be relieved by providing joints in the slab.
2. Slabs of different shapes expand and contract by different amounts. The larger the slab, the greater the movement will be. At the junction of unlike adjoining slabs, provision must be made for differential horizontal movement to prevent random cracking.
3. Slabs may experience some vertical movement because of heaving or settling of the subgrade. The amount of movement may vary at different areas of the slab because of lack of uniform subgrade bearing pressure, unequal changes in loading conditions, changes in moisture content of the soil, and freezing and thawing of the subgrade.
4. Slabs may also require joints to accommodate construction schedules and unforeseen causes. Frequently, concreting must be stopped before an entire floor is placed to allow for finishing.

There are three types of joints that will provide the necessary control. Although the same type of joint has often been given different names, the following names are herein used to identify the three types of joints recommended for slab-on-ground floors.

1. *Control joint.* This allows differential movement only in the plane of the floor.
2. *Isolation joint.* This allows differential movement in all directions.
3. *Construction joint.* This allows no movement in the completed floor, when not coinciding with the location of control joints. This is difficult to achieve, and when joints are required by construction schedules they should coincide with control joints whenever possible.

Control Joints

Vertical loads on one side of a control joint are transferred to the adjacent slab across the joint without significant differential vertical movement. Adequate provision must be made for transferring moving loads from one slab across a joint to the adjacent slab, and control joints are installed to allow for contraction caused by drying shrinkage. If no joints were used, random cracking in the floor would occur when the drying-shrinkage tensile stresses within the concrete were less than the resistance of the subgrade.

Similarly, the control joint eliminates random cracking resulting from thermal volume changes. Cracking due to thermal shrinkage may occur during dropping temperatures. Its effect is comparable to the shrinkage described above. On the other hand, thermal expansion is rarely a problem on interior slabs because the drying shrinkage of concrete is greater than any subsequent expansion within the normal temperature range.

For example, before expansion becomes a problem, the concrete must experience about a 100° F rise in temperature to equal the total shrinkage from drying. This means that the concrete volume will rarely, if ever, be as large as it was at the instant concrete was placed. For this reason, expansion joints, widely used in the past in floor construction, have not been necessary and are being rapidly replaced in new construction by control joints.

In general, joint spacing to control drying and thermal shrinkage should be from 12 to 25 ft in unreinforced or lightly reinforced concrete floors. Variation in spacing is the result of differences in local conditions, such as the concrete itself, climate, construction practices, and type of soil. Experience in some localities has indicated that joint spacing can be increased up to 30 ft without cracking. In general, however, this should be the top limit. Also, the joint spacing should be chosen so that the panels are approximately square.

There may be cases where joint spacing as great as the above recommendations cannot be uniformly maintained. Consider the case where an abrupt change in slab thickness occurs, or where the plan of the floor is irregular, such as in an L-shaped building. In such situations, it may be difficult to maintain joint spacing of 12 to 25 ft for all control joints. The importance of proper joint location and spacing cannot be overemphasized. Joints should be planned in advance and should be indicated on the drawings.

Control joints are made by purposely creating a vertical plane of weakness in the slab. The cracking then occurs at this weakened plane rather than at random locations in the slab. Control joints may be made in several ways, but one of the most widely used methods is to saw a slot in the top of the finished slab. Cracking occurs beneath the sawed slot when the shrinkage stresses exceed the tensile strength of the concrete. The configuration of the crack (the meshing of the aggregates in their sockets) provides an interlock across the crack, allowing vertical load to be transferred to the adjoining slab without differential vertical movement. Because the joint is made after the concrete is placed and finished, sawed joints do not interfere with concreting operations. The saw cut should be made as early as possible prior to drying shrinkage and preferably during rising temperatures. Sawing during dropping temperatures may result in random diagonal cracking directly ahead of the saw. Therefore, sawing is generally done during early morning, after the concrete has been placed. When slabs are not under cover, it may be necessary to saw at night.

When a limited amount of sawing is required, an electric-powered handsaw guided by a board held in the proper position can be used to give a straight cut. On larger jobs, heavier self-propelled equipment should be used with a chalk line as a guide. For early sawing, a carborundum blade is used. If the concrete has hardened appreciably or if extremely tough aggregate has been used, a diamond blade may be required. The width of the cut should not exceed $\frac{3}{16}$ in and the depth should be one-fourth to one-fifth the thickness of the slab. Immediately after sawing, the joint should be flushed with water or air under pressure to remove sawing residue.

After the joints are sawed, they should be sealed to prevent infiltration of foreign material. However, it is advisable to delay sealing the joint as long as possible for two reasons: first, to allow the concrete to gain strength; and second, to allow some shrinkage to occur so that the joint opening will be nearer its maximum width when sealed. After the usual curing period of at least one week, the floor should be allowed to air dry for at least one week to permit some initial shrinkage to occur.

The operation of joint sealing can then be done, after the joint has been cleared of all foreign material. Sealing is accomplished by filling the joints with either a bituminous joint sealer or lead.

Control joints can be formed in other ways to function in the same manner as the sawed joint. Premolded joint material, such as metal strips similar to those used in terrazzo work or premolded asphalt strips ($\frac{1}{8}$ in. wide and 1 to 2 in. deep), may be used. These strips are inserted when concrete is plastic, prior to finishing, and while concrete is being placed. Experience indicates that the top edge of the premolded material should be flush with the concrete surface to eliminate spalling at joints under heavy traffic. An edging tool with a maximum radius of $\frac{1}{8}$ in. should be used where rounded concrete edges are desired along the premolded material. As with sawed joints, the premolded material should extend into the slab at least one-fourth to one-fifth the slab depth.

There is another method of forming a control joint. Instead of creating a weakened plane, the joint is constructed completely through the slab. Such control joints are often used to aid in minimizing the effect of shrinkage by following a concrete placement sequence commonly termed "checkerboarding"—placing concrete in alternate squares as if it were being placed in the squares of one color on a checkerboard. In this way, partial shrinkage of half the parts occurs prior to placing the alternate ones.

The resulting joint, between the previously placed concrete and the freshly placed concrete, should not be bonded. It is usually simpler to use the nonbonded, keyed control joint shown in Figure 10-10. Breaking the bond between the new and previously placed slabs by spraying or painting with a curing compound, asphaltic emulsion, or oil allows the freedom of horizontal movement necessary for crack control. The key provides the same type of vertical load transfer achieved by the aggregate interlock in the sawed control joint, and prevents unequal settlement of the slabs. Keyways may be formed by a sheet metal form, by premolded mastic $\frac{1}{8}$ in. thick, by composition hardboard, or by a wood bulkhead with a strip of metal or wood attached to form the groove. If wood strips are used to form the keyway, they may swell and cause cracking of the concrete lip above the keyway. Therefore, they should be well oiled and grooved with a saw cut on the back side to minimize swelling. Experience has shown that the tongue-and-groove need not be as deep as has been often used in the past. A 1×2-in. keyway with beveled edges is sufficient for slabs of 5 to 7 in. thickness. Larger keyways, such as 2×2 in., should be avoided. If metal is used to form the keyway, it should be thick enough to withstand handling. One obvious advantage of the metal keyway form is ease of cleaning for reuse.

As is shown in Figure 10-10, the joint should be sawed. However, the saw cut need not be as deep as that made for the sawed control joint because its only function in this case is to provide space for the joint-sealing material.

If wire mesh or reinforcing steel is used in the slab, it should be stopped or eliminated at a control joint, regardless of the joint construction method. Wire mesh reinforcement is best reduced by cutting out alternate wires where the joint will be. If reinforcing bars are used, lapping should not occur at the joint, and half the number of bars should be eliminated across the joint. Reinforcing steel is sometimes used to increase the distance between joints. However, the greater the joint spacing, the more the joint will open. Where joints open wide, jolting by vehicles may result and increased joint maintenance will be required. The easiest joint to maintain is one that is narrow and straight.

Isolation Joints

Isolation joints separate or isolate concrete slabs from columns, footings, or walls. These joints permit horizontal movement of the slab caused by drying shrinkage, but they also permit vertical movement that invariably

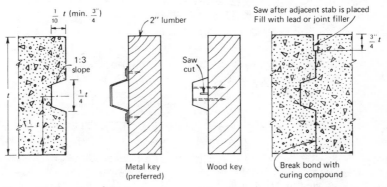

Figure 10–10   Keyed joints.

occurs because of differences in unit soil pressures under floors, walls, columns, and machinery footings. Because the isolation joint is used to allow freedom of movement, there should be no connection across the joint by reinforcement, keyways, or bond.

Columns or footings should be boxed out as is illustrated in the details of an interior and exterior column in Figure 10-11. The rectangular or

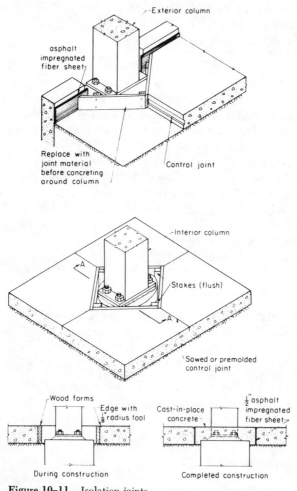

Figure 10-11   Isolation joints.

square form is placed so that its corner point at the control joints along the column lines. This is preferred because it eliminates notching the slab corners that might cause diagonal cracking.

An isolation joint between a floor and wall may be made by attaching an asphalt-impregnated sheet, not more than ¼ in. thick, to the wall prior to placing the concrete. If watertight isolation joints are desired, they can be formed with beveled wood strips held from the wall by wedges. The wedges and strips can be easily removed after the concrete has hardened, and the space is partially filled with sand or other granular material. The top of the joint can then be filled with calking compound or other resilient material to provide the watertight joint.

Construction Joints

Construction joints are used at temporary stopping places in concreting. They may be necessary in large projects because of the difficulty of placing and finishing large areas of concrete in one continuous operation.

A construction joint that does not coincide with the location of a control joint should constitute neither a plane of weakness nor an interruption in the homogeneity of the concrete. Therefore, every effort must be made to insure bonding of the newly placed concrete to that previously placed. To assure bond, the hardened concrete should be thoroughly cleaned and dampened before the fresh concrete is tamped or vibrated into place. To avoid the possibility of breaking the bond after the floor is in use, tiebars are frequently used to assist in vertical load transfer across the joint.

Floor Joint Checklist

1. Have the exact locations of all joints been shown on the plan?
2. Have control joints been placed along all column lines?
3. Have intermediate control joints been placed where the column spacing between columns and walls exceeds 25 ft?
4. Has a control joint been provided where a change of floor section occurs, either in thickness or width?
5. Have isolation joints been provided between floor slab and walls?
6. Have isolation joints been provided around all columns and footings?
7. Have isolation joints been provided around machinery foundations, floor pits, pipes, and the like?
8. Have adequate detail sections of each required type of joint been shown on the plan?

Suggested Spacing of Control Joints

Factors affecting joint spacing include: (1) type of coarse aggregate, (2) total amount of water in concrete mix, (3) temperature and humidity ranges experienced, and (4) subbase restraint. Occasionally, field experience indicates that spacing of joints can be greater than is indicated in the following table; however, the table should be used except when reliable data are available to indicate that more widely spaced joints are feasible (Table 10-A).

Table 10-A

Type of Coarse Aggregate	Joint Spacing in feet
Crushed granite	25
Crushed limestone	20
Crushed flinty limestone	20
Calcareous gravel	20
Siliceous gravel	15
Gravel less than 3/4 in. size	15
Slag	15

WALL PANELS

In addition to adding visually to a building facade, the functions of joints are:

1. To connect similar or dissimilar structural materials.
2. To accommodate changes in temperature, moisture, and structure.
3. To correct any variations in alignment or dimensions between plan and building site.

A joint must seal out water, air, and dust from the building, and the insulation properties of the wall must not be impaired by the joint.

There are two main types of joints now in use. One is the open joint which allows water to enter the joint, where it is channeled into some sort of drain. The other, called a closed joint, creates a seal that keeps all water and air out.

Open Joints

Open joints are easier to construct and, since they do not require sealing, there is no maintenance over the years. They accommodate changes in temperature, moisture and structure more easily than closed joints. There are two kinds of open joints—lap joints and vertical joints.

Lap joints are either side by side or overlapped the way tiles are laid. They work well for precast concrete curtain walls where the weather is not severe. Heavy, driving rain can cause problems, however.

Vertical joints are more of a problem and usually require a system of baffles and drains to collect the moisture that gets inside the joint. The cost of studying weather conditions and doing the necessary planning often makes this an impractical method of jointing.

Closed Joints

Closed joints are sealed for the purpose of keeping all moisture out. The greatest problem they pose is that, while they must be weathertight, they must also allow for movement of members on either side. This is an expensive type of joint, requiring a lot of labor and maintenance; however, it is the most widely used.

Joint Shapes

There are two main categories of joint shapes from which the architect can choose. Integral joints need no additional parts, while accessory joints require other pieces to complete them.

Choosing the type of joint suitable to a particular structure depends on many factors affecting the movement and stresses that the building must endure. The joint should be chosen to protect the joint seal and should be easy to install and replace. The number of different joint parts and shapes should be kept to a minimum.

Integral Joints. A butt joint is the simplest kind of joint. Weatherproofing is dependent on the seal, which is entirely exposed to the elements.

With the lap joint the whole panel can overlap or the panels can be shaped to overlap each other partially. Either method gives better protection to sealants than the butt joint does. It may be difficult to get the members lined up to make a smooth surface when using lap joints, however.

Another integral joint is the mated joint. This forms an interlocking joint which may be difficult to seal, but which provides a little better protection than lap joints (Fig. 10-12).

Accessory Joints. A spline is a loose accessory that fits either between two grooved members or around two tongued members. The spline must be inserted from the end of the panel after adjacent panels have been placed.

A cover is put on the face of a joint and can be removed easily for joint

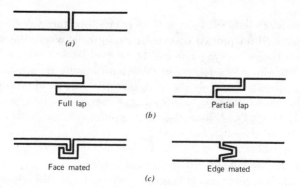

Figure 10–12 Integral joints (*a*), (*b*), (*c*).

maintenance. It can be attached to one or both members, to a frame on the back, to clips, or to another cover.

The frame and stop joint is the most expensive means of jointing. However, it is easy to dismantle and replace. Materials of different thicknesses can be used on either side of this joint. The drawbacks are that there are two joints to seal and the metal passing through from front to back allows for heat conductivity. It is a useful shape for a joint, and one of the most common means of holding glass in a wall (Fig. 10-13).

Sealing Joints in Curtain Walls
Movement in and around the joints caused by thermal changes, changes in dimension of the concrete itself, structural changes, and the like, must

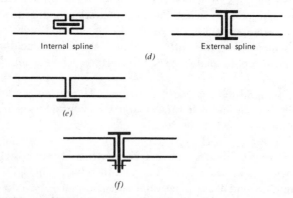

Figure 10–13 Accessory joints (*d*), (*e*), (*f*).

be allowed for in the planning of a wall. Oleoresinous (oil based) calking compounds have not proved elastic enough to seal joints adequately in normal temperature ranges. During thermal contraction their adhesion to the joint surfaces is not adequate. A later development in joint sealants is proving quite successful, however. The basic compound of the new sealant materials is a polysulfide liquid polymer. It is a two-part chemical compound which usually comes in two containers. One holds a pastelike accelerator and the other a syruplike base compound. When mixed, they become thick enough to apply and later thicken to a solid resembling rubber. Pigment can be added for permanent coloring.

This sealant is a significant improvement over previously used materials because it is flexible enough to stretch to, at least, double its original size. It adheres to concrete in spite of weather and is durable. Tests on polysulfide sealants show that a low-maintenance service life of at least 20 years can be expected. However, the initial cost of polysulfide compounds is so high that it is usually used in combination with a premolded sealer to reduce the amount required.

These sealants change shape during extension and contraction, but they do not change their volume. Depth-to-width proportion determines joint efficiency. The shallower the joint, the more volumetric changes it can withstand.

In extension a ½-in.-wide-by-1-in.-deep joint, extended ¼ in., increases the outer fiber's length by 94%. (See Fig. 10-14.) By simply reducing the joint's depths to ½ in. (for a ½ × ½-in. cross section), the strain on the outer fiber will be only 62% for the same extension—about a third less, while using half the material. Doubling the joint width to 1 in. (for a 1 × ½-in. cross section) reduces the strain on the outer fiber to 32% compared to the first section—a 65% reduction in strain with the same amount of material. As a hypothetical case, if the proposed sealant could resist a strain of 75% on the outer fiber, it would fail with ¼-in. movement in a ½ × 1-in. joint, but have a considerable factor of safety for the 1 × ½-in. joint.

These data apply to free-moving joints on two sides. The side opposite the exposed face should not bond to the inside of the joint slot, as this increases strains by more than 100%, as is shown in Figure 10-15. The use of a bond-breaker prevents bonding against the inside of a joint slot. Good bond-breakers include polyethylene, silicone-treated paper, and wax paper.

Strains on sealants during compression for the same three joint shapes shown in Figure 10-14 show outer fiber strain values are much greater for

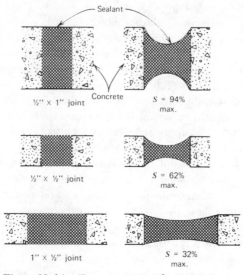

Figure 10–14 Extension strain of ¼-in. movement on the extreme fiber of a sealant of different shapes.

the same movements than for extension. (See Fig. 10-16.) In addition, all elastomers are subject to the phenomenon of compression set. (When maintained in a compressed condition for prolonged periods of time, they may not fully recover their original characteristics.) In designing elastomeric joints for compression, their movement should be limited to 25% or less, to insure that the elastomer will recover as the joint expands.

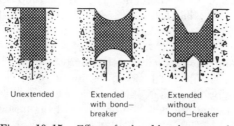

Figure 10–15 Effect of a bond-breaker on sealant performance. Without a bond-breaker the sealant cannot deform uniformly, therefore the strain is increased more than 100%.

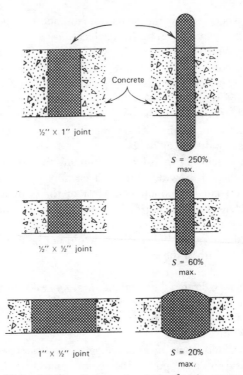

Figure 10–16 Compression strain of $\frac{1}{4}$-in. movement on the extreme fiber of a sealant of different shapes.

Application. Since application is not easy, some firms have specialized in mixing, storing and applying polysulfide sealants. For small applications the two sealant components can be mixed with a slow drill (300 rpm) with paddles attached. For larger jobs, it is preferable to mix them in a shop or plant.

After mixing, the compound is loaded in a polyethylene cartridge and quick-frozen to $-70°$ F. After it is frozen, it is stored at -10 to $-20°$ F. It can be kept at this temperature for up to a month. The cartridge should be transported to the job in dry ice and removed 20 min before it is to be used.

Joint surfaces should be clean and free from oil, dirt, and loose material. A primer specified by the sealant manufacturer should be used, although it is not always called for in specifications.

Sealant is applied with guns—either hand-operated calking guns or air guns. The air guns are speedier and easier to handle.

Hot, humid weather shortens the "pot life" and curing time of the sealant. "Pot life" is the length of time a sealant is workable. On cold, moist days with temperatures below 40° F., condensation is apt to prevent the sealant from setting. Good results can be obtained at 50 to 80° F., with best conditions at 70° F.

Because the sealant is expensive, a minimum amount should be used. Completely filling a 1-in.-wide joint in a curtain wall which is 4, 5, or 6 in. thick would not be economical. Most of the joint is usually filled with a premolded elastic material, and the polysulfide sealant is used only to fill ¾ to 1 in.

Less than ¼ in. of polysulfide sealant should not be installed. It should fill at least ½ the width of the joint.

Premolded Sealants. Premolded sealants are generally used when joints are wider than 1½ in. This occurs when curtain wall panels are large, when the building is subject to severe movement, or when other structural features require it.

Premolded sealants come in either sheets or tubes. They are made of neoprene or butyl rubber. Epoxy adhesive or a nonsag, field-molded sealant is used to bond them to the sides of the joint slot. They should be designed to bend and flex, but never to stretch, since they will pull away from the edges and cause leakage. They should be pitched for drainage.

Tubes should be compressed during installation so that they can recover their original shape during extension. They should never recover completely, or the adhesive binding them to the joint slot will stretch and peel.

Joint sealing is often done with a combination of field-molded sealants. This is sometimes the most economical method.

Gaskets. Many joints are sealed with gaskets of various kinds. Architects are continually designing new types of gaskets, and there is almost no limit to their possibilities. Complicated shapes can be made because die costs are low.

Close contact with joint surfaces is assured by pressure applied to the gasket, either by adjacent surfaces or by a special inlock strip that fits into the seal itself to hold it in shape. The inlock strip may be made separately or as an integral part of the gasket. Either way it is "zipped" into a groove in the gasket by a special tool.

Some gaskets are designed to have structured as well as sealing properties. These are subjected to great pressures over the years. There has been no long-term study to show how well synthetic rubbers behave in structural capacities for long periods. However, neoprene exposed to weathering for 30 years has held up well (Fig. 10-17).

JOINTS IN RESERVOIR LININGS

Spacing

The spacing of joints in reservoir linings is affected by other structures such as bridges, inlet and outlet pipes, roof columns, and integral stairways. Except roof columns, most of these items affect the joint location of only a few panels in their immediate vicinity. Column footings to support the roof slab are usually spaced over the entire reservoir area at some

Locking strip

Gasket before locking strip is inserted.

Figure 10-17   Sealing by resilient gasket: pressure applied by locking strip.

uniform interval; therefore, they have a very important influence on the joint layout.

Kinds of Joints

Three kinds of joints are used in reservoir linings: construction, contraction, and expansion.

A construction joint is one that is required for construction purposes. It may later become a contraction or expansion joint.

Contraction joints are usually of the weakened-plane type formed by constructing a groove in the top portion of the slab. They are used at predetermined locations to localize the shrinkage of concrete resulting from volume change (due to temperature or moisture). In nonreinforced concrete slabs, the distance (approximately 10 to 12 ft) between contraction joints should be such as to reasonably control cracking. In reinforced concrete slabs, the spacing (approximately 20 to 30 ft) may be varied as desired, although in most instances it will be found that there is a maximum practical limit.

The purpose of expansion joints in concrete pavement slabs is to prevent rupture of concrete by reducing compressive stress caused by an increase of temperature above that at the time of placement. If compressive stresses can be kept within 50% of the ultimate strength of the concrete, the possibility of rupture is remote. The shrinkage of concrete during hardening and the plastic flow under compression will, under ordinary conditions, reduce the compressive stress resulting from a 100° F. increase in temperature over that at the time of placement so that the resulting stress in the concrete will be no more than 1500 psi.

This stress would be less than 50% of the ordinary strength of most concrete used in reservoir linings. Therefore, expansion joints in reservoir linings are not ordinarily required except at junctions with fixed structures or under other extreme conditions. Experience has shown that the use of expansion joints has invariably resulted in greater opening of contraction joints. This is particularly objectionable in reservoir linings because of the increased difficulty of maintaining a watertight seal at joints.

Where a reservoir lining intersects a rigid structure, such as a bridge pier or column footing, an expansion joint should be provided to allow movement without danger of cracking the concrete slab. Although certain types of premolded rubber, cork, or asphalt expansion materials have been used, strips of redwood or other soft wood sealed with a joint filler are the most desirable for this purpose.

Joint Fillers

In contraction or construction joints, the groove is filled with a bituminous or rubber-latex compound. In expansion joints a similar filler is used to seal the joint space left above the expansion-joint material. A good filler is one that will adhere well to the concrete and will not become brittle in winter or soft enough to run in summer.

Waterstops

Concrete is practically impermeable. Therefore, the watertightness of a concrete lining depends primarily on the watertightness of the joints. Where it is impossible to maintain a watertight joint by use of a suitable joint filler, some type of waterstop must be used.

The waterstops must maintain their efficiency under stresses set up by shrinkage or expansion of the adjacent slabs and under any differential movement that might occur at the joint. They are constructed of various materials, such as copper, rubber, or plastic, and are generally of two types:

1. The diaphragm or membrane type that spans the joint and is embedded in the concrete for some distance on each side of the joint. Copper and rubber-dumbbell waterstops are representative of this type (Figs. 10–18a and 10–18c).
2. The keyed or integral types (Fig. 10–18b).

JOINTS IN HYDRAULIC STRUCTURES

The performance of hydraulic structures depends on the durability of the concrete of which they are made. Construction joints are vulnerable points for the start of deterioration in hydraulic structures. Construction at horizontal joints, in particular, should be given careful attention to see that bleeding does not take place. The durability of concrete at such joints is affected by the quality of concrete immediately below the joint and by the care taken to prepare the joint surface before fresh concrete is placed for the next lift.

Preparing Joint Surface

When concrete is placed in deep sections, there is a tendency for water, as well as the silt and clay present in the aggregate, to rise to the surface

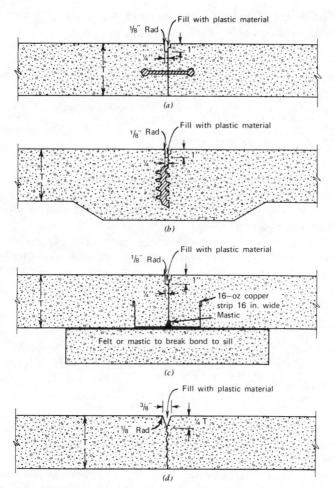

Figure 10–18 Typical contraction and construction joints. (*a*), (*b*), (*c*), (*d*).

and form a weak, porous surface layer. In air-entrained concrete of proper design, workability, and slump, this tendency is greatly reduced and there will be little, if any, such material on the surface. If a layer of inferior concrete exists, it is good practice to sweep the top surface with a fairly stiff broom just before the concrete becomes thoroughly hard. This will remove the soft mortar from the surface and will uncover the coarse aggregate which, if adequately anchored, will provide good bond for the next lift.

Placing High Lifts
In case of high lifts, such as for retaining walls, concrete may be placed to a level about 1 ft below the joint location on top of the wall and allowed to settle for about a half hour. Then, concrete of the same proportions, except with somewhat lower slump, is placed to complete the lift. Thorough vibration of this top layer of concrete is essential.

Preparing Hardened Concrete
The hardened concrete that forms the surface of a construction joint can best be prepared for continued concreting by sandblasting and washing the surface immediately before placement of the next lift. Sandblasting should be limited to the removal of weakened paste (laitance) since further cutting and roughening may weaken the joint.

Placing New Concrete
In all cases, the new concrete should be placed on a layer of cement mortar spread evenly over the joint surface. When possible, the mortar should be scrubbed into the surface of the joint with wire brooms. This may be followed by placement of a 3- to 5-in. layer of concrete from which a portion of the coarse aggregate has been omitted on the fresh mortar layer. This will help prevent rock pockets at the bottom of the new lift.

Vertical Joints
Properly made horizontal construction joints are as impervious to moisture as adjacent areas of concrete. Vertical construction joints, however, are frequently designed to function as contraction joints. Vertical joints in concrete dams almost always are grouted to unite the several blocks into one structural mass. Whether or not grouting takes place, waterstops should be installed in vertical joints. The size and shape of the waterstop and the material of which it is made will be governed by the hydraulic pressure at the joint, the maximum expected movement, and economic considerations. Careful placement of the waterstop and consolidation of the concrete in its vicinity are the most effective safeguards against subsequent leakage.

Subgrade pools should be designed to withstand the water pressure from within and to resist the pressure of the earth when the pool is empty. To prevent the possibility of cracks forming as a result of temperature changes and shrinkage, it is necessary to provide sufficient reinforcement and correctly designed expansion joints in the walls and floor. Adequate curing of high-quality concrete will reduce the possibility of cracks.

Practically all pools contain either construction joints or expansion joints or both. The proper location, design, and construction of these joints are very important in obtaining a watertight pool. Construction joints, or control joints, are intended to be rigid, while expansion or contraction joints are intended to allow for the movement caused by changes in temperature and moisture content.

Construction Joints

Construction joints should be avoided as much as possible by planning the work so that a complete section between expansion joints may be placed in one continuous operation. However, it will generally be necessary to make a construction joint at the base of the walls. Such horizontal joints can be made satisfactorily by simply bonding the new concrete to the hardened concrete. At such joints, it is good practice to key the two sections by forming a longitudinal slot in the center of the first section before the concrete hardens.

Vertical construction joints in walls should be keyed together and a waterstop used; or extra dowel bars should be used equivalent in area to 0.015 times the cross-sectional area of the wall and extending 30 bar diameters on each side of the joint.

Construction joints are seldom necessary in floors, but if required, should be made the same as suggested for vertical construction joints in walls. Screeds must be removed as soon as possible and the space should be filled with concrete worked into the adjacent portion of the slab.

Expansion Joints

The expansion joints in walls should be keyed and made watertight with waterstops. Reinforcement should never extend across any expansion joints.

Expansion joints in floors should be made over a beam or footing, or some special type of joint used that will keep the adjacent slabs in line as well as being watertight. The most common type of joint is made with a metal waterstop and mastic fill. Where the ends of the slab rest on beams or footings, the waterstop is frequently omitted, and dependence for watertightness is placed entirely on the filling and mastic between the slab and the footing or beam. The use of asbestos-fibered mastic, or multiple layers of open mesh burlap mopped with mastic, will permit differential movement and assure that the mastic remain in place. Strips of soft, clear-grain wood may be used as filler between slabs. A layer of bentonite, about ½ in. thick, placed under the mastic in floor joints has been very effective in preventing leaks in both old and new pools. Off-setting the expansion joint in the floor about 1 ft from wall joints will aid in making tight joints.

The proper location and spacing of expansion joints must be determined for each job, and although there are no fixed rules for this determination, some general comments will be helpful. They should be placed where there will be the greatest tendency to crack, such as at changes in section or direction of members. For pools of regular shape, such points will be at the junction of floor and walls and where there is a sharp change in grade of the floor. The distance between joints should ordinarily not exceed 60 ft. However, a number of successful small and medium-sized pools have been built without expansion joints.

SEALANTS

Selecting the proper sealant for joints is very important and not a simple task. There is a wide variety of new sealing products on the market that have not been around long enough to be field tested. Many times the manufacturer's claims are overly optimistic.

Some specifications call for sealing joints with premolded asphaltic materials, while others call for resin-type sealers. Foamed plastics are used, sometimes alone and sometimes in conjunction with other types of sealants. Even lead is effective in special conditions. The options are varied and often confusing. However, some characteristics that all seal-ants should have are resistance to sagging in joints, nontoxicity in normal use, and ease of application. Materials presently on the market differ in: price, ease of application, adhesion and cohesion, pliability, resistance to weathering, life expectancy, and behavior in cold weather.

Some joint sealants are primarily rubberlike, and tend to return to their original shape after being stretched or compressed. Others are perma-nently plastic (puttylike) and, when deformed, undergo "plastic flow" and remain deformed until subjected to some new force that again changes their shape. Most joint sealants fall somewhere between these two extremes, and many of them change their characteristics somewhat because of aging or weathering. A perfectly elastic sealant placed in a joint at 70 to 75° F. could become compressed during hot weather. When the temperature falls and the joint opens to its original width again, the sealant should return to its original shape. Likewise, if the temperature drops below 70° F. and then warms up again, the sealant should stretch and then return to its original shape.

Because most sealants are not perfectly elastic, each one will be kneaded into a narrower shape when compressed. The next time the joint opens it will have to stretch proportionately farther, and this is more likely to cause failure than if the sealant had not undergone plastic flow. (Actually, if the stretching is sufficiently slow, there can be plastic flow in the reverse direction, which may, at least, partially make up for the plastic flow in compression.) Sometimes the ability of a sealant to undergo plastic flow will decrease, the sealant will become harder, and it will become more likely to fail from the same amount of movement that it could formerly withstand.

Joint sealants are installed in various ways, depending on the material. Some require heating just prior to installation while others simply require careful mixing. Some are performed and ready to insert in the joint. The liquid types are usually installed by pumping or with a compression gun.

If the manufacturer recommends a primer it should be used as directed. Usually it is furnished as part of the sealant package. Timing is crucial when using a primer. If the sealant is placed in the joint before the primer is dry, the primer will hinder adhesion.

Regardless of the type of sealant used, joints must be thoroughly cleaned of oil, dirt, or any material that might prevent bond with the sides of the joint. This can be done by sandblasting and blowing the joint out with oil-free air. If there are honeycombed areas on the sides of joints, they should be patched before the sealant is placed. Sandblasting the joint surface is always necessary to prepare it for bonding.

Sealants usually require a backup material to prevent bond to the concrete at the bottom of the joint, and to insure correct sealant depth. Compressible nonabsorptive backup materials, such as closed-cell neoprene, polyurethane foam, or polystyrene foam, are used to maintain the shape of the joint. When it is only necessary to prevent bond, silicone-treated paper, polyethylene tape, or even strips of newspaper or wax paper can be used.

Thermoplastics

Thermoplastics are materials that become soft when heated. Materials like rubber asphalts, pitch coal tar and rubber-coal tar, are used primarily in highway, airport, and street paving. Some are designed for jet-fuel resistance. These may be mixtures of rubber and tar with plasticizers and stabilizers that jet fuel does not affect. They are melted in a double boiler to temperatures not in excess of 400 to 450° F. At higher temperatures there is a tendency for these materials to stiffen and become solid in the kettle, making them unsuitable for use in the joint.

Some thermoplastics, like vinyls and acrylics, contain 20 to 30% solvent, so they may be applied without heating. These are convenient to handle, but they undergo large shrinkages from solvent evaporation.

Sandblasting followed by blowing compressed air into the joint slot is necessary to prepare the joints. Thermoplastics must be poured or pumped into dry, clean joints, so they are usually used only in joints in horizontal surfaces.

Cold-pour Thermoplastics

These rubber and asphalt materials are not true thermoplastics because they undergo permanent changes in properties with prolonged or repeated heating. They do resemble thermoplastics, however, so they are included in this category for convenience.

They come in two types: single-component materials ready for immediate use, and two-part materials that require mixing. The single-component materials contain a flammable solvent; for this reason, they should be kept away from open flames. The solvent slowly evaporates after the sealant is in place, causing some shrinkage. The sealant also undergoes some curing in the joint.

These materials have greater elasticity and resilience than mastics, but they are not as effective as elastomers. They have low resistance to weathering and oxidize rapidly causing the sealant to harden. Installed and used carefully, they will perform very well for some time, however.

Butyl Rubbers

Butyl rubber materials are made from polybutene. They are mixed with resins, oils, fillers, solvents, and pigments to make a wide variety of sealing compounds. The butyl rubbers are nonoxidizing and almost noncuring, since they remain tacky for an extremely long period of time. In this respect they resemble the nonhardening mastics. These sealants are used more as calking compounds than as joint sealants, but they do fall under the general heading of sealants.

It is possible to compound the butyl rubbers into a skinning or semi-drying material by adding a drying oil. This makes a better calking material than the old, oil-based products that have been on the market for many years. The butyls will not tolerate much movement but they have some flexibility.

Elastomers

Polysulfide systems are used primarily for vertical joints in buildings. They are sold either as one liquid or as two liquids which, when mixed, cure by chemical reaction. Polysulfide elastomers are made from polysulfide polymers that cure by a linking of molecules from end to end. The polysulfide polymer starts out as a long molecule. When many of these long molecules are linked together by means of the curing agents, they become a rubber with the required elasticity. A rod of this rubber, $\frac{1}{2}$ in. in diameter, can be stretched up to 8 times its original length, depending on how it is formulated. However, field installations are subjected to a wide variety of influences that are not present in laboratory tests. Therefore, joint design should limit movement to a maximum of 25% extension and 25% compression.

It is recommended that joint shape be twice as wide as it is deep to get the best results. Manufacturers may differ here, and some may recommend that the joint width equal the depth. Such recommendations are based not only on the properties of the sealant but on the distance between joints and the climate. Prices vary depending on how much filler is used in the sealant. Obviously, this reflects differences in quality, so the buyer should take great care in choosing a sealant.

Two-part polysulfide sealants have service records dating back to 1955, but extensive use of one-part sealants became common only recently, so they have relatively short performance records by which to evaluate them. The one-part polysulfide sealants are usually sold in calking cartridges, ready for use. They are cured by the humidity in the air. When they start to cure, the sealant will skin over. Total curing takes from 14 to 180 days depending on the atmospheric conditions. Traffic and air pollutants can damage and discolor the materials while they cure. After they are cured, one-part polysulfide sealants have physical properties close to those of two-part compounds. Since there is no mixing, it is easier to use one-part sealants. Calking cartridges can be loaded on the scaffold, or a pail of the sealant from which calking guns are loaded can be supplied to the workmen.

Two-part sealants have to be mixed on the ground, and a man has to get them up to the workman filling the joints. Mixing equipment is required at the site, which usually means another man has to be employed to operate it. Any material that is not promptly removed from the mixer will cure there and will cause a cleanup problem. Proper proportioning is essential to good quality. This is not always easy at the job site.

For horizontal joints, where poured polysulfide joint sealants can be used, the two-component systems are usually easier to handle. Since they cure faster, the area can be opened to traffic earlier.

Silicones cure by contact with moisture in the air. In 48 to 54 hr they produce a flexible sealant. Other characteristics of silicone sealants are: very low shrinkage, lack of toxicity, ability to withstand service temperatures ranging from -200 to $+500°$ F., good strength, and excellent resistance to the effects of sunlight, ozone, and ultraviolet light. They also have good resistance to organic solvents, salts, both oxidizing and nonoxidizing acids, as well as to water.

Silicones function almost entirely by the mechanism of elastic deformation rather than plastic flow. The silicones, therefore, are in the same class as the polyurethanes that exhibit full recovery after either compression or extension.

A minor disadvantage is that silicones generate a small amount of acetic acid during curing. This acid attacks concrete and certain metals. However, they have a long service life and are tougher and harder than polysulfide sealants.

Acrylic sealants are of the solvent or emulsion type. They set when the water portion of the emulsion evaporates. Acrylics are rather new and seem to have found favor more as calking compounds than as joint sealants. This is true because they do not hold up well when compressed after being extended. They behave well under extension but their plasticity is low, so they cannot accommodate movement by means of plastic flow. They have fair-to-poor adhesion, some shrinkage, a service temperature range of -35 to $+150°$ F., fair strength, low dynamic properties, good weather resistance, and good-to-excellent resistance to water, salts, and nonoxidizing acids.

Several grades of polyurethane are available: a nonsag type, a hard one for foot traffic, and a fast-curing type for joints in bridge decks, highways, and runways. Polyurethane sealants are among the toughest of the elastomeric polymer compounds available today. They are nonstaining, durable, and flexible. They have high tensile strength, are tear resistant and highly abrasion resistant, and they recover well from elongation and compression. Hence, they are a good choice for sealing where there is heavy traffic and high stress. They give good service in temperatures ranging from -45 to $+275°$ F. They resist weather aging, compression set, water immersion, dilute acids, dilute alkalis, and solvents. They adhere excellently.

Polyurethane sealants have excellent resistance to compression set and plastic flow. They will last through many cycles of compression and elongation, usually returning to their original $\frac{1}{2}$-in. width of their own accord.

The polyurethanes are also resistant to increased hardening as they grow older. They do not undergo additional oxidation reactions during long-term exposure and, therefore, retain their original hardness over a much longer period of time than most other sealants.

Polyurethane might appear to be the perfect sealant, but actually there is no entirely perfect sealant. Like all other products, improper formulation may make some polyurethane sealants sensitive to moisture or may produce an improper balance of physical properties. The user must choose only products with proved quality and service life.

Neoprene seals are now widely used in pavement joints and structures. They are the only type of sealant that can stand up under the movement of bridge decks. They are extruded strips of neoprene rubber designed so their width is greater than that of the joint at its maximum opening. The strip is mechanically compressed just before inserting into the joint slot. In highways it is positioned just below the surface of the pavement. Once it is in place, its force exerts pressure against both faces of the joint and results in a positive seal that stops the entry of noncompressibles such as sand and silt. These noncompressible materials could cause joint cracking with thermal expansion and contraction of the pavement slab. As in all sealing systems, the joint must be sandblasted, swept, or blown free of all solid debris and water.

Relatively simple equipment is needed to install these strips. There are no mixing, heating, or temperature limitations for installing them, but a coating of neoprene adhesive should first be applied on the joint faces.

Proper installation requires that the seal be properly sized for joint extremes. Proper size is determined by the type and behavior of the joint into which it is to be inserted. Joint edges must be properly formed. Stretching of the seal during installation should be avoided. If a neoprene seal becomes damaged it can be replaced easily but it should give many years of maintenance-free service. Sometimes subgrade restraint may cause one joint in a pavement to do the work of adjacent joints. When this happens, it will open excessively and the neoprene seal may come out. In such circumstances it should be replaced with a wider seal.

Compressible polyurethane foam is impregnated with a waterproofing polybutylene to produce a strip of spongelike material that is quite tacky. The polyurethane must be of the open-cell type and must be thoroughly impregnated with a fluid composed of a polybutylene base plus a catalyst and a drying agent. The polybutylene serves a dual function: it provides the pressure that controls the rate of recovery after compression, and it helps to make the compressed strip impervious to liquids. The polybutylene does not serve as an adhesive agent. The strip is installed in the joint under compression in much the way neoprene strips are.

Polybutylene-filled urethane strips can be used as the only sealant or they can be used as backup materials below other sealants.

Polybutylene-filled strips are slightly tacky to the fingers. They are very easy to install compared to gun-type sealants. They are not adhesive in themselves, but they can be bonded with a contact cement. Polybutylene becomes slightly less extensible on weathering, but this change is not likely to affect these seals because they are continuously in compression. The seals are impermeable to water and can be installed in any weather (Table 10-B).

Mastics

Mastics are composed of asphalts and nondrying oils. They are heated before being placed, so heating equipment is needed at the job site. Their service temperature range is -35 to $+175°$ F., but their extensibility is extremely low at low temperatures.

They have poor adhesion qualities and poor strength in tension. They are excellent materials to use where joint movement is small but where joints are exposed to large quantities of water; consequently, they are recommended for joints in reservoirs, water tanks, and basements. Priming is advisable before application.

Waterstops

A waterstop is usually a section of flexible, waterproof material cast in place across a joint in concrete to prevent the passage of water. Early waterstops were made of rigid or semiflexible metal. These types relied

Table 10-B
SEALANT CHARACTERISTICS

Characteristic/Type	Mastics	Thermoplastics	Elastomers				One Part Polysulfide and Polyurethanes	Performed Neoprene Compression Gaskets
			Polysulfides	Silicones	Acrylics	Polyurethanes		
Composition	Asphalts, nondrying oils, etc.	Rubber asphalts, Coal tar	Long chain polymers	Silicone polymers	Acrylic acid monomers and polymers-emulsions		Same	Neoprene rubber
Care in surface preparation	Fair	Fair	Critical	Critical	Moderate	Critical to moderate	Same	Joint clean
Cure characteristics	Check manufacturer	Solvent evaporation or by cooling	Catalyst curing agent and 50°F +	Humidity pick up. Temp. not critical	Water evaporation. Temp. not critical	Catalyst curing	Humidity pick up	(Not applicable)
Cure time	When cool	Varies	4 to 7 days			3 to 6 days	PS 20 to 30 days PU 5 to 10 days	(Not applicable)
Handling	Needs equipment. Must be heated	Needs equipment	1 or 2 part 2 part needs equipment	1 part	1 part	1 or 2 part 2 part needs equipment	—	Equipment preferable
Storage	Indefinite	Indefinite	6 months	Limited	Limitless if sealed. Damaged by freezing	Limited	Same	Indefinite
Adhesion	Poor	Fair to excellent	Good	Excellent	Poor	Excellent	PS Good PU Excellent	Function of seal
Low shrinkage	Poor	Poor to excellent	Fair to good	Good	Fair	Good	Same	None
Toxicity—uncured	Moderate	Moderate	Moderate to high	None	Moderate	Moderate	Same	(Not applicable)
Service temperature range	−35°F to 175°F	−35°F to 250°F	−60°F to 250°F	−200°F to 500°F	−35°F to 300°F	−60°F to 250°F	Same	−50°F to 200°F

| Characteristic/Type | Mastics | Thermoplastics | Elastomers | | | | One Part Polysulfide and Polyurethanes | Performed Neoprene Compression Gaskets |
			Polysulfides	Silicones	Acrylics	Polyurethanes		
Strength—tension	Poor	Poor to excellent	75-300 psi.	150-600 psi.	75-150 psi.	150-600 psi.	PS 100-200 psi. PU 300-500 psi.	2000 psi.
Dynamic properties	Fair to poor	Poor to excellent	Good	Excellent	Fair to poor	Excellent	Same	Excellent
Aging	Fair	Fair to excellent	Fair to good	Good to excellent	Fair	Excellent	Same	Excellent
Weathering	Poor to ultraviolet	Fair to good	Fair—some good	Good	Good	Good	Same	Excellent
Resistance to:								
Water	Excellent	Excellent	Good	Excellent	Excellent	Excellent	Same	Excellent
Solvents (organic)	Poor	Poor	Poor	Fair	Fair	Poor	Same	Excellent
Salts	Excellent	Excellent	Good	Excellent	Excellent	Good	Same	Excellent
Alkalis	Good	Good	Good	Poor	Poor	Good	Same	Excellent
Nonoxidizing acids	Good	Good	Excellent	Good	Excellent	Excellent	Same	Excellent
Oxidizing acids	Good	Good	Poor	Poor	Poor	Poor	Same	Excellent
Recommended uses:								
Highways	No	Yes	Yes	No	No	Yes	Same	Yes
Overpasses and Bridge Decks	No	Yes	Yes	No	No	Yes	Same	Yes
Buildings—metal to metal	No	No (6)	Yes	Yes	Yes	Yes	Same	Yes
Buildings—concrete to joints	No	Yes (1)	Yes, with care (3)	Yes	No	Yes	Same	Yes
Buildings—precast panels	No	No	Yes, with care	Yes	Yes (3)	Yes	Same	Yes
Buildings—caulking	No	No	Yes	Yes	Yes	Yes	Same	Yes
Buildings—floor joints	No	Yes	No	Yes (5)	No	Yes	Same	Yes

Table 10-B cont.
SEALANT CHARACTERISTICS

			Elastomers				One Part Polysulfide and Polyurethanes	Performed Neoprene Compression Gaskets
Characteristic/Type	Mastics	Thermoplastics	Polysulfides	Silcones	Acrylics	Polyurethanes		
Underwater—such as:								
Reservoirs	Yes	Yes	Yes (4)	No	No	Yes	Same	Yes
Sewage plants	Yes	Yes	Yes (4)	No	No	Yes	Same	Yes
Water tanks	Yes	Yes	Yes (4)	No	No	Yes	Same	Yes
Basements	Yes	Yes	Yes	No	No	Yes	Same	Yes
Airstrips—Jet resist.	No	Yes, if special (2)	No	No	No	No	Same	Yes
Taxiways—Jet resist.	No	Yes, if special (2)	No	No	No	No	Same	Yes
Runways	No	Yes	Yes	No	No	Yes	Same	Yes
Swimming pools	No	Yes	Yes (4)	No	No	Yes	Same	Yes
Sidewalks	No	Yes	No	Yes (5)	No	Yes	Same	Yes
Recovery	Poor	Excellent	Moderate	100%	Very poor	100%	PS moderate PU 100%	Excellent
Abrasion and puncture resistance	Poor	Fair	Fair to Poor	Excellent	Poor	Excellent	Same	Excellent
Hardness increase—Age	Poor	Poor to excellent	Poor	Good	Poor	Good	Same	Excellent
Hardness increase—Low temperature	Poor	Poor to good	Poor	Good	Poor	Good	Same	Excellent
Priming	Recommended	Varies, ask manufacturer	Recommended			Recommended	Recommended	Adhesive lubricant

Notes
1. Use only if color is not important.
2. Use only special "jet resistant" thermoplastics.
3. If movement is within capability of sealant.
4. Only if code requirement of exposed area of joint surface per million gallons of water is met—curing agent lead dioxide is toxic.
5. If movement is within sealant capability—also cost is high in large quantities.
6. Thermoplastics have excellent adhesion to metal but they are not recommended here because they cannot be applied with a caulking

solely on a good bond between the metal and concrete for their effectiveness. They could accommodate only limited movements.

A fully flexible material, like natural rubber or neoprene, is resilient enough to stay locked in the concrete for some time without reducing joint efficiency from expansion, contraction, or settlement (Fig. 10-19).

The kind of rubber used for waterstops should not become brittle when exposed to alkali, sunlight, or continuous flexing. Many installations require the rubber to be resistant to chemicals, petroleum products, and sewage.

Polyvinylchloride (PVC) is the most popular plastic used to make waterstops. PVC performs about as well as rubber but is more rigid and less elastic. Some waterstops are made of two materials in combination, such as metal and neoprene, or flexible rubber and plastic. The flexible rubber and plastic waterstops have many advantages over those with metal. They can be purchased in long, coiled lengths (usually 15- to 125-ft rolls), which reduces the need for frequent splicing. They can be bent to accommodate irregular contours. Finally, they are resistant to impact and other mechanical damage and to chemical attack.

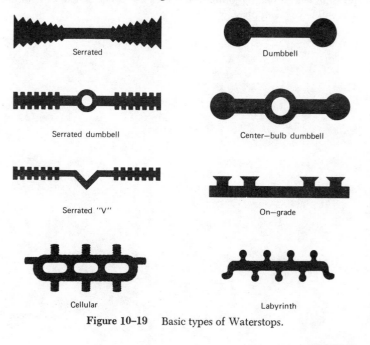

Serrated   Dumbbell

Serrated dumbbell   Center—bulb dumbbell

Serrated "V"   On—grade

Cellular   Labyrinth

Figure 10–19   Basic types of Waterstops.

There are several essential requirements for proper installation of waterstops:

1. The center bulb, or other provision to accommodate joint movement, must be positioned directly at the joint opening. With strip-type waterstops, adequate support is usually provided by splitting the form. Lashing the waterstop to the reinforcement with wire provides additional permanent support. Labyrinth waterstops can be nailed in place. Formwork must always be tight fitting.
2. The waterstop must be clean to achieve a good seal with the concrete.
3. The concrete must be properly compacted, preferably by vibration, to insure continuous intimate contact with the waterstop. A waterstop must always be embedded in concrete of high quality. Waterstops are designed to provide an effective waterproof barrier across construction, contraction, and expansion joints. They are not intended as a remedy for porous concrete.
4. All joints must be sealed properly. The joint sealant will prevent entry of dirt and foreign material. Joint sealing is essential where freezing and thawing may be a factor or where edges of the joint need to be protected against mechanical damage. It is better to wait as long as possible before sealing joints so the greatest amount of drying shrinkage will have taken place.

CHAPTER 11
CURING

When cement hydrates, chemical reactions take place between the cement and the mix water. Concrete does not harden because it dries out, but rather, it hardens because it does *not* dry out. The development of tobermorite gel and other products resulting from the reactions between water and cement causes the hardening of the concrete. Water is actually entrapped as the concrete cures. If water is constantly available to the cement particle, this chemical reaction will continue until all the cement has been combined. Hydration can continue in concrete for a long time. In a massive structure of solid concrete, like a gravity dam, hydration can continue for years. As long as it does, the concrete will continue to gain in strength.

The process of curing concrete is merely one of keeping the concrete moist and at a temperature of about 50° F. In some instances, this is relatively easy to do; in others, it is more difficult. The total effect of temperature and moisture on concrete depends to a great extent on the total mass and thickness, the type of cement used, and the age of the concrete. Under proper conditions, concrete will continue to cure until total hydration is achieved.

Cubes of cast concrete with sides of 8 and 12 in., when kept moist, showed much greater strength at 11 years than they had after four weeks of curing. It is also true that if a piece of concrete has been improperly cured, it can be restored to some extent by the resumption of curing. It may not achieve the total ultimate strength it could have had if proper curing had been sustained from the beginning, but a certain increase in strength and decrease in porosity can be gained.

REASONS FOR CURING CONCRETE

Reducing Shrinkage

If the surface of concrete is allowed to dry out before final set occurs, the surface may actually shrink, causing unsightly cracks in the piece that may sometimes be harmful. These cracks are called "plastic shrinkage cracks" because they occur before final set in the concrete. They can be easily identified because, unlike cracks from other causes, they appear in the slab as it starts to set. These cracks may have considerable depth and sometimes will have a defined crow's-foot pattern. Plastic shrinkage cracks, unlike other defects, appear as the concrete sets but do not get worse.

The major cause of plastic shrinkage cracks is high ambient temperatures (above 90° F.), low relative humidity, and winds of over 10 mph in clear weather, together causing rapid evaporation of mix water from freshly placed concrete. When the surface of the concrete has developed some rigidity, it still is not strong enough to withstand any rapid volume change or tensile stress. It is at this point, or soon thereafter, that the plastic shrinkage cracks will appear on the surface if moisture is not retained in the slab.

As soon as concrete is placed, the solids tend to settle, forcing the mix water to rise to the top. This is called "bleeding." If the rate of evaporation of water from the surface of the slab exceeds the bleed rate, then shrinkage and cracking are likely to occur. This is also the point when curing should start.

Another less dramatic result of shrinkage is the presence of fine pattern cracks in the surface layer of the concrete. This also is caused by shrinkage resulting from a too-rapid loss of moisture.

Ordinary shrinkage can cause deeper cracks that extend sometimes entirely through and across the slab.

Accelerating Strength Gain

The most convincing argument for curing concrete is the chart showing the strength of concrete cured for various amounts of time.

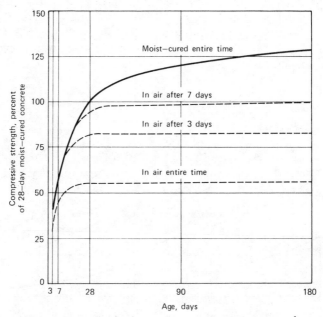

Figure 11-1 Strength of concrete continues to increase as long as moisture is present for hydration of cement.

The solid black line in Figure 11-1 indicates the laboratory standard of the curing arrangements: cured at 100% relative humidity, 70° F. Note also that this black line passes 100% of compressive strength at 28 days and actually ends up at 125+%. The 100% figure is the norm, the 28-day strength of moist-cured concrete: all other strengths of concrete shown here are based on percentages relative to this.

Concrete that has been moist-cured for 7 days and has been in the air thereafter will achieve almost 100% strength by the end of 180 days. Three-day-cured concrete will attain about 80% strength, and uncured concrete will have only a bit more than half of its possible strength. These strengths are approximately the ultimate that will develop unless moisture is reintroduced.

This is vivid visual proof that curing is important—especially when the ultimate strength of the concrete is a vital concern.

Figure 11-2 shows compressive strengths (measured in pounds per square inch) of moist-cured cylinders at various ages. It is interesting to

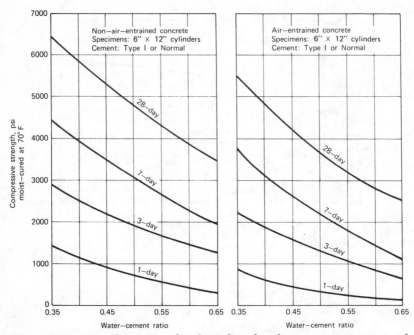

Figure 11-2 Typical age-strength relationships based on compression tests of 6 x 12-in. cylinders right, for air-entrained concrete: left, for non-air-entrained concrete).

note that a 0.35 water-cement ratio, non-air-entrained concrete, achieves almost 1500 psi the first day. At the end of the 28-day curing period, this mix will develop about 6300 psi.

These two charts graphically demonstrate two important principles—the effect of the water-cement ratio and the effect of curing on the ultimate strength of concrete. In both air-entrained and non-air-entrained concrete (using Type I cement), the strength at the end of 28 days of curing is almost five times the 1-day strength.

We can easily draw these conclusions: it is important to keep the water content of the mix as low as possible and to extend the period of curing as long as is practical.

Minimizing Creep

For some time, engineers and scientists have known that when concrete is subjected to heavy and sustained loads or pressures, it will deform slightly with time. This deformity is known as "creep." Creep is believed to be caused by the imperfect crystallization of some of the hydration products, causing these molecules to be both viscous and elastic.

Creep is important in many phases of construction and precasting, but it is of prime concern to the prestressed industry. There is great need to control the deflection and change of shape of a prestressed beam. Even a slight shortening of the concrete beam could negate the prestressing, causing complete or partial loss of tension in the steel cables. For this reason engineers and designers have had to pay much more attention to the fact that concrete can deform or creep. It has been found, however, that if concrete is properly cured, creep will be less.

Figure 11-3 illustrates the fact that high-pressure steam-cured concrete has much less creep deformation than seven-day moist-cure.

Improving Durability

Durability includes resistance of concrete to freeze-thaw cycles and wet-dry cycles, and stability in the presence of chemicals. Concrete is expected to have a long life together with low upkeep. To achieve this, concrete must have durability, and to have durability when exposed to deicing chemicals and freeze-thaw cycles, it must have entrained air and be properly cured.

In tests conducted to determine curing requirements for scale resistance of concrete, the results in Table 11-A were obtained. At intervals during the scaling test, the surfaces were examined, rated as to extent and depth of scale, and assigned a numerical rating.

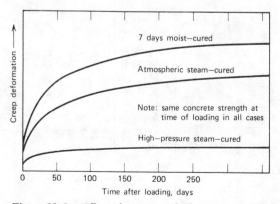

Figure 11-3   Effect of curing method on magnitude of creep for typical, normal-weight concrete.

Table 11-A
RESULTS OF SURFACE SCALING AND STRENGTH TESTS—
TYPE I CEMENT

Type I Cement—Lot 18868. No CaCl$_2$.
Cement content of all concretes—6 sacks per cu yd.
Neutralized Vinsol resin added at mixer for air-entrained concrete.
All specimens cured continuously moist for times indicated.
Net W/C: Non-A/E concrete—4.8 gal per sack. A/E concrete—4.5 gal per sack.
Air content (pressure): Non-A/E—1.60%. A/E—4.70%

Days of Curing	Curing Temp F	Compressive Strength psi 6- × 12-in. Cyl	Scale Ratings at Indicated Number of Cycles									
			5	10	15	25	50	75	100	150	200	250
			Non-Air-Entrained									
1	73	960	2	3	4+	(16)[a]						
3	73	2700	1−	2+	(15)	3						
28	73	6270	0	0	0+	3	(50)					
2	40	330	2	3	(15)							
4	40	1120	3+	4+	(13)							
6	40	1940	1+	3+	4+	(16)						
8	40	2960	1	3+	(15)							
12	40	4020	1	3	4+	(17)						
19	40	4880	1+	3	4+	(19)						
30	40	5540	1	4−	5−	(16)						
60	40	6410	0+	3+	4+	(23)						
9	25	440	0+	2	(15)							
18	25	560	0+	1	2	(16)						
28	25	560	0+	(10)								
40	25	520	1−	1+	3	(17)						
60	25	680	0+	1+	(15)							

1	73	1050	1+	2	2	2+	3+	4−	(95)			
3	73	2720	0	0	0	0+	1−	1	1+	2−	2−	2−
28	73	5500	0	0	0+	0+	0+	0+	0+	0+	0+	0+
2	40	420	3−	3+	4−	4+	(35)					
4	40	1960	2−	2	3+	4−	(40)					
6	40	1780	0+	0+	1−	1−	3−	3	3+	5−	5−	5−
8	40	2730	0	0+	0−	1−	1+	2	2+	4+	5−	5−
12	40	3460	0	0	0+	1−	1+	2−	2−	2+	3+	4−
19	40	4360	0	0	0	0+	0+	0+	0+	1−	1+	2
30	40	4900	0	0	0	0+	0+	0+	0+	1−	1	1−
60	40	5680	0+	0+	0+	0+	0+	0+	0+	0+	1−	1−
9	25	540	0+	1−	2+	3+	(31)					
18	25	720	0+	1−	1	(22)						
28	25	900	0	1−	1−	1−	1−	1−	1+	(106)		
40	25	1030	0	1−	1−	1−	1−	1−	1−	3	3	(167)
60	25	1010	0+	1−	1−	1−	1−	1+	1+	2	3	(220)

[a] ( )—Number of cycles at which the scaling test was discontinued at a rating of 5.
At intervals during the scaling test, the surfaces were examined, rated as to extent and depth of scale, and assigned a numerical rating as follows:

0—no scaling
1—very slight scaling
2—slight to moderate scaling
3—moderate scaling
4—moderate to bad scaling
5—severe scaling

These tests demonstrate the importance of curing for the development of scale resistance. Non-air-entrained concrete cylinders cured for 28 days showed only moderate scaling after 25 freeze-thaw cycles. The same concrete cured for 1 day showed from slight to moderate scaling after 4 cycles and severe scaling at 16 cycles. Specimens cured at 40° F. displayed greater scaling at all points of curing from 2 to 60 days.

Reducing Efflorescence

Efflorescence is the presence of soluble salts—usually calcium hydroxide —on the surface of the concrete. One of the causes of efflorescence is the crystallization of calcium hydroxide from the cement-water paste. This crystallization occurs at or near the surface when water passes through concrete and evaporates. Proper curing prevents mix-water evaporation. Efflorescence is unsightly and points to possible serious trouble within the concrete.

Improving Abrasion Resistance

Curing develops strength in concrete, and strength indicates a high resistance to abrasion. To develop strengths sufficient to produce a hard and durable surface, it is necessary to cure adequately.

Concrete with 6000 psi strength displays excellent resistance to abrasion. This is extremely important to the durability of paving, industrial floors, spillways, and other structures that must withstand a great amount

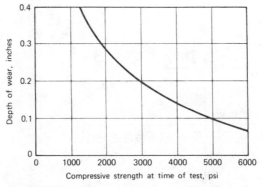

Figure 11–4 Effect of compressive strength on the abrasion resistance of concrete. High-strength concrete is highly resistant to abrasion.

of frictional wear. Concrete with a compressive strength of only 2000 psi had a 0.29-in. groove worn into it at the end of one test period (Fig. 11-4).

Improving Impermeability

There is a direct relationship between the method of curing a piece of concrete and its ultimate watertightness. If concrete is made with non-porous aggregate, and if it is properly cured, hydration will continue until all cement particles have been used up or until there is no space left for the formation of gel. Billions of air cells are trapped in air-entrained concrete, but these cells are not interconnected and they are microscopic in size—from 0.001 to 0.003 in. in diameter.

In Figure 11-5, it can be seen that two factors affect the watertightness of concrete: water-cement ratio and curing time. Concrete with a water-cement ratio of 0.50 when moist-cured for 7 days will be watertight. Poor quality concrete (water-cement ratio of 0.80) never will achieve water-tightness even with 28 days of moist curing. So, while correct curing is vital, it will not transform poor quality concrete into good quality concrete.

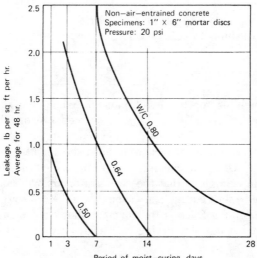

Figure 11-5 Effect of water-cement ratio and curing on watertightness. Note that leakage is reduced as the water-cement ratio is decreased and the curing period increased.

VARIABLES AFFECTING CURING

As concrete hydrates, there are three variables that must be considered: the concrete must be kept wet (at or near 100% relative humidity), the temperature should not be under 50° F., and both temperature and moisture must be maintained for 3 to 7 days.

Temperature variations affect the compressive strengths of concrete. Concrete cured for the first 28 days at 55° F. attained the highest strength at one year—more than 130% of the normal 28-day strength. Concrete cured at 40° F. will do about as well at the end of a year, but the early strengths of both are very poor. Comparing strengths at 7 days, concrete being cured at 73° F. is almost twice as strong as that being cured at 40° F. At lower temperatures the rate of strength gain in concrete is appreciably slowed down until, at some point below freezing, hydration stops.

On the other hand, the process of hydration creates heat of its own, and it is possible that concrete placed at a temperature of 75° F. would reach internal temperatures much higher during the curing process. This would be an important factor during hot weather especially in massive sections. Higher temperatures increase the rate of set and make the concrete extremely difficult or even impossible to finish. While intense heat can be applied during curing (as in steam curing or autoclaving), this is only done when the humidity can be carefully controlled.

THE TIME FOR CURING

There is no timetable for the exact moment when a curing method should be applied. Generally speaking, a selected process should start when the surface of bleed water has disappeared. This rule of thumb cannot be applied to air-entrained concrete, however, for little or no bleed water will appear. Once the concrete has been finished or settled in the forms and is firm enough so that the curing medium will not harm the surface, it is safe to start curing. Care should be taken to start the curing procedures before the surface of the concrete shows any signs of drying out. On a hot and windy day drying can occur very soon after placing.

LENGTH OF CURING TIME

The length of time that concrete should be protected against loss of moisture is dependent on the type of cement used, mix proportions, required strength, size and shape of the concrete mass, weather, and future exposure conditions. This period may be a month or longer for lean concrete mixtures used in structures such as dams; conversely, it may be only a few days for richer mixes, especially if Type III (high-early-strength) cement is used. Since all the desirable properties of concrete are improved by curing, the curing period should always be as long as practical.

The curing period should be prolonged for concretes made with cements possessing slow strength-gain characteristics.

For most structural uses, the curing period for cast-in-place concrete is three days to two weeks, depending on conditions such as temperature, anticipated loads, cement type, mix proportions, and the like. More extended curing periods are desirable for bridge decks and other slabs that are exposed to weather and chemical attack.

The minimum curing period for adequate scale resistance to chemical deicers generally corresponds to the time required to develop the design strength of the concrete. A period of time for air drying should be allowed after curing and before deicing chemicals are used. This enhances resistance to scaling. Because membrane curing during the late fall may seal the concrete and prevent drying, it should not be used in such circumstances. This drying period should last at least one month if at all possible. If use of deicers is required within one month after the curing period, a surface treatment of linseed oil may be helpful in sealing the concrete surface against penetration by water and deicer mixtures. Boiled linseed oil may be used for faster drying.

Generally, job specifications will state the type of curing procedure to be used and the length of curing time. On the usual job there is rarely sufficient time to make test cylinders and cure them adequately to determine the best length of curing time. Also, site conditions rarely correspond to laboratory curing conditions.

METHODS OF CURING

Curing is basically the process of assuring hydration by maintaining high relative humidity within the concrete. This can be done by supplying additional water to the surface of the concrete or by retaining the mix water that is in the concrete. While total immersion in warm saturated-lime water is perhaps the ideal method of curing, we have seen that properly mixed concrete does contain more than enough water to complete hydration.

Adding Water

When needed, however, additional water can be supplied to the concrete surface by ponding, sprinkling, soaking, or fogging.

Ponding is one of the best methods of curing concrete slabs. A dike of earth can be constructed around the slab so that a couple of inches of water covers the entire surface of the concrete. The water will also serve to keep the concrete cool during hot weather. There are disadvantages, however. It is easy for a careless workman to disrupt the earth dike and allow the water to escape, or the water may seep into nearby foundation excavations.

Water can also be applied by sprinkling the hardened concrete or by allowing a slow stream of water from a soaker hose to run over it. Both of these methods are good if carefully controlled. In both cases the entire surface of the concrete should be kept constantly wet. Care should be taken to ensure that the water running over the concrete does not stain it and does not wash away the subgrade undermining the concrete. Unless there is a provision for drainage, both sprinkling and soaking can turn the surrounding area into mud and can create a bad work situation.

Fogging the surface of the concrete will result in good curing conditions—providing the wind does not deflect the spray. Care must also be taken during the first 12 hr of curing to make sure that the fogging equipment does not harm the surface of the slab.

Water application should be maintained for at least 72 hr. This is often very difficult to do, and so these methods are sometimes combined with other curing methods. The surface of the slab is ponded or sprayed for 24 to 36 hr and then covered in some way to retain the water. This combination of curing methods works quite well.

Earth, sand, hay, straw, burlap, sawdust, or most any other inexpensive and absorptive material can be spread on the concrete slab and saturated with water. These curing methods are less apt to create muddy conditions at the job site, and they provide a good curing atmosphere for the concrete. They must be kept saturated with water, however, and on a hot, dry day this can keep a workman—or several workmen—busy. Special curing blankets of burlap with waterproof plastic backing, cotton bats, treated tarpaulins, and other types of curing coverings are manufactured (Fig. 11-6).

Wet moisture-retaining fabric coverings should be placed as soon as the concrete has hardened sufficiently to prevent surface damage. Care

Figure 11-6 Wet-curing with burlap.

should be taken to cover the entire surface, including the edges of slabs of pavements and sidewalks. The coverings should be kept continuously moist so that a film of water remains on the concrete surface throughout the curing period.

Wet coverings of earth or sand are effective for curing, but in recent years they have been largely discontinued because of their high cost. The method often is useful on small jobs, however. Moist earth or sand should be evenly distributed over the previously moistened surface of the concrete in a layer about 2 in. thick. It should be kept continuously wet.

Moist hay or straw can be used to cure flat surfaces. It should be placed in a layer at least 6 in. thick and kept continuously moist. A major disadvantage of moist earth, sand, hay, or straw coverings is the possibility of their discoloring the concrete. The effect of discoloration on surface appearance should be considered.

Retaining Water

Waterproof paper, plastic sheets, or sprayed-on membrane-forming curing compounds are used with greater frequency than are water-supplementing methods to cure newly placed concrete.

Where it is important to maintain an extremely smooth floor surface, it is wise to use waterproof curing paper. Not only does this do a good job of curing, but it also warns workmen that the floor is not yet ready for full use. In hot weather it is often necessary to supply additional water to the slab for the first 24 hr before putting the paper down. When used out-of-doors it is necessary to anchor the sheets of paper so that the wind cannot blow them off the slab. The use of kraft paper or roofing felt should be strictly avoided for curing because these materials are not waterproof (Fig. 11-7).

Waterproof curing paper is very often a sandwich of paper-asphalt-paper. Paper-backed plastic is common, too. To make curing papers punctureproof, some have a three-directional pattern of glass fibers between the paper layers. This type of curing paper is strong even when wet.

Figure 11-7 Waterproof curing paper is efficient in preventing evaporation of surface moisture.

Although these reinforced curing papers are more expensive, they can be reused. It is recommended that seams be overlapped and glued with a nonstaining mastic or sealed with tape.

Polyethylene film also does a good curing job. It is light and completely impervious, can be reused, is flexible, and can be used for other things on the job site such as covering window and door areas and providing temporary windbreaks.

Polyethylene sheets can be used on vertical as well as horizontal shapes and no cutting is required to fit the sheets around corners or odd shapes. The weight is constant whether wet or dry. If required, the plastic can be welded to form a watertight seam. Plastic sheeting is available in widths up to 32 ft. (Fig. 11-8).

In some instances, the use of thin plastic sheets for curing may cause discoloration of the hardening concrete. This may be especially true if the concrete surface has been steel-troweled to a hard finish or when calcium chloride admixture has been used. When such discoloration is objectionable, some other curing method is advisable. The use of plastic sheeting on colored concrete should be avoided unless the plastic can be kept from touching the surface of the concrete. White-pigmented plastic sheets are available to cut down on heat absorbed from the sun.

Figure 11-8 Curing with plastic sheeting. To prevent discoloration of the slab, plastic must be kept flat on the concrete surface.

Curing Compounds. Liquid membrane-forming compounds are frequently used. They retard or prevent evaporation of moisture from the concrete. They can be used not only for curing fresh concrete but also for further curing of concrete after removal of forms or after initial moist curing (Fig. 11-9).

Curing compounds come in several colors:

> clear or translucent
> white-pigmented
> light-gray pigmented
> black
> reddish-brown

and are made with various bases:

> synthetic resin (plastic)
> wax
> combination wax and resin
> chlorinated rubber

Clear or translucent compounds may contain a fugitive dye that fades out soon after application. This helps assure complete coverage of the

Figure 11–9   Spraying with a pigmented curing compound.

exposed concrete surface. During hot, sunny days, white-pigmented compounds are most effective because they reflect the sun's rays, thereby reducing the concrete temperature. Black compounds are rarely used on pavements because the color causes great heat to develop within the slab and can cause surface cracking.

Care should be taken to avoid getting the curing compound into expansion or contraction joints if they are to be filled with joint sealing compound. (These can be protected with tape before the compound is applied.) The joint sealing compound will not adhere to the curing compound. The edges of the slab, if not covered by forms, should also be sprayed with the compound.

Chlorinated rubber curing compounds not only form a film that protects the concrete from drying out but also fill in the minute pores in the surface of the slab. The surface film will wear off eventually. Chlorinated rubber is also resistant to the alkalies in concrete. It may be applied with a roller, brush, or by low-pressure spray units. Some contractors use inexpensive garden-type sprayers that can be discarded after use. Chlorinated rubber compounds offer the following advantages.

1. Immediate protection from a sudden rainfall—within 15 min. of application.
2. Good moisture barrier—retains the mix water to create good curing conditions.
3. Resistance to alkalies.

Concrete so cured also can withstand exposure to jet fuels, oil, and gasoline, making it a good curing agent for airport runways and parking aprons as well as for garages, driveways, and highway slabs. Chlorinated rubber curing compounds meet or exceed the water-retention requirements of ASTM C309-72, CRD-C300-72 (U.S. Army Corps of Engineers), and AASHTO M148. They are stable from -20 to $+105°$ F. Some chlorinated rubber compounds are mixed with plasticizing resins. The compound is slightly yellow in color and should not be used where discoloration of the slab would be objectionable. It is a highly volatile mixture and so should not be used near flames.

When curing compounds are used on architectural concrete, a fugitive dye is added so that coverage can be observed while applying the compound. This dye quickly fades and disappears. Black curing compounds are made from asphalt bases. They are extremely thick and sticky and, hence, give especially good protection to the concrete below grade. They also provide an excellent moisture barrier and are often used in a two-course basement floor slab for waterproofing. They can also be used on floor slabs that will ultimately be covered with linoleum or tile. This type of curing compound is also frequently used to cure soil-cement.

Application Methods. Curing compounds may be applied with hand or power-driven sprayer equipment. Application should be made as soon as possible after the water sheen has left the concrete. One gallon of compound is adequate for 150 to 200 sq ft. Care should be taken to completely cover the concrete so that maximum water retention is possible. Ordinarily, it is sufficient to apply one coating, but when two are necessary, it is wise to place the second coat at right angles to the first.

Large machines have been developed for the application of curing compounds on paving jobs

Disadvantages. Curing compounds should not be used where they will be worn off before the concrete has sufficiently hydrated. In very hot weather, a moist cure for the first day or two, followed by the application of a curing compound is desirable. When the concrete to be cured is to be subjected to ice-melting salts and other deicing agents, it is best not to use a curing compound. Concrete should cure for a period of time before being exposed to deicers. Curing compounds will keep all air away from the concrete until the membrane is worn off by traffic or by weather. When concrete is placed in the fall or winter in northern climates, it is best to cure it by some other method so that the surface will have an opportunity to air dry before deicers are applied. If the surface of the concrete is to be painted or covered with mastic, the curing compound must usually be removed to insure a tight bond between the concrete and paint or mastic.

Hot- and cold-weather conditions create special problems in regard to curing procedures, and these will be discussed in Chapters 12 and 13, respectively.

Steam Curing

Concrete block and other precast producers have found that curing by steam in chambers enables them to reuse their molds much more frequently—often each 24 hr. To maintain a competitive position in the market, it was necessary for precast producers to find a method of curing to attain high early strength in their product. In a factory situation it is not feasible to leave castings in forms for three to seven days—forms and molds are too great an investment and must be reused as often as possible.

Steam curing provides two especially favorable conditions for the curing of concrete and the attainment of early strength—high temperature and high humidity. This type of curing is usually done in some type of enclosure so that the temperature and the humidity can be carefully controlled.

The profile of a steam-curing cycle contains four specific curing periods:

1. Initial delay prior to steaming—3 to 5 hr.
2. Temperature increase period—2½ hr.
3. Constant temperature period—12 hr.
4. Temperature decrease period—2 hr.

Thus a good steam curing cycle would take from 19½ hr to 21½ hr.

Steam curing has been used in the production of concrete pipe, concrete masonry, and other small concrete products for many years. More recently, however, there has been a rapid rise of the precast concrete industry. Beams, columns, floor slabs, roof panels, bridge girders, and other large components are precast for easier and cheaper construction. Because of the capital outlay for production necessities such as forms, prestressing beds, and the like, it has been important to find a way of keeping this expensive equipment at a minimum. Steam curing has been an answer. It is much faster than moist curing, so forms can be reused more frequently. Many producers are able to keep their plants operating on a 24-hr cycle. In each 24-hr period, one set of castings is placed and cured and the forms are stripped. About 6 hr are required to strip the forms, clean them, reassemble them, and pretension any steel reinforcement. The remaining 18 hr are spent in concrete placement and in the presteam, temperature-rise, maximum-temperature, and steam-release periods.

The preset (or holding) period before steaming starts varies from 1 to 5 hr, depending on the type of concrete being cured and the temperature under which the curing will be carried out. Longer preset periods are necessary where higher temperatures are used. A 1-hr preset would only be adequate for low-temperature steam curing. This preset period allows the concrete time to start hydration at a slow rate.

Figure 11-10 illustrates the relationship between 18-hr strength and the delay period prior to steaming. On each curve the optimum delay time

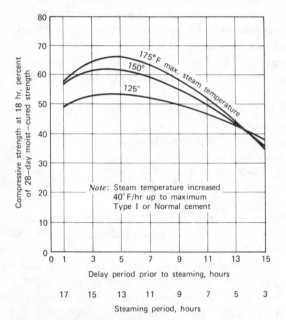

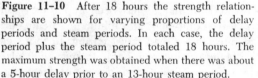

Figure 11-10 After 18 hours the strength relationships are shown for varying proportions of delay periods and steam periods. In each case, the delay period plus the steam period totaled 18 hours. The maximum strength was obtained when there was about a 5-hour delay prior to an 13-hour steam period.

seems to be about 5 hr. Also there is not much difference in compressive strength between concrete cured at 150 and 175° F (Fig. 11-10).

A delay period prior to steaming is an important part of the curing process. This present period has a significant effect on the ultimate compressive strengths of the concrete. This is an important factor to the operator of a prestressed plant where strengths of 3500 to 4000 psi are required of the concrete. Steam curing is used in the production of pretension and posttension components.

If concrete blocks are allowed to dry before steam curing, calcium carbonate efflorescence can result; therefore, it is especially important to avoid moisture losses in lightweight blocks. These blocks are unusually susceptible to water evaporation because they are made with porous aggregates and cast with thin walls. Care should be taken to assure their moisture retention during the presteam period.

When steaming starts, it is important that the temperature rise in the chamber be kept at a steady rate. This rate of increase of temperature can vary from 20 to 60° F per hr, depending on the product being steamed and on the length of the holding period.

Optimum curing results can be achieved by keeping the concrete below 120° F for 5 to 6 hr. This can be done in two ways:

1. A delay period followed by a rapid increase of temperature.
2. A continuous and slow rate of temperature increase.

The first method is usually used in manually operated chambers, while second method is most often used with automated equipment.

Once the maximum temperature has been reached, the steam is turned off and the kiln is allowed to return slowly to normal temperature and humidity. The insulation of the chamber should be sufficient to allow this to happen very gradually. When precast or prestressed pieces are being steam cured, it is considered best to continue the heat and steam until the maximum strength is achieved. Temperatures as high as 180° F are sometimes allowed, but high temperatures have a tendency to lower the ultimate strength of the concrete, although higher early strengths are obtained. The National Concrete Masonry Association recommends a maximum steaming temperature of 150 to 165° F for regular weight concrete products and 170 to 180° F for lightweight block.

There is no published information at this time that would indicate that rapid cooling is harmful to concrete being taken out of the steaming chamber. However, in winter weather, it would be wise to avoid taking the hot concrete out into below-freezing temperatures.

Steam curing does not have to take place in a chamber. The concrete can be cured by the construction of a temporary housing and the injection of steam heat into the area in cold weather.

Types I and III cements are usually used for concrete that is to be steam cured, although Types II and V have also been used on occasion. Either lightweight or natural aggregates can be used. Lightweight aggregates demand longer curing periods because concrete made from them takes longer to heat up. They also cool off more slowly. Admixtures may be used with concrete that is to be steam-cured but calcium chloride should not be used in prestressed concrete. The reactions of pozzolanic admixtures are speeded up by steam curing above about 180° F.

When designing the mix for concrete that is to be steam-cured for high early strength, the water-cement ratio is usually quite low, the mixture is rich in cement, and it will often contain a water-reducing agent. Small castings of concrete will gain 60% of 28-day strength in 24 hr when steam cured. Strengths of 4000 psi at 18 hr have been reported.

ACI Committee 517 report states:

"The recommended steam curing cycle for concrete block depends to some extent on the aggregate being used. For normal weight block, the presteaming period should be a minimum of 2 hours at 70° to 100° F. during warm weather, and 1 hour longer in cold weather. The temperature rise should not exceed 60° F. per hour, to a maximum of 150° to 165° F. The elevated temperature should be maintained for 12 to 18 hours, or until the required strengths are developed."

Features of a Good Steam Chamber. Steam chambers should be designed to include the following features:

1. Air-entraining agents are sometimes useful as an aid in compaction. essary curing temperatures.
2. A present period is more important to dense masonry than to light-
3. Water-tightness and vapor-tightness to insure against moisture loss.
4. Adequate instrumentation to record and control humidity and temperature.
5. Turbulent air motion within the kiln, usually supplied by a fan.
6. A method of exhausting the heat and steam at the end of the curing cycle.

Importance of Steam Curing to Masonry Industry. The discovery of steam curing probably did as much to strengthen the growing concrete masonry industry as any other single development. Before steam curing it was necessary to shut down masonry plants during cold weather. Some experimentation was made with stove-heated drying sheds, but steam curing produced much better masonry.

When steam curing was first introduced to the industry, temperatures usually hovered around 120° F. During the 1920s, however, a few producers were heating as high as 190° F. Early producers steam-cured for a period of time and then finished the cure by sprinkling the stockpiles of masonry with water. During the years, this low-temperature steam curing was gradually replaced by high-temperature curing. Temperatures of 170 to 200° F cut curing time and increased plant productivity. About

1945, some experimentation was done with steam curing and hot-air drying of masonry to try to reduce shrinkage to a minimum. It was found that by increasing the temperature of the steam during the curing process, the curing time could actually be shortened. More research indicated that a waiting period prior to steaming would result in stronger masonry. The findings of an eight year study begun in 1947 showed the following about steam curing concrete masonry.

1. Air-entraining agents are sometimes useful as an aid in compaction. In extremely wet mixes this will result in improved resistance to freeze-thaw exposure.
2. A preset period is more important to dense masonry than to lightweight masonry.
3. 40° F per hr of heat rise is best.
4. Temperatures above 190° F are harmful to concrete. A temperature of 165° F is recommended for dense aggregate and 180° F for lightweight aggregate masonry.
5. 100% humidity should be maintained.
6. A soaking period substituted for part of the steam curing resulted in decreased strength.
7. Kiln drying of newly cured masonry adds to early strength.

Steam Curing Prestressed Members. The report of ACI Committee 517, "Low Pressure Steam Curing," states that changing from moist curing to steam curing may eliminate up to 50% of the creep.

Problems

Some problems remain to be solved in the production of steam-cured products. It is not known why the ultimate strength of steam-cured concrete is below that of moist-cured concrete, or if this can be corrected. Masonry that is steam-cured sometimes expands or spreads and the reason for this is not known. During the soaking period, the heat and moisture tend to stratify in the kiln and the effects and causes of this are not known. More research needs to be done on the relative merits of the various types of kiln insulation.

Autoclaving or High-pressure Steam Curing. Autoclaving is curing with steam at temperatures of 325 to 375° F and pressures of 80 to 170 psi. The ACI published its first report on autoclaving in 1944. Since that time, the use of autoclaving in the concrete masonry and asbestos-cement industries has increased markedly. In 1965, autoclaving accounted for 15 to 20% of concrete block curing and virtually all of the asbestos-cement pipe curing (Fig. 11-11).

Autoclaving produces block that have these qualities:

1. Reduced change in size and shape.
2. Reduced cracking.
3. Permanent high early strength—1-day strength equivalent to 28-day moist-cure strength.
4. About half the volume change resulting from drying as found in other curing methods.
5. Improved resistance to sulfate action.
6. Reduction of efflorescence and leaching.
7. Reduction of popping or spalling.
8. Lighter color.
9. Lower moisture content.

The main reason manufacturers have selected autoclaving is that masonry and other precast items do not shrink nearly as much when they are cured in this manner. In moist-cured concrete, the process of hydration produces tobermorite gel, while in autoclave-curing, tobermorite (a more crystalline hydrous calcium silicate) is produced.

Figure 11-11 These large autoclaves cure concrete products with steam at temperatures of 325° to 375° F., and pressures of 80 to 170 psi.

In general, it has been observed that the shrinkage of tobermorite is approximately half that of tobermorite gel, and that the strength of tobermorite is approximately two to three times greater than that of tobermorite gel.

The concrete mix used for autoclaved products contains from 0 to 50 lb of powdered silica for every 100 lb of portland cement, depending on the type of aggregate.

There are various methods of autoclaving and several different schedules that can be followed. Which one is to be used by a plant will depend on the product being cured, the type and number of kilns in the plant, the labor scheduling, and the rate of production required.

The cast units can be placed in the curing chamber, and the steam can be injected into the chamber of atmospheric pressure for 1 to 4 hr. The contents of the chamber are then heated to 350° F in a period not to exceed 3 hr. Pressure is built up and maintained at 125 to 140 psi for about 5 to 8 hr. This is allowed to drop to atmospheric pressure in 30 min.

Concrete cured in this manner can be used 24 hr after casting.

Another method of autoclaving is to precure the block with low-pressure steam before placing it in the kiln. This will set the block sufficiently to guard against chipping or deforming during firing. In this manner the autoclave can be filled with more units per cycle, since the units may be placed closer together in the kiln.

Care must be taken in autoclaving to make sure the cast block or other precast pieces are supported out of contact with liquid. The kiln usually has a reservoir, and this water is vaporized by the action of the indirect heat within the kiln.

Either normal or lightweight aggregates can be used in the concrete to be autoclaved. The grading of the aggregate is extremely important, however. The quantity and variety of reactive fines in the aggregate should be carefully ascertained before the proper type of cement is selected. Types I, IA, III, and IIIA cements are suitable for autoclaving. Care must be taken to keep the cement content of the mix high enough to insure sufficient early strength so that the masonry will not deform during curing.

Colored masonry can be manufactured by using the oxides of iron, manganese, and chromium in the mix. The design of the mix to be used in autoclaved products depends on the following.

1. Type of cement.
2. Type and fineness of siliceous material.

3. Character and grading of aggregate.
4. Curing cycle.

Batching of the mix should be done by weight because of the greater accuracy possible.

When autoclaving lightweight cellular concrete products, most manufacturers stipulate a long preset time so that the entrained air cells within the piece are sufficiently strong to withstand the pressures of autoclaving.

Much time has been spent by scientists and technologists trying to ascertain at just what point the unmolded green masonry will start to deform during autoclaving. Masonry rapidly heated from 104 to 212° F was found to deform. The water in the concrete expands 3.5% during this temperature rise and exerts tremendous force. If concrete had nowhere for the water to go, the concrete would surely fail, but concrete does have air voids and channels of varying sizes to relieve the pressure. Nevertheless, the concrete must develop sufficient strength to withstand this pressure until it is relieved, and to force the water to find outlets other than through the concrete. To achieve this strength, it is necessary to either precure the concrete or to start autoclaving very slowly.

High-pressure steam curing makes the concrete more resistant to sulfate attack.

While autoclaving offers many advantages to the producer of precast concrete products, there are some drawbacks to the system. The bond between steel and concrete is partially destroyed during autoclaving. The absorptive qualities of the concrete and its permeability are likely to be increased. Autoclaved concrete is more brittle than moist-cured products, and its impact strength is not as high.

In 1935, C. A. Menzel reported in a published paper that flexural strength in concrete was improved by decreasing the rate of temperature rise and increasing the cooling time before the autoclave was opened and after curing was completed. Still another study indicated that strength in lightweight concrete is increased in autoclaving even though brittleness is also increased.

CURING SPECIAL MATERIALS

Cement Paint
Portland cement paint should be kept damp for 48 hr after application to insure good bond with the concrete or concrete masonry of the wall being painted. A simple garden sprayer with a fog nozzle is an excellent device

for doing this. The spray should be applied as soon as the paint has set enough so that it will not be washed away. When properly cured, portland cement paint becomes a part of the concrete to which it is applied. It seals the pores in open-textured concrete and makes it weathertight. If not properly cured, however, the paint will dust and chip.

Cement Stucco and Cement Plaster
These materials must also be cured for the development of strength and durability. The vast exposed surfaces and the relative thinness of the applications make curing essential.

The curing process is best accomplished by fogging the surface of the stucco or cement plaster with a fine spray of water. Care should be taken to be insure the spray is not so intense that it erodes the surface and that it is applied after plaster has reached sufficient hardness. The finish coat, however, is not cured. Sufficient water in the base coat allows cement hydration. It is necessary only to protect the surface from rapid water loss. The walls should also be protected from hot and dry winds. This can be done by draping the well-dampened wall with tarps or plastic sheeting.

Stucco and cement plaster should not be applied when the temperature is likely to fall below freezing, unless some protection against the cold is provided. It is more difficult to keep an outside wall warm in cold weather than it is to maintain some heat in a horizontally placed piece of concrete. It is also important that stucco and cement plaster never be applied to surfaces containing frost. By uniformly curing the plaster or stucco, an attractive, strong, and durable finish will result.

Bonding Cement Plaster to Concrete. In bonding portland cement plaster mixtures to concrete, it is important to prevent the early drying of the new mortar; otherwise, it has a tendency to shrink from the old surface. The new work should be kept constantly moist for several days and, if practical, curing should continue for one day for plaster scratch and for seven days for plaster brown coat. When patching or stuccoing is to be done in warm, dry weather, it is advisable to work under tarpaulins, or at least to cover the work as quickly as possible with tarpaulins, wet burlap, or heavy waterproof paper.

Soil-cement
When soil-cement is compacted and finished, it contains sufficient moisture for adequate cement hydration. A moisture-retaining cover is placed over the soil-cement, as soon after completion of the compaction as pos-

sible, to retain this moisture and permit the cement to hydrate. Until the cover is placed, the soil-cement should be kept damp. In recent years most soil-cement has been cured with bituminous material, but other materials such as waterproof paper, or moist straw or dirt, are entirely satisfactory.

At the time of application of the bituminous materials, the soil-cement should be free of all dry, loose, and extraneous material. Bituminous cover materials are applied to a very moist soil-cement surface, and in most cases, water is applied immediately ahead of the bituminous application. When traffic is to be maintained, the bituminous material is sanded to prevent its pickup by traffic.

SECTION THREE
SPECIAL CONDITIONS
FOR CONCRETE

CHAPTER 12
HOT WEATHER CONCRETING

Among plastic concrete's worst enemies are heat and drying. Hot weather provides climatic conditions of both heat and drying. The heat of the component parts of concrete added to the heat of hydration can create volume changes within the concrete that may later cause serious fractures when the structure changes temperature. Hot, dry weather evaporates any surface moisture quickly. This is detrimental to the proper curing of concrete.

The ideal day for placing concrete would be a dark, windless, humid day with a temperature in the low 60s. These precise weather conditions are rare in any climate and are almost nonexistent in some areas. It would be impossible to restrict the placement of all concrete to days when the weather conditions were most favorable. Although extremes of temperature and other weather conditions may add to the difficulty of placing, it is possible, with precautions and planning, to achieve excellent results in the heat of summer or in winter's cold.

In the past few decades, the problems of placing concrete during hot or cold weather have increased because more and more work is being done under such temperature extremes. Large jobs cannot wait for a good day. Modern-day cements are more finely ground than previously, and this contributes to the increased heat of faster hydration. Higher and higher compressive and flexural strengths are required of concrete that, in turn, demands richer mixes. Thinner concrete sections, that contain a greater percentage of steel reinforcing, are being designed. These sections generally have a greater surface area and are much more apt to dry out before hydration is completed. Contractors have tighter schedules to meet and often find it necessary to remove forms and shoring members as soon as possible, exposing the concrete to potential drying. The use of larger capacity batch plants and mixer trucks has made possible the delivery to the job of larger amounts of concrete at one time. Much of this concrete has low slump, which is more difficult to place in forms—especially in hot weather.

Modern construction demands more of concrete, and these demands call for tighter scheduling and greater attention to the exact timing of each step. In hot weather, it is especially true that advanced planning and the taking of a few simple precautions can save money in the long run. Once hot weather has damaged plastic concrete, it is almost impossible to restore it.

PROBLEMS

Concrete at 60° F will set up in about 2½ hr and will be completely firm in about 6 hr. Concrete placed at 100° F may set up in 45 min and may be completely firm in 3 hr or less. This accelerated setting, of course, greatly increases the difficulty of finishing the 100° F concrete. The chances of cold joints developing during placement are greatly increased because there is not enough time to strike off, float, and trowel the surface.

For each 10° F rise in temperature, an additional 7 lb per cu yd of water may be required to produce the same slump. If additional water is added to the mix, an additional measure of cement must also be added to maintain the water-cement ratio. If this is not done, strength and durability will be impaired. The addition of more water also can result in greater volume change in the concrete. The addition of the extra cement will, however, further increase the total heat of hydration. As concrete temperatures increase, rate of slump loss increases and the water requirement increases sharply (Fig. 12-1).

The chapter on curing emphasized the necessity of maintaining sufficient water in the concrete to allow for complete hydration of cement in the mix. With an increased amount of mix water present in concrete at higher temperatures, this problem becomes increasingly critical. If the mix water is allowed to evaporate, plastic shrinkage cracks may develop.

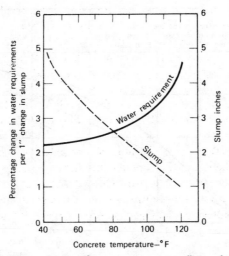

Figure 12-1 The representative effect of concrete temperature on slump and on water required to change slump.

If concrete cools too rapidly, thermal shrinkage cracks may form. The increased danger of possible volume changes within both the freshly mixed and the hardened concrete is further implemented by increased temperatures and high evaporation rates. In a hot, dry summer day, it is much more difficult to protect concrete from drying. When concrete has been hot, it is more apt to build up tensile stresses as it cools. Evaporation causes efflorescence and leaching that can mar the surface.

After 24 hr, concrete placed at 120° F may have almost twice the strength of concrete placed at 73° F. But the 73° F casting achieves the same strength at 28 days that it takes the 120° F piece a year to achieve (Fig. 12-2).

Hot weather and the resultant increase in concrete temperatures make the control of entrained air in concrete more difficult. More air-entraining admixture is needed at higher temperatures to obtain a required percentage of entrained air. It is, however, difficult to determine in advance the exact amount of additional admixture that will be needed for a given mix at a given temperature.

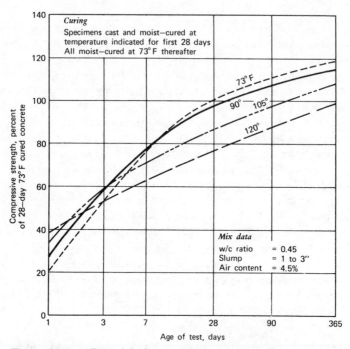

Figure 12-2 Effect of high temperatures on concrete compressive strength at various ages.

The durability of concrete placed at high temperatures may be affected. We have seen that durability is usually a function of entrained air and of strength. Concrete placed during hot weather without taking precautionary measures is often less resistant to freeze-thaw and wet-dry cycles.

The appearance of the concrete is sometimes affected by hot weather. Cracking and crazing are apt to occur. Other surface damage and discoloration can be caused by undue evaporation of mix water and rushed finishing.

This is a long and rather grim list of all the things that can go wrong if hot weather effects are ignored. Yet a great percentage of the concrete placed is done so during some of these adverse conditions. But there are realistic precautions that can be taken.

PRECAUTIONS

We speak of "hot" weather, but that is not a very accurate description of the weather conditions that necessitate the precautions for hot-weather concreting.

In addition to temperature, the wind velocity and humidity must also receive consideration. The rate of evaporation is highest when the relative humidity is low, when concrete and air temperatures are high, when the concrete temperature is higher than the air temperature, and when the wind is strong over the surface of the concrete. When the relative humidity changes from 90 to 50%, for example, the evaporation rate may increase ninefold. In summer, the relative humidity in the morning hours is often much higher than at noon. When both concrete and air temperatures increase from 50 to 70° F, evaporation may double; if temperatures further increase to 90° F, evaporation may quadruple. The evaporation rate may be four times greater when the wind velocity is increased from dead calm to 10 mph, and nine times greater when increased to 25 mph.

When the evaporation rate reaches 0.1 lb per sq ft per hr, it is wise to take some precautions; when it rises above that, it becomes increasingly necessary to take measures to insure that mix water will not evaporate and that the concrete is not too warm (Fig. 12-3).

One way to avoid the effects of hot weather and the rapid evaporation rate is to place concrete that is cool and to be sure that it remains cool. This can usually be done without too much trouble.

Before any concrete is ordered from the ready mix producer or from the mixing plant, it is essential that preparations be made at the job site to assure that the delivered concrete will be placed immediately, and that enough labor and tools will be on hand to consolidate and finish it. It is not always possible to add workmen to the crew, but sometimes a foreman can arrange the day or night schedule to allow the shifting of men from one site or area to another.

A list should be made of the available vibrators, floats, darbys, trowels, and the like, so that there is no waiting for the proper tools. Allowance should also be made for a possible breakdown of equipment, which should be test-run before it is needed.

Placing of concrete on a day when the weather conditions are such that rapid evaporation will take place should never be attempted if there is a lack of labor or equipment that will significantly delay the placing or finishing of the concrete.

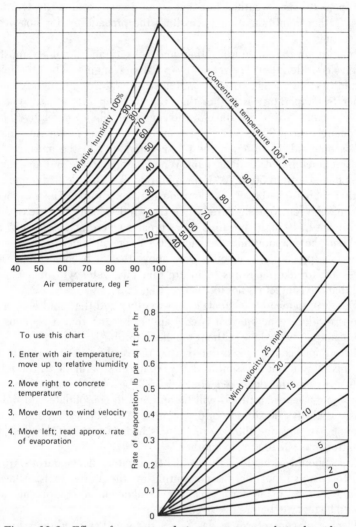

To use this chart

1. Enter with air temperature; move up to relative humidity

2. Move right to concrete temperature

3. Move down to wind velocity

4. Move left; read approx. rate of evaporation

Figure 12–3 Effect of concrete and air temperatures, relative humidity, and wind velocity on the rate of evaporation of surface moisture from concrete.

The delivery of mixed concrete should be planned so that there will be no waiting for trucks to arrive at the job, and so that trucks will not be lined up waiting to unload.

When temperatures and evaporation rate are especially high, it might be advisable to schedule the placing of concrete on large projects after sundown or before sunrise, so that as much protection as possible can be given the concrete during the cooler hours.

Before the concrete is to be placed, it is good practice to dampen the subbase. Just before placing, the work area can be fogged to dampen forms, cool the air, and raise the humidity. Sunshades and windscreens should be erected if these are considered necessary. Equipment, subbase, and forms can be covered with damp burlap to maintain moisture (Fig. 12-4).

It is also necessary to have curing supplies immediately at hand so that the curing operation can be started as soon as the concrete surface can withstand the curing medium without being marred. White-pigmented curing compounds are widely used in hot-weather applications.

For more massive structures, in which significant heat is generated

Figure 12–4   Before placing concrete in hot weather, it is good to dampen the subbase and forms. This not only cools these surfaces and the surrounding area, but dampened forms and subbase will not leach moisture out of the mix.

through hydration of cement, a mix temperature of not more than 60° F is desirable. For other structures, a maximum of 90° F is allowed for the mix. However, difficulties may be encountered with concrete at a placing temperature approaching 90° F, and every effort should be made to place at a lower temperature.

HOT WEATHER CONCRETING MATERIALS

Each material in a concrete mixture affects the temperature of the whole mix.

Aggregates

Aggregates compose 60 to 80% of the total weight of a concrete mix, and thus the temperature of the aggregates has a great influence on the temperature of the mix. Stockpiles should be shaded and sprinkled to keep them moist. Evaporation is one good way to keep aggregates cool with a minimum of effort. This is especially useful when the relative humidity is low. When it is necessary to cool aggregates to temperatures lower than can be obtained by shading or sprinkling, the aggregates can be refrigerated or immersed in chilled water. Care should be taken to be sure that the water is continually circulated and that enough time is allowed for the aggregates to cool. It is important to drain all excess water from the aggregates at the end of the chilling period. This is difficult with sand, so usually only coarse aggregates are cooled with water. A temperature log should be maintained on all materials.

A vacuum process has been successfully used to cool aggregates on large projects. On the Green Peter Dam in Oregon, the U.S. Corps of Engineers specified a mixed concrete temperature of 50° F. The contractor constructed a vacuum tank in which the damp aggregates were placed. As the relatively warm surface water of the aggregate boiled, the latent heat was carried away, resulting in a chilling of the aggregate. Such methods are commonly used only on projects such as gravity dams, where the masses of concrete being placed are so great that they produce tremendous heat from hydration.

Water

Water is perhaps the easiest of the concrete ingredients to cool. By cooling water (which has a specific heat $4\frac{1}{2}$ to 5 times that of cement or aggregate), it is possible to effect a great reduction in the final heat of the mix. In hot weather, care should be taken to make sure that mixing water comes from a cool source and is permitted to remain as cool as possible. Water storage tanks and pipelines should be shaded, insulated and/or painted white. Often it is advisable to bypass the storage tanks on a hot day and obtain the water directly from the city supply, which is generally much cooler than water that has been stored in a tank.

Water can be drastically cooled by adding ice to it. Crushed or flaked ice works very well in cooling the concrete mix, but it must be considered as part of the mixing water. One pound of ice absorbs 144 btu. in melting. If 75 lb of ice were substituted for each cubic yard of water in a mix with a temperature of 90° F, the temperature of the mix would drop to about 75° F. Care should be taken, however, to be sure the ice is in pieces small enough to be completely melted by the time the concrete is discharged from the mixer (Fig. 12-5).

When the 8-ft-thick foundation pad of the One Shell Plaza Building was placed in Houston, Texas, the specifications called for a maximum concrete temperature of 58° F. By using ice for a great percentage of the

Figure 12-5 Charging crushed ice into a transit mixer can greatly reduce the temperature of the mix.

mix water, it was possible to meet this specification. Three days after placing, the concrete temperature had risen to 170° F, demonstrating the tremendous heat released during hydration. Liquid nitrogen has been used successfully to cool concrete to even lower temperatures than those obtained with ice.

In constructing the dam at Hells Canyon on the Snake River, at the northern end of the Oregon-Idaho border, the difficulties of avoiding heat in a gravity dam were compounded by summertime heats that regularly reached 110° F. Aggregates were sprayed with 35° F water as they traveled on conveyors to the batch plant. Ice also replaced up to 95% of the mix water.

Cement

Because cement represents such a small percentage of the mix and has such a low specific heat, its temperature is not too significant a part of the final heat of the mix. It is true, however, that when cement is ground it is very hot and because of the fineness of the grind and storage conditions, it loses this heat very slowly. Thus it is often delivered to the job site while hot. Because of this, some specifications limit the temperature of the cement to less than 150° F.

If cement has been ground at too high a temperature, it sometimes will cause a premature "flash" set. Flash set occurs during mixing, and is more common in summer than winter. Extra heat is generated and additional water and mixing are required to regain plasticity. However, the concrete will not have the same final properties. Fortunately, it is extremely rare. Another effect called "false" set is caused by dehydration of the gypsum due to grinding at too high a temperature. Such concrete can normally be retained in the mixer a little longer, and mixed back to a plastic condition without adding water.

The concrete technologist should be fully aware of the major types of cement and their properties. Cement type and source should be selected with care.

From the graphs in Figure 12-6, it can be seen that the type of cement used has a great effect on the amount of heat from hydration that will be produced within the concrete. Type III cement should seldom be used during hot weather.

Formulas

Several formulas have been devised to calculate the temperature of the mixed batch of concrete when the weight and temperature of ingredients

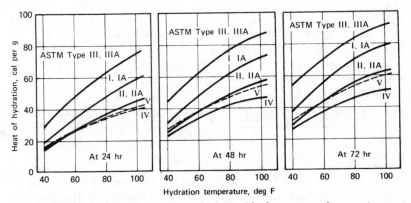

Figure 12-6 Effect of temperature on the heat of hydration at early ages. Average heat of hydration of the different types of portland cement was determined by a conduction calorimeter.

(in pound and degrees Fahrenheit) are known. Two such formulas follow.

$$T = \frac{0.22W_cT_c + 0.22W_aT_a + W_fT_a + W_wT_w}{0.22W_c + 0.22W_a + W_f + W_w} \text{ where}$$

T = temperature of the fresh concrete
T_c = temperature of cement
T_a = temperature of aggregate
T_w = temperature of mixing water
W_c = weight of cement
W_a = weight of aggregates
W_f = weight of free moisture in the aggregates
W_w = weight of mixing water

Another equation allows for the addition of ice to the mix to cool it down:

$$T = \frac{0.22W_cT_c + 0.22W_aT_a + W_fT_a + W_wT_w - 112 I}{0.22W_c + 0.22W_a + W_f + W_w + I} \text{ where I = weight of ice}$$

At other times it is desirable to calculate how much cooled water is required to cool the batch sufficiently (to a temperature, T):

$$W_{ics} = \frac{0.22W_c(T_c - T) + 0.22W_a(T_a - T) + W_f(T_a - T) + W_w(T_w - T)}{T - T_{ics}},$$

where

$$W_{ws} = \text{weight of water from cooled supply}$$
$$T_{ws} = \text{temperature of water from cooled supply}$$

Or it may be desirable to be able to find out the amount of ice necessary to cool the batch:

$$I = \frac{0.22W_c(T_c - T) + 0.22W_a(T_a - T) + W_f(T_a - T) + W_w(T_w - T)}{112 + T}$$

Nomograph

The basic formula has been adapted for a nomograph. When trying to ascertain the temperature of a mix containing 450 to 600 lb of cement, it is possible to approximate it without going to the trouble of working through the formula by using the nomograph of Figure 12-7.

Instructions for Using a Nomograph (see Fig. 12-7)

1. Take temperatures of aggregate, cement, and water.
2. Draw a straight line from temperature of aggregate (1) to temperature of water (2).
3. From intersection of this line with vertical line X (3), draw a straight line to cement temperature (4).
4. Read temperature of concrete where this line intersects concrete temperature line (5).

For the example in Figure 12-7:

$$\text{Temperature of aggregate} = 80°$$
$$\text{Temperature of water} = 50°$$
$$\text{Temperature of cement} = 140°$$
$$\text{And, temperature of concrete} = 81°$$

It is important to remember, however, that the formulas and nomograph do not allow for heat added to the mix by the sun or by the physical friction of mixing.

Admixtures

Another way in which the properties of the plastic concrete can be controlled is by the judicious use of chemical admixtures. Several types can be used, provided that preliminary trial batches are made and evaluated prior to use.

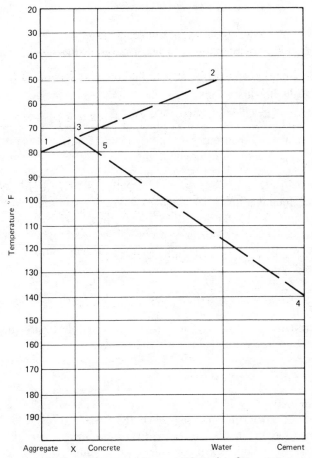

Figure 12–7 Temperature Nomograph is for determining approximate temperature of concrete, applied to normal mixes containing 450 to 600 lbs of cement/cu. yd., based on temperatures of ingredients.

A water-reducing retarding admixture (ASTM C494 Type D) is a material that is used for the purpose of retarding the setting time of concrete and reducing water content. High temperatures of fresh concrete (85 to 90° F and higher) are often the cause of a high rate of loss of slump, which makes proper placing and finishing difficult. Water-reducing retarders do not decrease the initial temperature or alleviate slump loss of the mix, but they do:

1. Offset the accelerating effect of hot weather on the setting of the concrete.
2. Delay the initial set of concrete or grout.
3. Generally provide a plasticizing effect on the concrete, giving greater workability at a lower slump.

Retarders may also entrain some air in concrete but, in general, are not considered as air-entraining agents. Generally, some reduction in strength at early ages (1 to 3 days) accompanies the use of retarders, but the 28-day strength is somewhat greater than nonadmixtured concrete.

Water-reducing admixtures serve to increase the slump without increasing the water content of the mix and, thus, may be used to offset slump loss, even though not preventing it. These admixtures should be used in liquid form, and the liquid should be added to the mix with the water.

Calcium chloride should never be added to concrete during hot weather. The use of high-early-strength cements (Type III) is usually not advisable. If specifications call for Type III cements and special precautions against temperature rise are taken, it is possible to use them with the addition of a retarding admixture to slow down the initial set.

COOLING EQUIPMENT AND SUBGRADE

Another effective way of lowering the temperature of the concrete as it is being mixed (or of maintaining a lowered temperature) is to cool the equipment used to handle the mix. Mixers, trucks, hoppers, bins, conveyor belts, water tanks and lines—all can be painted white to reflect as much heat as possible. Very often it is possible to shade and insulate storage and mixing equipment to reduce the heat. In extremely hot weather, it is sometimes necessary to cover the equipment with damp burlap, utilizing the cooling effect of evaporation to lower the temperature.

Before placing concrete, forms and subgrade should also be sprinkled to cool them and to avoid the possibility of dry forms and subgrade quickly drawing mix water from the concrete. It is usually wise to sprinkle the subgrade the evening before placing so that there will be no standing water.

TRANSPORTING, PLACING, AND FINISHING CONCRETE IN HOT WEATHER

Transporting of concrete should be done as rapidly as possible on a hot day. The concrete should not be allowed to stand in the mixer and should not be overmixed. If delays occur on the job it is best to stop the mixer and agitate intermittently. In general, concrete should be discharged within 90 min after the start of mixing, except in hot weather when it may be necessary to reduce the limit. Delays in mixing and placing contribute to a loss of slump and increase the temperature of the concrete. Mixing delays also increase the difficulty of controlling the air content.

As we have pointed out, enough workmen should be on hand to place, consolidate, strike off, and finish the concrete. Great care should be taken to avoid segregation of the mix and to place rather than pour the mix into the forms. Areas to be worked should be small, and layers should be shallow to avoid cold joints. Each layer should be well integrated with the previously placed concrete. Proper and sufficient vibration equipment should be at hand, with extra equipment available in case of breakdowns. Sufficient water supply and spraying or fogging equipment should also be at hand. Sunshades and windbreaks should be erected beforehand, if necessary.

When one layer of concrete is placed, it is advisable to cover it with wet burlap while waiting for the next load.

If plastic shrinkage cracks start to form while the concrete is still plastic, it is sometimes possible to eliminate them by revibration.

As soon as the concrete is placed, it should be struck off and then screeded and bull floated or darbied. When the surface water has disappeared, floating or troweling should be started. On a hot day this frequently occurs very quickly. Wet finishing must be avoided because it mixes water with surface paste, resulting in a poor-wearing surface. If the start of finishing is delayed too long, a ridged and wavy surface may result.

It is wise to cover the slab between the various floating and troweling operations.

CURING

We have seen that curing is one of the most important steps in concrete construction. This is especially so in the summer when all weather elements are against the proper hydration of concrete. Curing should be started as soon as the final finishing is completed. Moist curing is the best method since it does the double job of keeping the concrete cool while keeping it moist. Often, the concrete is moist-cured for 24 hr, and then a curing compound or other water-retention curing method is applied. If a curing compound is used it should be of the white-pigmented variety. If a 24-hr period of moist cure is not possible, a satisfactory cure can be achieved by fogging the slab and applying a white-pigmented curing compound directly to the wet concrete.

During hot weather, forms should not be relied on for a cure unless some method of running water over the forms is used. If the structure is to be back-filled, this should be done as early as possible.

TESTING AND INSPECTING

In hot weather it is wise to remember that more tests are necessary, but these tests may not always have the same results or compare normally with each other. Slump tests, for instance, may be misleading because, as concrete heats, it loses slump rapidly. Any testing should be done with greater frequency than during cooler weather. Test specimens are apt to dry out unless particular care is taken to avoid this. Test cylinders should be well protected. The cylinders should be treated carefully and should be transferred to the laboratory (or other suitable location) for curing in accordance with specifications.

As the temperature of the mixed concrete approaches the upper limits, it is wise to check the temperature frequently with a metal-sheathed thermometer. This "delivered" temperature is often taken immediately before the slump test is made. The temperature can also be taken "as placed" by inserting the thermometer into the placed slab or structure. Two or three readings from different spots should be taken and an average temperature determined. It is important to leave the thermometer in the concrete long enough to obtain an accurate reading.

Spring reaction-type apparatus for measuring penetration resistance can be used to determine the point at which it is no longer possible to vibrate the concrete—500 psi compressive strength. It is at this point of hardening, before the addition of another layer of concrete, that a cold joint may be formed. Retarding admixtures are able to extend this vibration limit. When the penetration resistance reaches 4000 psi, the compressive strength of the concrete has reached about 100 psi.

The Inspector

The role of the inspector in hot-weather concreting is an especially important one. The inspector should check to see that all preliminary precautions are taken to assure the prompt placement and finishing of the concrete. He should also be on hand constantly to check and record slump, temperatures, weather conditions, and curing operations.

Careful records should be made of the results of all tests and observations as well as climatic conditions. These records should become a part of the final job record.

CHAPTER 13
COLD WEATHER CONCRETING

While there are problems caused by placing concrete in cold weather, these problems are not, in some ways, as severe as those caused by hot weather. (For our purposes, we shall consider weather "cold" when the mean daily temperature drops below 40° F. When the temperature is above this level, it is necessary to give only a minimum of protection to plastic and "green" concrete.)

In areas where the yearly volume of construction is high, there is an increasing amount of concrete construction being done in the colder months. Generally, the areas with the coldest climates also have the shortest summer seasons—and the largest volume of winter placing of concrete. In Canada, for example, more and more construction is being done in the winter because the summer seasons are not long enough to take care of the construction needs of a growing nation. Construction goes on in heated enclosures when outside temperatures are −35 to −45° F.

With an increased demand for concrete in the winter, more and more ready mix producers are installing heating equipment in their plants, so that they are able to mix warm concrete even in subzero temperatures. With the development of new insulated forms, insulated bats and blankets, and other materials, it is possible to maintain the heat of hydration or supply additional heat to concrete at relatively reasonable costs. As with hot-weather concreting, placing concrete in cold weather demands careful planning and thoughtful attention to the smallest details. The results, however, can be rewarding.

Plastic concrete must not freeze. The freezing temperature of concrete is near the freezing point of water—32° F; when the mix water forms into ice crystals, the concrete is considered frozen. Protection should be given the concrete until hydration has used enough of the free water to reduce the degree of saturation and create enough strength to withstand stresses.

ADVANTAGES OF COLD WEATHER CONCRETING

There are several important advantages to placing concrete in cold weather. One of the most important is the more abundant supply of labor during the winter months. Construction work has traditionally offered seasonal employment: in the summer months all available crews are employed; during the winter this is not true. By working 12 months, it is possible to keep well-trained workmen employed full-time. This builds morale among the workers and promotes a loyalty to the company that might not exist if the employment period were of shorter duration. Also materials are often easier to obtain during the slack winter months, and delays for supplies are rare.

A contractor can amortize his equipment over a 12-month period when jobs are scheduled for winter as well as for the warmer seasons. It is often possible to maintain a smaller supply of the expensive equipment needed for construction by leasing the additionally required tools during the heavy summer months. As a result, overhead is reduced.

Big contracts can be completed sooner when work continues for 12 months. Schedules can be met more easily, and larger jobs can be bid.

There is another important advantage to placing concrete in the winter. Concrete placed at lower temperatures (but above 40° F) will ultimately gain higher strengths than concrete placed at 73° F, for example. (Concrete will obtain greatest ultimate strength when placed at 55° F.) Although the development of this strength is slower, cold weather provides a good slow-cure environment for concrete (Fig. 13-1).

With the advent of increased amounts of wintertime construction, advantages are being felt by the labor force, the contractors, and the economy of the community.

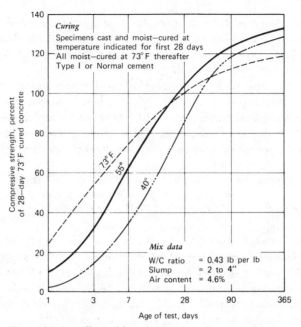

Figure 13-1 Effect of low temperatures on concrete compressive strength at various ages.

DISADVANTAGES OF COLD WEATHER CONCRETING

Placing concrete in cold weather, however, is not without problems. A workman's efficiency is cut by about 25% in extremely cold weather, and the efficiency of machines is also lowered. Breakdowns are more frequent when machines are run in cold weather, and usually it is more costly to run them because of the increased care required to keep them operative.

Extra equipment in the form of heaters, insulating materials, and tarps are needed to protect the workmen and the newly placed concrete. Fuel bills are high when temporary structures must be heated in extremely cold weather. Extra men are needed on night and weekend shifts to monitor the heating equipment.

The necessity of supplying heat to the concrete can create fire hazards that are a threat to the work area and to the workmen. When using heaters, care should be taken to prevent drying shrinkage cracks. Combustion heaters generate CO_2, which is detrimental to the surface of flatwork and can be dangerous to workmen; heaters should always be vented to the outside.

Often admixtures or premium cements are required in cold weather. In some cases, it is necessary to develop more heat from hydration by using a richer mix design to protect the concrete and to obtain a faster early strength gain.

A contractor's expenses are as much as 10% higher during cold weather; weather is uncertain and, hence, bidding is more uncertain. The variation of possible trouble is wide: there are many more problems to consider and many more risks to take.

Cold concrete has slower hydration rates. At 40° F, it takes longer to develop a given strength than it does as 73° F. When subsequent construction depends on a quick strength gain in the concrete, the delays caused by cold weather interfere.

If the concrete freezes before a compressive strength of approximately 500 psi has been developed, considerable damage is done. At best, the piece will have a damaged surface and will never attain watertightness; at worst, the freezing will cause such heaving and structural faults that it will lack strength and deteriorate. The critical point at which this damage is apt to occur is determined by the design of the mix, the type of cement used, whether or not the concrete is air-entrained, and the degree of hydration reached.

The use of as low-slump concrete as possible is recommended. The less mix water there is to freeze, the sooner a safe strength will be attained.

In cold weather the temperature variations from top to bottom of a slab can lead to thermal stresses within the slab that can cause cracking.

The first 24 to 48 hr after placing are the most critical for concrete. Once the level of compressive strength has reached 500 psi, it can survive several freeze-thaw cycles. Favorable curing conditions with good protection must continue, however, for as long as possible.

BASIC RULES FOR COLD WEATHER CONCRETING

Before any concrete is ordered, it is most important to make careful plans for receiving, placing, finishing, protection, and curing. All equipment and materials needed should be available. Provision should be made for possible equipment breakdowns on the job. An ample number of men should be on hand to do the work, and the supervisor should keep in mind that cold weather reduces human efficiency. It may be necessary to add one or more men to a crew to accomplish the work quickly in extremely cold weather.

The preplanning should include provision to thaw the subbase, if frozen, and to remove ice and snow from the area of placing. Concrete should never be placed in frozen forms, around cold reinforcing steel, or on frozen ground. Contact with the cold can freeze the concrete, and spring thaws of the subbase are apt to heave and shatter the concrete. The subbase can be thawed with steam, by building fires on the surface of the ground, or by applying $CaCl_2$. Care should be exercised to make sure that there is no opportunity for the subbase to refreeze after it has been thawed. Forms should be warmed immediately prior to placing, and this is especially important when using metal forms.

Some provision should be made for providing proper curing temperatures for the concrete. Generally, this is done in one of two ways: (1) insulation is provided to maintain the heat of hydration within the concrete by using straw, black polyethylene, and the like, or (2) a temporary structure is built around the forms and heat is provided to the structure (Fig. 13-2).

Figure 13-2   Protection from cold weather can be created by enclosing the construction area with a temporary shelter as was being done with this frame and tarpaulin.

The design of the concrete mix can be altered to obtain high early strength and quick-setting properties. It is often a good practice to heat the water or the aggregate, or both so that the concrete temperature, when placed, is between 55 and 65° F if the subgrade and air temperatures are below 50° F. Generally, it is wise to avoid using lean mixes in cold weather.

These special precautions add to the cost of cold weather construction, but they will insure good concrete, and the advantages of working all year compensate for the increased costs.

HEATING CONCRETE

One of the ways concrete can be protected from freezing in cold weather is to heat the ingredients of the concrete so that it is placed at a temperature above that of the ambient temperature. As the masses of concrete are greater, the temperature can be lower because the heat of hydration is greater. At temperatures higher than 70° F, it is hard to maintain slump, and the rate of heat loss of concrete at higher temperatures is much greater. If temperatures are too high, hydration may proceed sufficiently prior to placing the concrete to permit ultimate strength loss. With "as mixed" concrete temperatures above 70° F, the problems of hot weather concreting can be encountered in the dead of winter: it becomes more difficult to entrain air; additional mixing water (with the resulting dangers of shrinkage cracking and strength loss) is required; slump loss is increased; and evaporation rates increase markedly, which may cause plastic shrinkage.

A warm concrete mixture can be achieved by heating the mixing water, the aggregates, or both. The procedure to be followed depends on the equipment available, the temperature desired for the concrete, and the temperature of the air where the concrete is to be placed. It is desirable to produce concrete of a given temperature regardless of which ingredients are heated.

A formula has been developed, similar to the one in Chapter 12, for determining the temperature of the resulting mix when the temperatures of the ingredients are known:

$$T = \frac{0.22(T_a W_a + T_c W_c) + T_f W_f + T_w W_w}{0.22(W_a + W_c) + W_f + W_w} \text{ in which}$$

T = temperature, in degrees Fahrenheit, of the fresh concrete

T_a, T_c, T_f, T_w = temperature, in degrees Fahrenheit, of the aggregates, cement, free moisture in aggregates, and mixing water, respectively (usually, $T_a = T_f$)

W_a, W_c, W_f, W_w = weight, in pounds, of the aggregates, cement, free moisture in the aggregates, and mixing water, respectively.

Note: this formula assumes all materials to be free of ice.

Assuming the following facts:

$$T_a \text{ (aggregate temperature)} = 47° \text{ F}$$
$$T_c \text{ (cement temperature)} = 53° \text{ F}$$
$$T_f \text{ (temperature of the free moisture in the aggregate)} = 45° \text{ F}$$
$$T_w \text{ (mix water temperature)} = 83° \text{ F}$$
$$W_a \text{ (aggregate weight)} = 2800 \text{ lb}$$
$$W_c \text{ (cement weight)} = 600 \text{ lb}$$
$$W_f \text{ (free moisture weight)} = 100 \text{ lb}$$
$$W_w \text{ (mix water weight)} = 300 \text{ lb}$$

then

$$T = \frac{0.22\ (47 \times 2800 + 53 \times 600) + 45 \times 100 + 83 \times 300}{0.22\ (2800 + 600) + 100 + 300}$$

or

$$T = \frac{0.22\ (131,600 + 31,800) + 4500 + 24,900}{0.22\ (3400) + 400}$$

$$= \frac{0.22\ (163,400) + 29,400}{748 + 400}$$

$$= \frac{35,948 + 29,400}{1148} = \frac{65,348}{1148} \sim 57° \text{ F}$$

Several charts based on this formula have been developed for determining the temperature of the mixed concrete.

By using these charts, the necessary temperatures for mixing water and aggregates can be determined for various percentages of surface moisture of the aggregate and for various "as mixed" temperatures of the concrete.

If either the water or the aggregates are above 100° F, they should be combined in the mixer before the cement is added to alleviate the danger of a flash set caused by hot spots in the mixture. For some of the higher concrete temperatures, it may be necessary to heat both water and aggregates. If a mixed concrete temperature of 70° F is desired and a mix-water temperature of 140° F is available, it is necessary to have an aggregate temperature of 50° F.

It is often advantageous to be able to determine just how much heat will be needed to raise the water or aggregate temperature to necessary levels. Btu's can be determined by totaling the weight of material to be heated (in pounds), multiplying by the specific heat of the material (in British thermal units per pound per degree Fahrenheit), and multiplying this figure by the temperature rise needed (in degrees Fahrenheit). Each material must be calculated separately because the specific heat (British thermal units required to raise the temperature of 1 lb of material 1° F) of each material is different. The average specific heat of the solid materials in concrete (cement and aggregates) may be assumed to be 0.22 Btu per lb. per degree Fahrenheit, compared to 1.0 for water. Cement is usually not included because it is almost never heated in any way. If free moisture in aggregates is in the form of ice, it is necessary to calculate the amount of heat necessary to melt the ice and then to compute the heat required to raise the temperature of the 32° F water.

A chart (Table 13-A) has been created to determine the best temperature of concrete for a particular job and the prevalent weather conditions.

Table 13–A
RECOMMENDED CONCRETE TEMPERATURES FOR
COLD WEATHER CONSTRUCTION[a]
(Air-Entrained Concrete)

Line	Type of Temperature		Temperature, degrees Fahrenheit		
			Thin Sections	Moderate Sections	Mass Sections
1	Minimum temperature fresh	Above 30° F	60	55	50
2	concrete *as mixed* for	0° to 30° F	65	60	55
3	weather indicated	Below 0° F	70	65	60
4	Minimum temperature fresh concrete *as placed*		55	50	45
5	Maximum allowable *gradual* drop in temperature throughout first 24 hr after end of protection		50	40	30

[a] Adapted from Recommended Practice for Cold Weather Concreting (ACI 306-66).

Line 4 indicates the "as placed" temperature, and lines 1, 2, and 3 determine the ideal "as mixed" temperatures. An increase in these temperatures will result in a more rapid development of heat of hydration, heat loss from the plastic concrete, a waste of fuel to heat the ingredients, and an increase in the danger of thermal shock and plastic shrinkage to the concrete.

Figure 13-3 shows what the temperature of the mix water should be to achieve the desired concrete temperature if the aggregates are *not* frozen.

Mixing water is the most easily heated ingredient, and because it also has the highest specific heat, it can impart more heat to the final mix than any other ingredient. When air temperatures dip below the freezing point, however, it is often necessary to heat either the fine or the coarse aggregates, or both, to achieve the proper mix temperatures. They may have to be heated to thaw them in order to remove them from the storage bins.

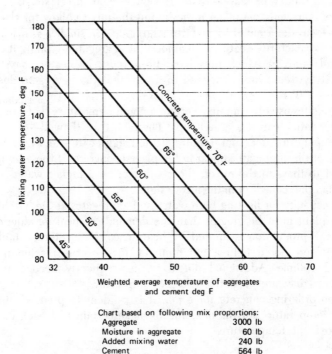

Weighted average temperature of aggregates
and cement deg F

Chart based on following mix proportions:
Aggregate	3000 lb
Moisture in aggregate	60 lb
Added mixing water	240 lb
Cement	564 lb

Figure 13–3 Temperature of mixing water needed to produced heated concrete of required temperature. Although this chart is based on the mixture shown, it is reasonably accurate for other typical mixtures.

It takes 30,000 Btu's to heat 300 lb of water 100° F. When dealing with above-freezing temperatures, 1 boiler hp (33.5 Btu's for 1 hr) will raise 1 ton of aggregate 60° to 65° F per hr. When aggregates are frozen, 1 boiler hp will raise the temperature of 1 ton of aggregate 30 to 40° F per hr. Computations can be made to determine the heating capacity of a boiler. Often the boiler has this information imprinted on the side.

Care should be taken to avoid heating the aggregate above 150° F. At this point the physical characteristics of some aggregates may change, and the aggregate can become extremely dry and hot, which may cause high slump loss and rapid stiffening.

Aggregates can be treated by stockpiling over steam coils. If the stockpiles are covered with tarps, the heat can be retained to warm the entire pile. A second method of heating aggregates is to inject the stockpile with live steam under pressure—75 to 125 psi. When direct steam is used, however, care must be taken in designing the mix to allow for the extra moisture (as determined by test) in aggregate. On smaller jobs it is possible to satisfactorily heat a small amount of aggregate by piling it on top of metal pipes in which fires have been built, or heaters have been installed. Extreme care is needed to insure that the aggregate does not heat over 150° F.

Aggregates usually contain a certain amount of moisture that can freeze in cold weather causing large lumps. These must be thawed—even if the mix temperature does not require heated aggregate—to avoid having pockets of the aggregate appear in the placed concrete. If these lumps are allowed to thaw in the mixer, the result can be a higher water-cement ratio than that called for in the mix design.

The capacity for heating both water and aggregates at the batching or mixing plant must be greater than the demands of peak production. In fact, peak production is limited by heating capacity. Storage tanks for warm water should hold a minimum of 75% of the maximum daily demand volume. Additional hot water is frequently needed in cold weather to thaw and clean the delivery equipment.

When ordering concrete for a job, it is prudent to predetermine the proper temperature for the mix, and to make certain that each batch of concrete is this temperature.

PLACING AND FINISHING

Once the concrete is delivered to the site, work should proceed quickly to retain as much "as mixed" heat as possible. Short hauls of large loads should be arranged. If an enclosure has been built to protect the new concrete, it is best to unload within the enclosure. If conveyors, pumps, or chutes are to be used to transport concrete, these should be warmed, enclosed, or insulated.

Low-slump concrete should be used to avoid long delays waiting for bleed water to evaporate before finishing can start. The evaporation rate in winter is very low, and if concrete contains too much water, untimely delays can result. The use of air-entrained concrete is strongly advised. Not only does air entraining contribute to the freeze-thaw resistance of the concrete, but it reduces the bleeding and allows finishing to proceed more quickly.

PROTECTING PLACED CONCRETE

Once concrete has been placed and finished, it still must be offered some protection from the cold. The amount and type of protection depends on the structure, the type of concrete used, weather conditions, and the relative cost of the various methods.

Wooden forms provide some protection and may be sufficient if the weather is not too cold. Metal forms, however, offer little protection from heat loss unless they are insulated in some manner.

Covering the concrete with straw gives some slight protection—especially when the straw is covered with tarps. Commercial curing blankets offer the best insulating protection to flat work. Insulating board, sawdust, lumber, damp sand, and dead air space are all used and afford varying degrees of protection.

Electric heating blankets have been developed for the protection of new concrete in winter. These have been used with great success. Some types are designed for reuse; others are for one-time use and are to be discarded. The latter type can be used underneath a slab on grade to maintain warmth from the bottom.

Plastic sheets placed a few inches above the concrete have been used with success in temperatures down to 18° F. When using plastic sheeting in the winter, it is advisable to order the clear type instead of the white-pigmented variety so that every advantage can be taken of solar heat during the day.

Of all the types of protection that can be offered concrete during cold weather, perhaps the best—and most expensive—is the construction of a temporary shelter around the work. This is often built of lumber covered with sheet plastic or tarps. Some use has also been made of plastic bubbles that inflate into a dome over the work area.

When erecting shelters over the concrete, care should be taken to make sure that a free circulation of warm air is possible around the entire surface of the slab and that corners and edges—especially when near the wall of the structure—are sufficiently protected. It is sometimes necessary to add some further protection, such as insulating materials, to these especially susceptible points.

The easiest method of curing exterior walls is to use insulated forms and to leave them on to protect the outside of the concrete while the interior of the building is heated. Insulated forms can be used repeatedly and, hence, are often an economical investment for a contractor who does a sizeable amount of winter work (Fig. 13-4).

No matter what method of insulating the concrete is used, the principle remains the same—the heat of hydration is kept within the concrete. This heat continues to build up, but 30 hr after placing, the temperature

Figure 13–4 If the cold weather is not too severe, the heat generated by the hydration of cement is usually sufficient to maintain the concrete at an acceptable temperature provided that heat can be contained. Insulated forms is the easiest method of retaining the heat of hydration.

has usually reached its peak. One cubic yard of concrete releases a total of 80,000 to 90,000 Btu's. This heat is often sufficient to protect the concrete and provide a suitable curing temperature.

If metal reinforcing bars or other metal pieces extend from the concrete to the cold air, it is wise to cover them to avoid heat loss through radiation. Metal is a very good conductor of heat and can lower the local temperature appreciably if left exposed.

SUPPLYING ADDITIONAL HEAT

In extreme cold additional heat often must be supplied to the concrete. This heat is best applied in the form of steam, which supplies moisture while furnishing heat. When cold air is heated it becomes dry, and unless moisture is added, the relative humidity may go down to 1 or 2%. Mixing water will evaporate from the concrete unless some protection is supplied. Membrane-forming curing compounds are often sprayed on the concrete to contain the moisture.

Ponding is an excellent way of curing concrete in slabs, but in recent years research has indicated that concrete cured with water remains saturated and thus, is more likely to be subject to postcuring damage when it is exposed to below-freezing temperatures than is concrete cured with other methods. Water-curing methods are often difficult to implement, and there is the risk of freezing the curing water and the concrete.

There are two ways in which heat can be introduced into the temporary shelter; moist heat (steam) or dry heat. Steam, with its built-in humidity, is preferred. Higher curing temperatures are possible when using steam and strength gains are proportionately faster. Care should be taken, however, to make sure that the steam outlets are not directed at the concrete.

When dry heat is used it is necessary to make sure that the heat circulates and does not warm the concrete unevenly. Crazing will result when water loss is too rapid.

Heaters should be vented to the outside air to avoid a buildup of carbon dioxide, which is harmful to fresh concrete, and carbon monoxide, which is harmful to people. Care should also be taken to keep the temperature as constant as possible. This can be achieved in part by regulating the amount of heat produced and by making the structure as airtight as possible while still well ventilated (Fig. 13-5).

Shelters can be constructed of wood, canvas, building board, plastic sheets, or waterproof paper, or these materials can be attached to the

Figure 13–5 Portable heaters are useful provided they can be adequately ventilated.

existing framework of the building being constructed. Extra care should be given the windward side of the structure. The more airtight the structure, the less money will be needed to heat it, and the better the protection offered the concrete.

The air temperature within the structure is not always the temperature of the concrete. It is most important to take readings frequently to check the exact temperature of the concrete. This can be done by placing thermocouples in the still-plastic concrete or by placing thermometers on top of the concrete long enough to get an accurate reading.

When heating an enclosure other than by electricity or steam, great care should be exercised to avoid fire. Flameproof materials should be used wherever possible and a 24-hr watch service should be maintained as long as heating units are in operation.

The placing of concrete floors on grade before a building has been enclosed should be avoided when possible. Protection can be provided more easily to a floor after the building is under roof and when the walls can be enclosed in some manner. If, however, the floor must be placed before the building can be sealed, the insulating protective materials should be placed a few inches above the slab to avoid marring the surface of the floor.

In Canada, where a great deal of winter construction takes place, several interesting methods of providing cold-weather protection have been developed by contractors.

In Manitoba, a contractor working on the construction of a small dam covered the whole work area with a temporary structure. Because of wind conditions prevalent at the working site, great care had to be taken to make the structure strong enough to withstand the wind that blew down the river valley. Space heaters provided adequate temperatures for the curing of the concrete and for the comfort of the workmen. In the construction of the shelter, polyethylene film was stretched tightly and attached to the wooden support pieces. It was found more satisfactory to stretch the film the long way and avoid the use of heavy-gauge film which becomes brittle in very cold weather. A 6-mil film was used. The film was reinforced across the width with slats before it was attached to the framework. Rough lumber was used for the supporting members because this gives a better grip on the film (Fig. 13-6).

A great amount of new construction in Montreal was scheduled for completion just prior to the opening of Expo '67 in that city. On a bridge-approach project, work continued through a very cold winter to make possible the opening of a highway to facilitate the flow of traffic to and from the fair. The contractor set up a central boiler plant and delivered steam to enclosures erected over the various work sites. In this way small enclosures could be used, saving the amount of heat needed, and the

Figure 13–6 Polyethylene plastic sheet enclosing a building.

central plant was safer than individual space heaters, which are prone to tip or run out of fuel and can be fire hazards.

Tarpaulins can also be used to cover temporary structures and to cover insulation placed on top of concrete. There are several disadvantages with tarps, however. It is hard to make them windproof when used to cover a shelter. They are heavy to handle and, hence, structures must be constructed with some degree of strength to hold them. They are extremely hard to manipulate when frozen. Because tarps do not allow sunlight through, they make for dark working quarters and permit no solar heating during the day.

With the use of temporary enclosures it is possible to provide the concrete with better curing conditions in winter than are available in the summer. Temperature and humidity can be controlled quite easily, and if care is exercised the winter-placed concrete can be stronger and more durable than that placed at almost any other time of the year.

CHANGING THE MIX DESIGN

We have seen that by warming the concrete and by offering protection to the newly placed concrete, the possible damages caused by cold weather can be avoided. The amount of protection required and, therefore, the expense associated with it, can be reduced by changing the design of the mix or by the addition of admixtures to the mix to change the physical properties of the concrete.

Generally, do these three things:

1. Use Type III cement.
2. Lower the water-cement ratio and increase the cement content of the content.
3. Use chemical accelerators in the mix.

During cold weather, high-early-strength cement (Type III) can be used advantageously to permit early reuse of forms and removal of shores, to effect savings in cost of artificial heat and protection, to achieve earlier finishing of flatwork, and to allow earlier use of the finished structure. In the winter, or when time is short, Type III cement can be used economically in concrete for pavements and bases. High-early-strength concrete is often desirable at important intersections, on busy streets where satisfactory detours are not available, for the last few days of construction on long stretches of highway, and for patching.

Type III concrete fabricated and cured for 3 days at 40° F gains about 35% of 28-day strength, while Type I concrete gains only a bit more than 10%. At 7 days the difference in strength development is still significant: about 68% for Type III and 33% for Type I. For concrete fabricated and cured at 55° F, the early strength development is much higher in Type III, but by 7 days Type I strength has almost caught up.

Type III cement costs more than Type I, however, and sometimes the value of the construction does not warrant the extra cost. Another method of achieving high early strength is to decrease the water-cement ratio of the mix. The rule of thumb for adding extra cement is that by increasing the amount of Type I cement by a factor of one-third, the early strengths of Type III cement are achieved. By adding extra cement to the mix it is possible to obtain high early strengths, and the heat of hydration is increased in the richer mix; however, this is usually more expensive than using Type III.

A third way of adapting concrete to cold weather is to add chemicals to the mix that change the physical properties of the concrete. Calcium chloride is the chemical most frequently used to accelerate setting and early strength gain. The calcium chloride used should meet ASTM specifications D 98 and should be sampled and tested in accordance with D 345. If in a crystal or powdered form, it should be stored in airtight containers to avoid lumping. Calcium chloride is frequently sold in solution, and it is in this liquid form that it should be added to the mixer. (Remember to include the water of the solution in the total mixing water to avoid increasing the water-cement ratio.) The calcium chloride should be added near the end of the mixing. It takes only 20 to 30 revolutions of a mixing drum to thoroughly blend the solution into the mix. By adding it at the end of the mixing cycle, the chance of a flash set being caused by some calcium chloride coming into direct contact with dry cement is almost eliminated.

Most trade-name accelerators on the market are composed mainly of calcium chloride and so should be handled in the same manner. However, non–calcium chloride accelerators are available. Extreme care in selection is necessary, since it is important to know what is being added to concrete.

Calcium chloride should not come into contact with air-entraining or water-reducing admixtures, since they can be precipitated out of solution. The amount of calcium chloride used should never exceed 2%.

Calcium chloride should not be used in the manufacture of prestressed concrete members or when the concrete is to be used with steel reinforcing and aluminum conduit or other nonferrous metals placed over metal for insulation purposes. Corrosion of the metals is likely to take place, causing a weakening or failure of the concrete. Concretes made with calcium chloride are less resistant to sulfate attack and should not be used when the presence of sulfates is likely. The addition of calcium chloride to the concrete is also apt to cause dark spots in the new concrete. This occurs most often when the temperature of the mixed concrete is too low. These spots can sometimes be removed by washing the surface with a solution of sodium hydroxide and, 24 hr later, with clear water.

Great care should be taken when using accelerators with heated concrete. The combination of chemicals and heated concrete can sometimes cause a flash set. By adding ingredients to the mixer in the proper order, this can be avoided. (The order should be aggregates and water first, cement next, and any accelerators after the other ingredients have been blended.)

Calcium chloride does not lower the freezing temperature of the mix significantly but, instead, acts as a catalyst to start the cement setting more quickly. There are, however, so-called "anti-freeze" compounds on the market that claim to actually lower the freezing point of the concrete. If they are used in sufficient quantity to do this, the properties of the hardened concrete can be adversely affected. These compounds should be strictly avoided!

In no case should the change in mix design, the use of Type III cement, or the addition of admixtures to the concrete take the place of proper protection and curing. These are helpful tools, but they do not in themselves solve the problems of placing in cold weather.

There is no protection that contributes as greatly to the durability of concrete as air entrainment. We know that air-entrained concrete can withstand freezing and thawing much better than non-air-entrained concrete. It is a good general rule to always insist on the use of air-entrained concrete when placing concrete that will be exposed to cold weather.

By using concrete with high early strengths it is possible to reuse forms sooner, remove shoring sooner, remove protective shelters and insulation, and use the structure or slab sooner. The reduced time can represent money saved for the contractor.

LENGTH OF PROTECTION TIME

The length of time that concrete should be protected depends on several factors: weather conditions, type of concrete used, strength demanded of the concrete, and type of protection given the concrete. Generally, the additional strength needed and the weather conditions to which the concrete will be subject after the removal of protection are the determining factors.

Concrete should be kept at the temperatures shown in line four of Table 13-A (page 410) for the length of time shown in Table 13-B.

Table 13-B
RECOMMENDED DURATION OF PROTECTION FOR
CONCRETE PLACED IN COLD WEATHER[a]
(Air-Entrained Concrete)

Degree of exposure to freeze-thaw	Normal concrete[b]	High-early strength concrete[c]
No exposure	2 days	1 day
Any exposure	3 days	2 days

[a] Protective for durability for concrete placed at temperature indicated in line 4, Table 13-A. Adapted from Recommended Practice for Cold Weather Concreting (ACI 306-66).
[b] Made with Type I, II, or normal cement.
[c] Made with Type III or high-early-strength cement, or an accelerator, or an extra 100 lb of cement.

Non-load-bearing concrete that will not be subjected to freeze-thaw cycles (such as foundation footings) should be protected from freezing for two days. Concrete that is non–load bearing (such as sidewalks and curbs) but will be subjected to freeze-thaw cycles should be protected for three days if air-entrained concrete is used and five days if non-air-entrained concrete is used. Concrete that is required to carry a load shortly after the first cure but will receive an adequate curing atmosphere after protection is removed should be protected for seven days or for enough time to develop the necessary strength. Examples of this type of structure are bridge decks and piers.

Structures that will be required to withstand stresses at the end of the protection period and that will be in a poor curing atmosphere after protection is removed should be protected for at least 14 days. Examples are columns and floor slabs. In cases where great strength or durability is ultimately required of the concrete, it is wise to make test cylinders to determine the length of the protection period. If strength-time curves have been established, they can be used to determine the protection time much more accurately than any general chart or graph. By using your own charts, the chance of error is greatly reduced.

As a general rule it is advisable that thin sections be kept for three days at 55° F if Type I or Type II cements are used. If Type III cement (or an accelerator or extra cement) is used, the 55° F temperature should be maintained at least two days for durability and four days for partial loading. Non-air-entrained concrete usually needs about twice as much time under protection.

Forms should be left in place and protection should be given the concrete as long as possible. Unless there is some reason to remove the forms (for reuse, for example), it is better to leave them on the concrete. If insulating blankets are not needed at other places on the job, they should be left to protect the concrete.

Many job specifications limit the removal of forms to not before the time when minimum strengths are attained. This time is best determined by making test cylinders on the job. Strengths for this purpose should be measured on test cylinders made from the job concrete and cured under job conditions. Job specifications often specify curing times.

When protection is no longer necessary, the temperature of the concrete should be lowered gradually to reduce the possibility of thermal shock. The temperature drop per 24-hr period is determined from Table 13-A, line five (page 410).

The cooling process of a structure can often be accomplished by merely turning off the heat and allowing the temperature within the enclosure to drop gradually. When the concrete is covered with insulating materials, these can sometimes be removed in layers or, in the case of an insulating blanket, they can be replaced with lighter insulation.

TYPES OF PROTECTION

Between the time of the first frost in fall and the time when the average temperature falls below 40° F, only moderate protection is necessary for concrete. Sometimes, just leaving the forms in place is enough. The forms need be left on for only 48 hr if air-entrained concrete is used. Non-air-entrained concrete should be protected for four days.

For temperatures between 30 and 40° F, it is wise to use air-entrained concrete placed at a maximum of 70° F. Between 20 and 30° F additional protection should be given the air-entrained, warmed concrete in the form of insulation and/or a heater shelter. Below 20° F total protection should be given.

ADDITIONAL CURING

Once protection has been removed no further curing is necessary so long as the air temperature remains below 50° F. An exception to this would be in a region where the relative humidity is extremely low—in such a case, additional curing (usually with a membrane-curing compound) would be required. If, in normal-humidity areas, the temperature rises above 50° F for more than 12 hr, some form of curing should be started. While it is true that at temperatures of 50° F and below there is a very low evaporation rate, still the quality and durability of the concrete will be improved if a curing compound is applied. If this is done, there is no need for watching the temperature rise and fall and checking on the condition of the concrete.

TESTING AND RECORD-KEEPING

When making test cylinders, the ACI standards suggest that one strength test be made for each 100 cu yd, or fraction thereof, of concrete placed. These cylinders, cast to test the quality of the concrete as delivered, should be made and laboratory cured in accordance with ASTM C 31. If job-cured cylinders are to be tested, additional specimens from the same sample should be made and field cured.

In addition to test cylinders, the inspector or job supervisor should insist that a careful check be made of the temperature of each batch of concrete at the end of mixing and that the "as placed" temperature of the concrete also be checked. After the concrete has been placed (or during placement on large jobs) the temperature should be checked again. These temperatures, together with the associated times and locations, should be entered on a record sheet that becomes part of the job records.

Any admixtures used are also noted.

The type of protection offered the concrete should be a part of the record as should ambient weather conditions.

A form should be designed for the recording of these facts. It is recommended that a separate sheet be prepared for each day's work. The following should be included.

1. Date.
2. Job location and particular work site.
3. Inspector's name.
4. Note of any pertinent factors such as labor shortages, machinery breakdowns, and the like.
5. Weather conditions, including temperature, relative humidity, wind direction and speed, and precipitation.
6. Batch number of mixed concrete.
7. Delivery time to job site.

8. Time of discharge from mixer or transporting carrier.
9. Mixed temperature of batch.
10. Temperature of concrete after placing, including a chart to indicate where readings were taken.
11. Indication of change in weather conditions during day.
12. Maximum and minimum temperatures.
13. Mix design data as available.
14. Temperature of curing concrete.

CONCRETE MASONRY CONSTRUCTION IN WINTER

When building with concrete block in cold weather, there are certain precautions that must be taken to insure the quality of the work.

The supply of block should be stored in a protected place so that it is kept clean, dry and at a temperature at or above 40° F. If the block is below 40° F, it should be warmed with some type of heater. This can usually be accomplished by covering the pile of block with a tarp and by placing a salamander under the tarp.

When making mortar in cold weather, it is advisable to warm the sand and the water to 160° F, being careful not to scorch the sand. Unused mortar should be kept warm—but not hot. Calcium chloride should not be used in making mortar.

The completed masonry should be protected with windbreaks, enclosures, and, if necessary, heaters until the mortar has reached final set.

CHAPTER 14
VOLUME CHANGES AND CRACK CONTROL

The value of understanding when and why volume changes occur in concrete is in minimizing cracking caused by this phenomenon. Cracking is a problem that hurts the reputation of concrete because many people feel that cracking is a sign of defective concrete. This is not necessarily true. Cracks rarely affect the structural functioning or the durability of concrete significantly but, since they are unsightly and admit water more easily, they may accelerate weathering or rusting in some instances. Therefore, minimizing them is important. Although it is difficult to isolate the cause of any given crack, this chapter discusses cracks caused by the shrinkage of concrete during and after hardening, shifting of the concrete before it hardens, and several other types of cracking.

SHRINKAGE

Drying shrinkage in concrete takes place in two stages: (1) while the concrete is plastic, that is, before it has taken its initial set, and (2) in the hardened concrete. Both types of shrinkage may cause cracking.

Plastic Shrinkage
Immediately after placement, concrete in the plastic state settles in the forms and bleed water migrates to the top, where it is lost by drainage or

evaporation. In addition, some of the water is absorbed by the cement and aggregate. These effects combine to produce a reduction in the volume of the concrete. In extreme cases, this shrinkage may amount to 1% or more. However, since the concrete in this stage is in a plastic or semiplastic state, it is free to move. Therefore, no appreciable stresses result from these volume changes, and major cracks are rare. Small cracks, however, called plastic shrinkage cracks, may occur within the first few hours after placement. These cracks may have considerable depth and are usually relatively long without any definite symmetry or pattern. They may occasionally have a crow's-foot pattern (Fig. 14-1).

The main cause of this type of shrinkage cracking is excessively rapid evaporation of water from the concrete surface. It usually occurs in slab and pavement construction when the surface is exposed to the sun and drying winds. Corrective measures are all directed toward reducing the rate of evaporation or the total time during which evaporation can take place. The likelihood of a plastic crack is greater whenever the rate of evaporation exceeds the rate at which water bleeds or rises to the surface. This condition usually occurs in summer when winds are strong, temperatures are high, and humidity is low. However, it may also be experienced during cold weather, especially when the concrete temperature is high in comparison to the air temperature at the time of placing. Table 14-A shows how various weather factors influence the rate of evaporation. The evaporation is highest under the following conditions: when the wind is

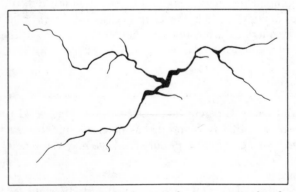

Figure 14–1 Plastic shrinkage cracks may occur within the first few hours after placement.

Table 14–A

EFFECT OF VARIATIONS IN CONCRETE AND AIR TEMPERATURES,
RELATIVE HUMIDITY, AND WIND SPEED ON DRYING TENDENCY
OF AIR AT JOB SITE

Variables	Case No.	Concrete Temperature (°F)	Air Temperature (°F)	Relative Humidity (%)	Dew Point (°F)	Wind Speed (mph)	Drying Tendency (lb/sq ft/hr)
	1	70	70	70	59	0	0.015
	2	70	70	70	59	5	0.038
1. Increase in	3	70	70	70	59	10	0.062
wind speed	4	70	70	70	59	15	0.085
	5	70	70	70	59	20	0.110
	6	70	70	70	59	25	0.135
	7	70	70	90	67	10	0.020
2. Decrease in	8	70	70	70	59	10	0.062
relative	9	70	70	50	50	10	0.100
humidity	10	70	70	30	37	10	0.135
	11	70	70	10	13	10	0.175
3. Increase in	12	50	50	70	41	10	0.026
concrete and	13	60	60	70	50	10	0.043
air tempera-	14	70	70	70	59	10	0.062
atures	15	80	80	70	70	10	0.077
	16	90	90	70	79	10	0.110
4. Concrete at	17	100	100	70	88	10	0.180
70° F; decrease	18	70	80	70	70	10	0.000
in air	19	70	70	70	59	10	0.062
temperature	20	70	50	70	41	10	0.125
5. Concrete at	21	70	30	70	21	10	0.165
high temper-	22	80	40	100	40	10	0.205
ature; air at	23	70	40	100	40	10	0.130
40° F and	24	60	40	100	40	10	0.075
100% R. H.							
6. Concrete at							
high temper-	25	70	40	50	23	0	0.035
ature; air at	26	70	40	50	23	10	0.162
40° F; variable	27	70	40	50	23	25	0.357
wind							
7. Decrease in							
concrete tem-	28	80	70	50	50	10	0.175
perature air	29	70	70	50	50	10	0.100
at 70° F	30	60	70	50	50	10	0.045
8. Concrete and							
air at high	31	90	90	10	26	0	0.070
temperature,	32	90	90	10	26	10	0.336
10% R. H.;	33	90	90	10	26	25	0.740
variable wind							

blowing over the surface of the concrete, when the relative humidity is low, when the temperature of the concrete and the temperature of the air are both high. The procedures that reduce the rate of evaporation were covered in Chapter 12, "Hot Weather Concreting."

Shrinkage in Hardened Concrete

Effects of Restraint. Shrinkage in concrete after initial set is generally more serious than while it is plastic. If the concrete were always free to move, volume changes would be relatively unimportant; usually, however, movement is restrained by a subgrade foundation, steel reinforcement, or by adjacent concrete. As the potential movement is restrained, stresses will be developed that may rupture the concrete. This is particularly true when tension is developed since concrete's tensile strength is much lower than its compressive strength. If concrete is dried or cooled without restraint, it will simply contract and no stress will be developed. If, however, concrete is restrained during drying or cooling, tensile stress develops. If at any time the tensile stress becomes greater than the tensile strength, the concrete will crack and the stress will be relieved.

Creep is a phenomenon in concrete that tends to relieve the stress in concrete in tension, especially when the concrete is reinforced. Creep causes a gradual reduction in the load on the concrete and a corresponding increase in the load on the steel. In various structural elements such as continuous beams or slabs, creep relieves some of the stress in the most highly stressed portions and increases the stress in adjacent portions of the concrete; finally, the stresses are more uniform throughout the member. This relieving of higher stresses serves to reduce the tendency toward cracking, even though the process may take several years.

Causes of Shrinkage

Shrinkage is a decrease in volume caused by drying and chemical changes, and is a function of time but not one of temperature or stress due to external load. The exact chemical and physical mechanics of shrinkage are not thoroughly understood. Some authorities believe that it results from stresses induced by loss of capillary water and absorbed water surrounding the particles of cement gel, resulting in compressive strains that cause a closer compaction of the gel particles. Others believe shrinkage is due to a change in the gel itself or, in technical terms, collapse of the gel lattices as the interlattice "zeolitic" water is removed by drying. Despite controversy over the causes of drying shrinkage, such shrinkage is an ever-present factor in concrete work. Its effects (drying shrinkage cracks) are quite apparent even to casual observers.

In general, however, shrinkage is caused by two conditions: drying and carbonation. The most important single factor affecting drying shrinkage is the amount of water per unit volume of concrete. Consequently, concrete with a wetter consistency will shrink more than one with a dry or stiff consistency.

Cement Paste. The cement paste is the principal material responsible for all volume changes. Cement paste in concrete, if it were not restrained by aggregate, would shrink from 5 to 15 times as much as the concrete. The quality of the paste is primarily a function of the water-cement ratio and the composition and fineness of the cement. A paste with a water-cement ratio of 0.56 by weight shrinks about 50% more than one with a water-cement ratio of 0.40. Finer cements generally exhibit slightly greater shrinkage than coarse cements. In considering composition, it appears that a given amount of tricalcium aluminate contributes most to shrinkage and tricalcium silicate contributes least. Gypsum has a major effect on shrinkage, and for a given cement, there is an optimum amount that produces the smallest shrinkage. In general, a high cement content increases the volume of the paste and, therefore, has a potential for higher shrinkage. Hence, even though the high cement content substantially decreases the water-cement ratio and increases the strength of the concrete, it may lead to increased shrinkage. To reduce the likelihood of shrinkage, it is advantageous to specify a mix with as small a cement content as possible while still meeting strength requirements for the particular project.

Admixtures. Some admixtures have been found to increase the amount of shrinkage in concrete. Some of these are calcium chloride, some pozzolans, and some water-reducing admixtures. Air-entraining admixtures do not seem to affect shrinkage. In general, there is very little information available on exactly which admixtures affect shrinkage under what conditions, and which do not. For this reason, any admixtures with which one has no experience should be tested before use.

Aggregates. Aggregates also have an important effect on shrinkage. A well-graded aggregate is always desirable, not only from a strength viewpoint but also for shrinkage qualities. Good grading gives the desirable quality of a low percentage of voids, therefore, requiring less paste volume. Also, the larger the maximum size of aggregate, in a well-graded aggregate, the smaller the shrinkage. For example, considering mixes with equal water-cement ratios, the shrinkage of a sand mortar may be two or three times greater than that of a concrete with a ¾-in. maximum size aggregate, and three or four times that of a concrete with a 1½-in. maximum. This is because the larger maximum size has more aggregate in the mix to resist shrinkage.

A principal function of the aggregate is to control the shrinkage of cement paste. Particles of rock embedded in cement paste restrain shrinkage of paste to a degree that makes concrete a practical building material. Ordinarily, restraint is provided by the rock particles acting separately. Each is separated from its neighbors by paste. If all rock particles could be removed from hard concrete leaving the paste intact, we would find a spongelike skeleton structure that would shrink or swell the same amount as the paste itself. When the holes are filled with a solid material (aggregate) that does not itself shrink, shrinkage of the skeleton is restrained.

Shrinkage is not only dependent on the aggregate content of a mix but also on the kind of aggregate. Hard aggregates that are difficult to compress are more resistant to shrinkage of cement paste. As an extreme example, if steel balls are substituted for ordinary coarse aggregate, shrinkage may be reduced by 30% or more.

Table 14-B shows some experimental values of the coefficient of thermal expansion of concretes made with aggregates of various types. These data were obtained from tests on small concrete specimens in which all factors were the same except for aggregate type. The fine aggregate was of the same materials as the coarse aggregate.

Table 14-B
EFFECT OF AGGREGATE TYPE ON
THERMAL COEFFICIENT OF EXPANSION OF CONCRETE[a]

Aggregate Type (From One Source)	Coefficient of Expansion (Millionths per Degrees Fahrenheit)
Quartz	6.6
Sandstone	6.5
Gravel	6.0
Granite	5.3
Basalt	4.8
Limestone	3.8

[a] Coefficients for concretes made with aggregates from different sources may vary widely from these values, especially those for gravels, granites, and limestones.

Carbonation. Another cause of shrinkage in hardened concrete is carbonation. When concrete is exposed to air containing carbon dioxide, it undergoes irreversible carbonation shrinkage that may be, under some circumstances, about as large as the shrinkage due to air drying. The carbon dioxide reacts with lime in the cement gel, causing loss of water as in the drying process. Carbonation proceeds slowly and usually produces little shrinkage at very low or very high relative humidities. The maximum effect appears to occur when the relative humidity is about 50%.

Shrinkage occurring as a result of carbonation is indistinguishable from drying shrinkage except by sophisticated research procedures. Most laboratory measurements of "drying shrinkage" actually are measurements of length changes that are due to drying shrinkage plus carbonation shrinkage; the distinction is mainly of concern only to researchers.

The physical chemistry of shrinkage caused by carbon dioxide is not well understood, but its effects have been proved. For example, it was observed that a masonry wall, although carefully designed to prevent cracking from drying shrinkage, nevertheless developed cracks after several years of service. This is now believed to be the effect of shrinkage produced by gradual carbonation. Laboratory and field experiments are now underway to precarbonate concrete products, particularly block, as a means of preventing subsequent carbonation shrinkage.

Curing. The effect of curing on the ultimate drying shrinkage of concrete is complex and varies with the water-cement ratio. Generally, however, longer periods of moist curing reduce shrinkage.

Cumulative Effects. It has been observed and experimentally shown that the cumulative effect of the individual factors that cause shrinkage can be very large. The combined effect of these factors is the *product*, rather than the sum, of the individual effects. Consider a concrete where everything was done wrong: unfavorable construction practices were used; concrete discharge temperatures were 80° F rather than 60° F; 7-in., rather than 3-in., slump was used; a ¾-in. maximum size aggregate was used instead of a 1½-in. maximum; and mixing and waiting periods were too long. In addition, a cement with a high shrinkage characteristic, dirty aggregates of poor inherent shrinkage quality, and admixtures that increase shrinkage were used. The cumulative effect of these factors was shrinkage about 5 times as large as the shrinkage that would have been produced by the best choice of variables (Table 14-C).

Table 14-C
CUMULATIVE EFFECT OF ADVERSE FACTORS ON SHRINKAGE[a]

Factor	Equivalent Increase in Shrinkage (%)	Cumulative Effect
Temperature of concrete at discharge allowed to reach 80°F, whereas with reasonable precautions, temperature of 60°F could have been maintained	8	$1.00 \times 1.08 = 1.08$
Used 6- to 7-in slump where 3- to 4-in slump could have been used	10	$1.08 \times 1.10 = 1.19$
Use of ¾-in maximum size of aggregate under conditions where 1½-in size could have been used	25	$1.19 \times 1.25 = 1.49$
Excessive haul in transit mixer, too long a waiting period at jobsite, or too many revolutions at mixing speed	10	$1.49 \times 1.10 = 1.64$
Use of cement having relatively high shrinkage characteristics	25	$1.64 \times 1.25 = 2.05$
Excessive "dirt" in aggregate due to insufficient washing or contamination during handling	25	$2.05 \times 1.25 = 2.56$
Use of aggregates of poor inherent quality with respect to shrinkage	50	$2.56 \times 1.50 = 3.84$
Use of admixture that produces high shrinkage	30	$3.84 \times 1.30 = 5.00$

[a] Based on effect of departing from use of best materials and workmanship.

Reducing Shrinkage

Construction practices that tend to minimize shrinkage are: good mixing, placing, and curing. Mixing is of great importance in getting a uniform low-shrinkage concrete. Thorough mixing also gives increased workability, which will allow slightly larger quantities of aggregate for a given quantity of cement and water. Good placing practices should be followed. Care must be exercised to prevent excessive vibration that would cause materials to separate. The excess water that accumulates on the top should be worked to a low point and removed. Laitance should not be permitted. Curing is a highly important factor in obtaining crack-free concrete. Even if all precautions up to this point are observed but curing is neglected, some cracks will probably occur unless moist weather provides naturally favorable curing conditions.

If shrinkage cracks are to be held to a minimum, the size of the placement should be considered. The largest dimension between contraction joints should be not more than 40 ft. For large slab areas in a complicated framing job, considerable study may be necessary to locate the proper construction joints. The sequence of placing is also important. If the contractor wants to place adjacent areas at the same time, an open section should be kept between the areas. The section should be filled in a few days later. Another method of placing is by alternating areas in checkerboard fashion.

OTHER VOLUME CHANGES

Shrinkage is only one cause of volume changes in hardened concrete. There are other factors that act to increase or decrease its volume temporarily or permanently.

External Temperature Changes

All materials change in dimension when their temperature changes. Most increase in volume as their temperature increases, and this is true for concrete. The unit in which this change is measured is millionths of an inch per inch, per degree Fahrenheit. Concrete expands about 5.5 millionths of an inch per inch, per degree Fahrenheit. This average coefficient of thermal expansion or contraction is equivalent to a length change of 0.66 in. for 100 ft of concrete subjected to a temperature change of 100° F.

A fast rate of temperature change is more damaging to concrete than a slow change.

There are a number of thermal properties of concrete. These include *diffusivity, conductivity,* and *coefficient of expansion.* Diffusivity is a measure of the capability of the concrete to undergo temperature changes. Conductivity is the rate at which concrete conducts heat, and the coefficient of expansion is the rate at which concrete changes volume with changes in temperature.

A high coefficient of thermal expansion in concrete causes high stresses in the surface of the concrete undergoing rapid temperature changes. Coarse aggregates with a high diffusivity compared to the mortar will cause differential volume changes and resultant stresses. Both effects would adversely affect durability in environments with extreme and rapid temperature changes.

The coefficient of thermal expansion of concrete is very nearly the same as that for steel. It is this similar coefficient of expansion of the two materials that makes it possible to use reinforced concrete.

There are two ingredients of concrete that affect its thermal properties. One is the type of coarse aggregate and the other is the water content. Natural aggregates vary in coefficient of expansion between about 7 millionths per degree Fahrenheit, for quarts, and 1 millionth per degree Fahrenheit, or less, for some limestones. Sandstone, granite, and basalt fall about mid-range. Gravel, of course, may vary considerably in mineralogical composition and may be anywhere in this range.

In general, if the concrete is made with aggregates of good quality, it should give no trouble.

A high moisture content in hardened concrete can cause some increase in its coefficient of thermal expansion. In neat cement the coefficient of expansion may vary by as much as 100%, depending on its moisture content. In concrete, the aggregates minimize this variance. Concrete structures that may be subject to moisture should be carefully designed, with temperature changes taken into consideration. The temperature at which the concrete hardens determines its initial volume. Expansions and contractions due to future temperature changes will be based on this original volume. For this reason, cracking caused by temperature changes can be minimized by having the concrete harden at approximately the mean annual temperature to which the concrete is to be exposed.

Control of cracking due to external temperature changes is accomplished mainly with reinforcing steel and expansion and isolation joints.

Internal Temperature Changes

Heat is given off during the concrete hardening period. This is called the heat of hydration. This heat, if not dissipated, can build up and cause expansion. This effect is not serious except in massive structures such as dams. In these cases, it is important to minimize the amount of heat generated by hydration. This can be done by using low-heat portland cements, by using concrete mixtures with low cement content, by precooling the materials to lower the temperature of the concrete as placed and, in some instances, by embedding cooling pipes in the mass concrete.

Continuous Moisture Exposure

Concrete subject to continuous exposure to moisture, such as in pipe lines and submerged marine structures, begins to expand after the initial shrinkage. This expansion is continuous for several years, with a total amount usually less than 0.025%. Concrete immersed in water has an expansion equal to about ⅓ of the contraction of the same concrete air dried for the same period. Under such conditions, Type I cement expands slightly more than Types II, III, IV, and V; mortars and concrete containing pozzolanic material usually expand slightly more under continuous moist storage than comparable mortars. and concretes made without pozzolanic materials. The amount of cement in a given mixture is of much greater importance than the type of cement, since the expansion of a neat paste is about twice that of an average mortar and the expansion of the mortar is about twice that of an average concrete.

Mechanical Forces

When a load is applied to concrete, the concrete deforms. If the concrete is in compression, as is most structural concrete, the result is a reduction in volume of the concrete. This reduction in volume depends on the magnitude of the load, the age of the concrete at the time of load application, the rate at which the load is applied, and the length of time that the load is acting on the concrete.

Creep. Like all structural materials, concrete has a certain degree of elasticity. This means that the concrete will compress if subjected to compressive loads, and will expand when subjected to tensile loads. Elastic length changes are, by definition, changes in structural dimensions due to loading and are instantly reversible when the applied load is removed.

Concrete, however, is not a perfectly elastic material. In addition to elastic deformation, it undergoes very slow length changes as long as loads are applied. This effect is known as creep. When the load is removed after a long period of time, the concrete recovers immediately from the elastic portion of the volume change and then, with time, recovers gradually from the volume changes due to creep. It probably will never completely return to its initial length before loading. When the loads are removed and the concrete structure creeps back toward its initial length, the process is known as "creep recovery."

In newly placed concrete, the change in volume or length due to creep is mostly unrecoverable. However, creep that occurs in old or dry concrete is largely recoverable.

The amount of creep is dependent on (1) the magnitude of stress, (2) the age and strength of the concrete when the stress is applied, and (3) the length of time the concrete is stressed. It is also affected by other factors related to the quality of the concrete and conditions of exposure. Some of these factors are the kind, amount, and maximum size of aggregate, the type of cement, the amount of cement paste, the size and shape of the concrete mass, the amount of steel reinforcement, and the curing and exposure conditions.

Under continuous load, creep continues for many years, but the rate decreases with time. Within normal stress ranges, creep is proportional to stress. The ultimate magnitude of creep of plain concrete per unit stress (1 psi) can range from 0.2 to 2.0 millionths in terms of length, but it is ordinarily about 1 millionth or less.

Figure 14-2 illustrates the effect of compressive strength on the amount of creep during a period of one year. Creep tests from which these data are taken were made with 6×12-in. concrete cylinders. They were loaded continuously in compression to a stress of 600 psi after moist curing for seven days. The stress was maintained uniformly for one year, while length changes were measured periodically. Total shortening due to creep at the end of the year was 730 millionths for the 3000 psi concrete (specimen A) and 490 millionths for the 4500 psi concrete (specimen B).

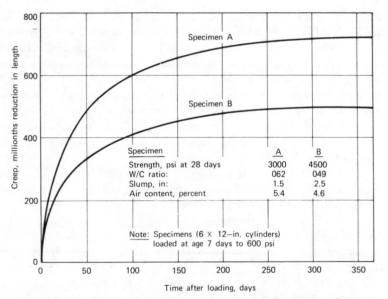

Figure 14-2 Relationship of creep, compressive strength, and period of load application for typical normal-weight concrete. Specimens were loaded at 600 psi. continuously after age 7 days.

An increase in strength results in a significant reduction in creep, as is indicated by these data.

Concrete specimens of equal strength, but of different ages, will have different creep characteristics. Those loaded at a late age will creep less than those loaded at an early age.

The method of curing prior to loading has a marked effect on the amount of creep of concrete. The effects of three different methods of curing on creep are shown in Figure 14-3. Note that very little creep occurs in concrete that is cured by high-pressure steam (autoclaving). Note also that atmospheric steam-cured concrete has considerably less creep than seven-day, moist-cured concrete. The two methods of steam curing shown in Figure 14-3 reduce drying shrinkage of concrete about half as much as they reduce creep.

Freezing and Thawing

Nondurable concrete subjected to freezing and thawing forces tends to expand slightly. This expansion is due to the freezing of water in concrete pores. It is the subsequent expansion of the freezing pore water that actually disrupts and expands the concrete surrounding it.

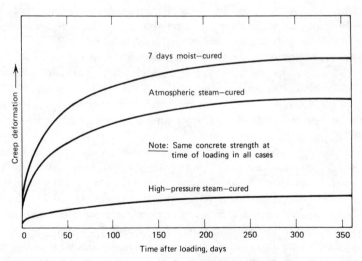

Figure 14-3 Effect of curing method on magnitude of creep for typical normal-weight concrete.

Exposure to Sulfates

Sulfate attack from exposure to sulfate soils expands concrete to the point where it may become disrupted.

Alkali-Aggregate Reactivity

Certain siliceous minerals (opal, tridymite, and chalcedony) react with the alkalis in cement to form silica gels. During the reaction, fluid pressure causes concrete made with a high percentage of reactive materials to expand disruptively.

Certain carbonate rocks may also react with cement alkalis. One of the current explanations for this is that the solid reaction products of this combination physically expand during the reaction, thereby causing disruptive pressures in the concrete.

MAGNITUDE OF VOLUME CHANGES IN CONCRETE

For convenience, the magnitude of volume changes in concrete is generally stated in linear rather than volumetric units. The volume changes that ordinarily occur are small, ranging in terms of change in length from a few up to about 1000 millionths.

Normal volume changes of concrete are caused by variations in temperature and moisture, and by sustained stress. Many factors affect the magnitude of volume changes, and reliable information is available on those factors that are the most important.

An average value for length change due to temperature variation is about 5.5 millionths (millionths of an inch per inch) per degree Fahrenheit, although values ranging from 3.2 to 7.0 have been observed. This average coefficient of thermal expansion or contraction is equivalent to a length change of 0.66 in. for 100 ft of concrete subjected to a rise or fall of 100° F.

Thermal expansion and contraction of concrete vary with factors such as aggregate type, water-cement ratio, temperature range, concrete age, and relative humidity. Of these, aggregate type has the greatest influence.

For ordinary conditions, an average thermal coefficient of 5.5 millionths per degree Fahrenheit is sufficiently accurate for most calculations. If greater accuracy is needed, tests should be made.

The thermal coefficient for steel is about 6.5 millionths per degree Fahrenheit, which is comparable to that for concrete. The thermal coefficient for reinforced concrete can be assumed as 6.0 millionths per degree Fahrenheit, the average for concrete and steel.

Concrete swells with a gain in moisture. If kept continuously in water, concrete slowly swells for several years, but the total amount of swelling is normally so small that it is unimportant (usually less than 150 millionths). Concrete that is not continuously wet is subject to water loss with resulting shrinkage.

Tests indicate that the unit length change due to drying shrinkage of small plain concrete (no reinforcement) specimens ranges from about 400 to 800 millionths when exposed to air at 50% humidity. This equals approximately the thermal contraction caused by a decrease in temperature of 100° F.

For many outdoor applications, concrete reaches its maximum moisture content during the season of low temperature. Thus the volume changes due to moisture and temperature variations frequently tend to offset each other.

The amount of moisture in concrete is affected by the relative humidity of the surrounding air. After concrete has dried to constant moisture content at one atmospheric condition, a decrease in humidity causes it to lose moisture and an increase causes it to gain. The cement paste, and the concrete of which it is a part, shrinks or swells with each such change in moisture content.

Alternate wetting and drying causes cycles of swelling and shrinking. The effects of these moisture movements are illustrated schematically in Figure 14-4. Specimen A represents concrete stored continuously in water from the time of casting, while specimen B represents another sample of the same concrete exposed first to drying and then alternate cycles of wetting and drying. For comparative purposes, it may be noted that the swelling that occurs during continuous wet storage over a period of several years is about one-third of the shrinkage of the concrete air dried for the same period. Note also that specimen B exhibits a net shrinkage after several cycles of wetting and drying. This is probably due to carbonation.

Shrinkage may continue for several years, depending on the size and shape of the concrete mass. The rate and ultimate amount of shrinkage are smaller for large masses of concrete than for smaller masses, although shrinkage continues longer for the large mass.

The shape of a structural member has an influence on the amount of drying shrinkage. Recent tests tend to indicate that shrinkage is a function of the volume to surface ratio of the members. The degree of corre-

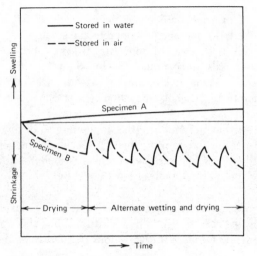

Figure 14-4 Schematic illustration of moisture movements in concrete. If concrete is kept continuously wet, a slight expansion occurs. However, drying usually takes place, causing shrinkage. Further wetting and drying causes alternate swelling and shrinkage.

lation found between volume to surface ratios and shrinkage and creep is satisfactory for purposes of practical design.

The rate and amount of drying shrinkage for 4×4×40-in. concrete specimens are shown in Figure 14-5. Specimens were initially moist-cured for 14 days at 70° F, then are stored for 38 months in air at 70° F and 50% relative humidity. Shrinkage recorded at an age of 38 months ranged from 600 to 790 millionths. An average of 34% of this shrinkage occurred within the first month. At the end of 11 months, an average of 90% of the 38-month shrinkage had taken place (Fig. 14-5).

MINIMIZING EFFECTS OF VOLUME CHANGES

There are two ways to control cracking caused by volume changes in hardened concrete: (1) minimize the volume changes, and (2) compensate for the effects of the volume changes. Shrinkage and creep are about the only volume changes over which control can be exercised and they are discussed above. This section considers what can be done to offset the effects of those volume changes that cannot be controlled.

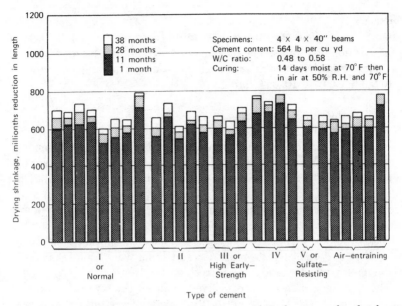

Figure 14–5 In these tests by the U.S. Bureau of Reclamation, the shrinkage ranged from 600 to 790 millionths after 38 months of drying.

If concrete survives the drying and curing stages without cracking, but develops cracks later, they may be because of improper design, jointing, reinforcement, or prestressing.

Design

Proper design of a structure is extremely important in minimizing cracking. If structural members are inadequate in load-carrying capacity, improperly placed or shaped, or the interaction of one member with another is not considered, cracking is likely to result, and complete failure is possible. Although structural design is beyond the scope of this course, design defects are often evident to even the untrained eye. Examples are a column improperly placed with respect to the beam it supports, lack of provision for proper drainage, and improper provision for expansion and contraction. Other examples of poor design are pavements and driveways that include narrow, feathered sections. These thin sections are not able to withstand the shrinkage and stresses that develop or the heavy loadings that they experience. In these instances, the best possible concrete and concreting practices will be of little help.

Jointing

Almost every concrete structure requires construction joints, control joints, or isolation joints. The design, construction, and location of these joints is extremely critical in minimizing cracking. This subject is covered in detail in Chapter 10, on jointing.

Reinforcement

The primary purpose of reinforcement in concrete is to increase the strength of structural members. Concrete is strong in compression, but weak in tension; while steel, on the other hand, is relatively weak in compression, but strong in tension. By combining the two materials, the best qualities of both are obtained. The use of reinforcement in structural members reduces cracking due to stresses. Although cracks still may develop, reinforcement limits the width of these cracks so that their effect on the structural member is negligible. Sufficient reinforcement should be used so that crack widths remain so small that water cannot penetrate to the reinforcing steel and cause corrosion.

The shrinkage of reinforced concrete is less than that for plain concrete, the difference depending on the amount of reinforcement. Steel reinforcement restricts, but does not completely prevent drying shrinkage. In reinforced concrete structures having normal amounts of reinforcement, the drying shrinkage is commonly assumed to be 200 to 300 millionths.

Engineering design manuals specify the amount of reinforcing steel required for various loads in all types of concrete structural members. In beams, girders, and columns, main reinforcing steel is placed parallel to the length of the member. This reinforcement minimizes cracking caused by loads on the concrete member, as well as cracks caused by volume changes in the direction of the reinforcement. However, volume changes occur in every direction; therefore, additional reinforcing is sometimes placed in the concrete to counteract these volume changes. Steel placed in concrete to counteract volume changes resulting from temperature variations is called temperature reinforcement (Fig. 14-6). This reinforcement should be placed near the surface, especially near edges where cracks may start. When only one side of the concrete is exposed to the elements, about 60 to 70% of the steel should be near the exposed side, unless the walls are thin. Special attention should be given to concrete aprons at doorways, to pavements of areaways, to sidewalks that adjoin foundations, to basement walls of placed concrete that are largely or entirely exposed to the atmosphere, as at the back of a

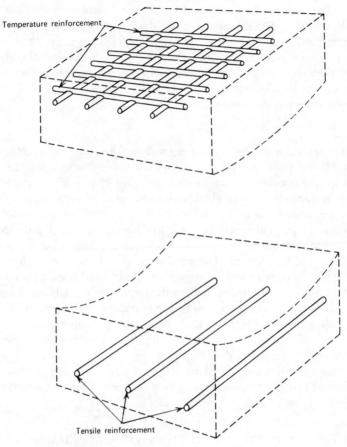

Temperature reinforcement

Tensile reinforcement

Figure 14–6

structure on a lot that slopes down toward the rear, and to long mono-lithic walls, particularly when they have a southern exposure in northern temperature zones. Structures that are narrow and high, such as piers, do not need steel to resist cracking due to temperature to the same extent as do long ones. They are free to expand or contract vertically, and their dead load will not let cracks open up.

Reinforcement to minimize cracking from drying shrinkage in concrete is a different problem. The drying shrinkage of the concrete, when rein-forced and after setting, is about equal to the volume change due to a drop in temperature of 30 to 80° F, but the reinforcement does not shrink as a result of drying. The shrinkage actually sets up compressive stresses in reinforcing rods so that if the percentage of steel is too great, the rods will be stronger in compression than the concrete is in tension, thereby producing the cracking that the reinforcement was designed to eliminate. The best way to handle drying shrinkage is to provide control joints to allow for the expected movement.

Prestressing

Prestressing is used to create very strong and slender concrete structural members. Highly tensioned wires or cables are utilized which, when released, place the concrete in compression and thus give beams a greater load-carrying capacity. This method is often too expensive to use merely for preventing cracks, but prestressed members are often free from cracking. Prestressing prevents cracking of lower fibers because it puts the entire concrete member in compression and resists volume changes and deformations due to loadings. The advantage of prestressing is that it places concrete in compression in areas that *would* have been subjected to a certain amount of tension, ordinarily requiring a relatively large amount of both steel and concrete to be structurally sound. A prestressed beam is more economical if it can be built thinner or shallower and still carry the same design loads. The prestressing force itself does not resist volume changes (creep and shrinkage)—in fact, these volume changes must be very carefully considered in design. There are fewer cracks in the lower fiber of a prestressed beam, solely because the precompression prevents tension from developing in the concrete.

CAUSES OF CRACKING OTHER THAN VOLUME CHANGES

The exact reason that a particular crack may have developed is often difficult to determine. There are a number of possible explanations. Once the crack has developed there is little that can be done about it, except to prevent the situation from becoming worse. However, if some of the causes of cracking are kept in mind during design and construction of the structure, potential cracking situations can be avoided.

Movement Before Final Set

While concrete is in the plastic state, shrinkage, expansion, and movement are not harmful. However, once it takes its initial set any movement is likely to cause cracking. This movement can occur for a number of reasons: unstable forms because of inadequate bracing and shoring, poor construction, miscalculating the loads that the forms must carry, inadequate provision for lateral pressure of freshly placed concrete, improper connections and bearings for a jointing formwork structure, inadequate bearing plates or mud sills, and premature removal of forms. These errors in forming often cause cracking or bulging of concrete and, sometimes, are responsible for the complete collapse of structures during construction.

Unstable Subgrade. Another cause of shifting before final set is unstable subgrade. This instability can be due to improper preparation because of inadequate studies of the soil conditions, miscalculation of the weight-supporting ability of the subgrade, hard or soft spots, or by conditions in the subgrade that did not exist at the time the building was designed.

Foundation designs are often based on widely separated test borings. During excavation and construction, the engineer should be at hand to check and confirm assumptions on which he based his design. If unexpected subgrade conditions are found, he should be ready to modify the foundation design. Unanticipated subgrade conditions have been the cause of cracking where footings were actually left suspended in soil without any bearing. Instead of providing support, the footing and the column above it acted in reverse, increasing the loads in surrounding bays and columns. This type of situation can cause complete structural failure. Another cause of cracking is movement of the subgrade. This can be the result of moving heavy construction equipment, moving near the forms, or bumping the forms themselves during construction of adjacent work.

Settling After Hardening

Subgrade settling is a common cause of cracking of structures after hardening and can be created by a number of conditions. Hard or soft spots in the subgrade below the structure, such as rock under one portion of the structure and sand under another, could cause uneven settling and possibly even collapse of the structure. Cracking can be caused by movement of soil and groundwater because of underground streams or springs, shifting of soil on slopes, or a variety of other causes. Expanding shales and clays also cause cracking of structures built on them. These materials expand in proportion to the amount of moisture they contain. If the subgrade on one side of a structure is kept dry by a driveway or adjoining pavement and the other side is subject to moisture, this can cause uneven expansion of the shale or clay underneath and can result in cracking. A large tree draws moisture from the soil surrounding it, and the effect of a large poplar tree, for example, can be noted as far as 35 ft from the trunk. It can draw moisture from soil on one side of a building and not on the other, thereby causing uneven settling and cracking.

Another cause of expansion of the subgrade is frost action. The formation of ice crystals, exerting pressures of several tons per square feet, is a cause of many problems. Frost penetration of a clay subgrade on which concrete footings are placed will heave the footings, if moisture is available to form ice crystals. Spring thaws will then bring settlement of the structure.

Rich Surface

An overrich surface is also a cause of cracking in concrete. Cracking from this cause can be identified by the size and configuration of the cracks. This type of cracking, called crazing, is identified by small cracks on the surface of the concrete arranged in an octagonal pattern. They usually appear in the first few days, and can deteriorate into a flaking surface. A rich crust can be caused by excessive early floating or troweling that brings water, cement, and dust from the aggregate to the surface, or by dusting dry cement on the surface of the plastic concrete. This wet, cement-rich material has a higher drying shrinkage than the underlying concrete. At this fine surface skin drys, it contracts extensively while the concrete below it contracts only slightly. The resulting tensile stresses create the characteristic surface cracking. Floating and troweling operations should be kept to a minimum to produce the required surface finish. These operations should be performed after the concrete has stiffened sufficiently so that excessive fines will not come to the surface. Dusting cement on the surface of the concrete to dry it prior to finishing should be prohibited.

Reinforcing Steel

Under certain conditions, reinforcing steel can cause cracks and spalls in the concrete over the steel. One of the causes of this cracking is rusting of the steel. The pressure exerted by this oxidation causes cracks in the concrete, particularly when the reinforcement is located near the surface. This is especially true when the concrete is exposed to seawater, to chemicals used for melting snow and ice, and when certain admixtures such as calcium chloride are used in excess. To prevent corrosion of reinforcing steel, an adequate cover of concrete over the steel should be provided. Ordinarily, a cover of 1½ or 2 in. is desirable for steel below large, flat surfaces exposed to air, 2 to 2½ in. at corners in air, and 3 in. when concrete is exposed to fresh water or moist earth. In seawater, 4 in. of cover is desirable. More detailed data are available in ACI 318-63, "ACI Standard Building Code Requirements for Reinforced Concrete."

When there is insufficient cover over reinforcing at changes in direction in the structure, cracks may occur. This condition concentrates stresses in the concrete near the surface, causing the cracking.

Reinforcing steel can also cause cracks in the surface when the concrete settles over it. This is because the reinforcing steel serves as a kind of pivot over which the concrete breaks as it settles. This can be prevented by using low-slump concrete, by vibration, and by making sure that the subgrade is stable or that the bar has more cover.

Stresses concentrate in the corners of structures, such as near openings for door and windows. Concrete should always be provided with reinforcement at corners to prevent such cracking. Again, adequate cover over the steel is important.

Nonferrous Metals

Nonferrous metals such as aluminum, copper, lead, and zinc are often used in construction and may come in contact with fresh and hardened concrete. When such materials are embedded, as in the case of conduits, pipe, and flashing, corrosion can be a serious problem.

Aluminum is almost sure to cause difficulties because it is corroded by alkalis present in fresh and unseasoned concrete. The corrosion is especially severe when aluminum conduit is embedded in concrete containing calcium chloride, and when steel is electrically connected to the aluminum. Conditions of this kind should be avoided. If aluminum must be embedded in concrete, it should be protected by a coating of asphalt, varnish, or pitch.

Copper is not affected by concrete unless admixtures containing chlorides have been used. It may be embedded safely under normal conditions.

Zinc is attacked by fresh concrete, but quickly forms its own protective film that prevents further corrosion. It is not affected by contact with dry, seasoned concrete.

Lead in contact with fresh concrete will corrode and cause severe localized retardation. It will also corrode in moist, hardened concrete when part of the lead is embedded and part is exposed to the air. Under such conditions, the lead should have a protective coating of asphalt, varnish, or pitch.

Aggregates, Near Surface

Large aggregate particles can cause cracks in the surface of the concrete in the same way as reinforcing steel placed too close can cause cracking. The phenomenon seems to occur when the aggregate particles form a skeleton through which the cement paste can settle and separate. The same preventive measures as for reinforcing steel are recommended.

Overloading

Excessive loading of concrete structures or pavements often increases the size of existing cracks as well as causing new ones. The remedies for this are to make sure that the structure is designed for any conditions it will subsequently encounter. In the case of old buildings that are converted for another use, often the original design limits are exceeded. In pavements or driveways, vehicles that exceed the design limits of the pavements are often allowed to pass. These overloads should be avoided if at all possible.

SHRINKAGE-COMPENSATING CEMENT

A new type of cement has recently been developed that has expansive properties. There are three types of expansive cement: Type K, which is an expansive cement containing anhydrous aluminosulfate blended with portland cement, calcium sulfate, and free lime; Type M, which is a mixture of portland cement, calcium aluminate cement, and calcium sulfate; and Type S, a cement containing a large amount of tricalcium aluminate and modified by an excess of calcium sulfate. In making concrete with expansive cement, all of the factors involved, including the composition of the cement, the proportions of the mixture, and the time and temperature of the curing, must be carefully controlled. Tests and evaluations of concrete made with expansive cements are currently being conducted. The data accumulated thus far indicate that the strength, volume change and creep characteristics, resistance to freezing and thawing, and abrasion resistance are all comparable to standard concrete. There is reason to believe that expansive cements will, in the near future, be used universally to overcome the chief fault of concrete—shrinking cracking. Expansive cement will never be a panacea and will not cover up poor design and construction practices. However, it will, when properly utilized, assist in the elimination of drying shrinkage cracks. The realization of the full potential of expansive cement will depend on the development of the necessary engineering technology and specifications.

Cracking of concrete is a highly complex subject. Cracks can be caused by a wide variety of conditions, and sometimes several contribute to the same crack. Prevention of cracking in concrete requires careful attention to detail and good concreting practices. A strong, attractive crack-free structure is a testament to the skill of all those who contributed to it.

SECTION FOUR
ESTIMATING CONCRETE

Chapters 15, 16, and 17 are intended as a complete guide to estimating the cost of concrete work. In addition, these chapters will serve as a reference for concrete estimators when actually compiling job costs. Pages 453 and 454 have two outlines of the three estimating chapters. Use the page references in these outlines to find the information you need when compiling an estimate. Page 453 should be used when making the quantity takeoff. The first page reference opposite each of the 20 work items is to the section in Chapter 15 that explains how to estimate the quantity of material required. The second page reference is to the section in Chapter 16 that gives man-hour estimates for the work described. Page 454 has an outline of Chapter 17 and will be useful when recapping and pricing each job.

Follow the order given on both pages 453 and 454 when estimating concrete work. Good estimating habits will prevent many estimating mistakes. Use the outlines as a checklist to avoid forgetting any part of the estimate. Of course, not every job will include all 20 items on the Quantity Takeoff Outline. Most jobs will have costs for all 7 items on the Recapitulation Sheet. Chapter 16 includes a series of estimating tables. Tables 16-A to 16-F give sample material prices and actual labor outputs per day for various types of formwork. The prices for materials are intended as samples only to help illustrate how an actual estimate might look. Convenient units have been selected to make understanding the tables easier. For example, these tables assume a cost of $500 per 1000 board feet of lumber and $1000 per 1000 square feet of plywood. These prices do not necessarily apply to any job. They are merely examples and should be replaced by actual material costs prevailing at the time the materials will be purchased. Likewise, a labor cost per day of $240 for carpenters and $200 for laborers has been assumed. Replace these figures with the actual cost for labor on the job you are estimating.

It may be convenient to modify the labor and material cost figures in tables 16-A to 16-F by a percentage to speed estimating. For example, if plywood actually will cost $900 per 1000 square feet, consider using 90% of the cost listed for plywood forms.

Table 16-G gives labor costs for placing concrete, setting screeds and finishing the surface. Both labor hours per unit and cost per unit based on assumed wage rates are shown. Again, modify the cost per unit to conform to the hourly cost of labor on your job.

Table 16-H is an example of a handy table for figuring the labor cost per unit for any kind of concrete work if the output per day is known. Several blank forms like Table 16-H are included so tables can be produced based on wage rates other than those given.

A sample estimate is included at the end of Chapter 17. This estimate has three pages of recapitulation and pricing and six pages of quantity takeoffs. Please note that the takeoff sheets cover only actual concrete quantities. Other takeoff sheets would be required to list the many other items that are included on the recapitulation sheet. Again, the prices given are intended to illustrate the procedure rather than provide actual prices for any area or time.

The information in this section was compiled by E. G. Le Jeune and is the product of over 20 years of experience in construction estimating. Much of the material in these chapters was originally published in *Concrete Construction* or was presented in lecture courses given by the author. The labor output per day figures are the product of the author's experience and many conversations over the years with concrete contractors. Naturally, every estimate in these chapters *is just an estimate* which may or may not apply on the job you are figuring. Only the estimator knows the conditions that exist on the job and can determine the correct labor output per day for each man on the job.

Quantity Takeoff Outline

Recapitulation Outline For Pricing

CHAPTER 15
QUANTITY TAKEOFF

To properly prepare a concrete estimate for bidding purposes the estimator must consider two facets of the estimate: (1) the quantity takeoff and (2) the pricing schedule. The quantity takeoff must be done by the same man who is to price the job or at least by someone who uses the same system of takeoff as the one who prices it. This is especially true in formwork quantities. Without this understanding between the "takeoff estimator" and the "pricing estimator" errors are introduced that can make the bid either too high (and therefore useless) or too low (and therefore costly to the successful bidder).

A quantity takeoff of all concrete items is the first step in preparing an estimate. These items are listed on the quantity takeoff sheets. Since the concrete yardage in a job is the most important item, this should be taken first along with related formwork, hand excavation, finishing and other items that can be gathered at the same time to expedite the takeoff. Most miscellaneous items can be worked out from the above quantities. No deduction from concrete yardage is made for reinforcing steel.

The next step is to prepare a recapitulation sheet. With a job properly taken off, the recap sheet can be prepared by either the "takeoff" or the "pricing" estimator since each understands the method followed. Items are listed on the recapitulation sheet under the following major headings:

1. Forms
2. Concrete
3. Finishing
4. Hand excavation and sand fill
5. Miscellaneous items

If reinforcing steel and mesh are to be included, add them as a sub-item.

The final step is the pricing of the work. This, of course, is *the* important step. Many estimators can be trained to do takeoff work. Very few are qualified *pricing* estimators. A qualified pricing estimator should be able to show how he arrived at every unit price he used. The unit price may be off to some degree, but the estimator should have a logical reason for using it.

If you are a small contractor and if all your employees are close friends, you can figure on a high rate of labor production. Those two "if's" are very important to your labor costs. Since you're getting much more work out of your men than the average contractor, you can do the work for a lower cost than the average contractor. Once you employ more than about a dozen men, you should figure work at an *average* rate of production because the average man will not give you maximum production.

The labor costs in this book are based on average labor output in units per day and an assumed local wage scale per day. With these figures (and experienced common sense) it is possible to figure labor costs for this year, next year, or any year.

The Quantity Takeoff Outline should not be unfamiliar to contractors. Here, it is in a form that logically follows all steps in taking off a job. For example, if you are figuring only slab work, you skip everything else on the outline that does not apply. There is no one job that has all 20 items on it.

The table on estimating wall forms up to 4 feet high shows that a man should be able to form about 300 square feet a day. If you know that you get 400 or only 200 square feet a day from your crew, you can still use this system. Simply apply your own figures and use your own labor rates to come up with the correct labor cost per square foot. The tables base all formwork erection costs on a carpenter's and half a laborer's rate. This assumes that you run the average crew with a ratio of two carpenters to one laborer and applies to all of your larger jobs and for many of your smaller jobs. However, you may use one carpenter to two laborers. You can still use this method. Simply add a carpenter's rate to two laborers and use his production. But remember that you may not get the same production out of a carpenter and two laborers as with two carpenters and one laborer.

The following method is highly recommended for checking extensions and additions. Whenever extending multiplication on a machine, always mentally calculate the approximate answer while the machine is getting the exact answer. This prevents a decimal point error. For instance, if the answer should be 225, you might get 240 if you punched the wrong number on the machine and you wouldn't know it. You couldn't get 24, because you would know in your head that you have an answer coming in the hundreds. In other words, if you are multiplying 158 by 112, in your head you know that 150 times 100 is 15,000, so you had better be somewhere close to that in the thousands, and if you get about 1,500, you know you are wrong.

When adding a column of figures, add the total down the column of figures and write the answer. Then add the total in reverse, or up the column of figures, and put a checkmark at the first answer if the totals agree. It is very difficult to have an error in addition using this method. Checking a column of figures against an adding machine tape introduces one more possibility for human error.

COLUMN OR BOX FOOTINGS

Start the job by looking for a column footing schedule. If given, list the description and dimensions from the schedule. Then turn to the footing plan and count the number of the footings. This makes sure that you include all of the footings and not just the ones listed in the schedule. This list should include all box footings which, in your judgment, would have to be located and staked out separately from the wall footings. It is important to list the total number of footings in the last column so that this information will be carried forward to the recapitulation sheet to aid in pricing.

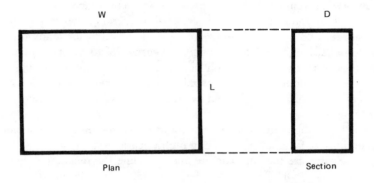

Figure 15-1 Column Footings or Box Footings—width, length, and depth.

Table 15-A
COLUMN FOOTINGS OR BOX FOOTINGS

Description	Dimensions	Square Feet Area	Square Feet Forms	Cubic Feet Concrete	Estimated Quantity
Column					
Footings:	W x L x D	WxL	(2W+2L) D	WxLxD	
F1	6 x 4' x 4' x 1'-0	96	96	96	
F2	3 x 3'-6 x 3'-6 x 1'-0	37	42	37	
F3	2 x 7'-6 x 7'-6 x 1'-6	113	90	170	
F4	2 x 6' x 8' x 1'-8	96	94	161	
Chimney					
Footing	1 x 10' x 8' x 2'-0	80	72	160	
		−2	−4		
Totals	14 Footings	420 ×̷	390 ×̷	624 ×̷	
				27	

Concrete = 24 C.Y.

Footing Forms (14) = 390 S.F. √

Hand Excavation Area = 420 S.F. ╲

The Outline begins with the heading "Column Footings: hand excavation area, forms, concrete." Column footings or box footings are individual footings, not anything that is continuous under a wall. Wall footings are covered under item 2. Under column footings pick up all the individual footings including a footing for a chimney. It might be 6 feet by 4 feet by 2 feet thick. Pricewise, this is an individual footing.

It should be pointed out here that the quantity takeoff is separated by items in order to get everything into the correct price bracket. In other words, a 4-foot by 3-foot by 1-foot column footing is in the same price bracket as a 6-foot by 4-foot by 2-foot chimney footing. The system is flexible because under the heading "column footings" you can include any other individual footings you wish to include in the same price bracket.

For the hand excavation area always take the exact area of the footing. That is, the machine excavator leaves the earth cut down roughly to plus or minus 1 inch of the bottom of the footing. Hand labor must be used to cut the bottom down to the needed grade. Always take the exact area, for the following reason: the width times the length is the footing area, and if you multiply this by the depth you have the concrete volume. You know that the laborers will hand excavate a little wider than the exact footing area, but you take this into account with your pricing unit. Concrete is not a simple thing to take off. You must break it down, step by step, as it would be built.

This method is not new or revolutionary. All good contractors follow this system in one form or another. But the point is this: the quantity takeoff should be simplified as much as possible in order to get the answers as quickly as possible.

To determine the formwork required for column footings, take twice the width plus twice the length times the depth of the footing. This gives the contact area of forms required. All formwork should be taken off and priced on the basis of contact area. This will make it easier to take off and recapitulate quantities, and will make for a better understanding between the takeoff estimator and the pricing estimator if they are separate persons.

Write down the number of column footings required of each size on the quantity takeoff sheet. From these figures extend the totals for hand excavation area, concrete volume, and formwork areas. These three totals are extended to your last column on the right and checked off thus

(√) to show that they are to be extended again later to your recapitulation sheet for pricing. Never list any quantity in the last column to the right unless it is to be moved to the recapitulation sheet for pricing. This will make these quantities stand out and prevent errors of omission on the recapitulation sheet. Always list the total number of column footings. It is easier to price the labor on column footings if you know how many footings are involved, and how many footings of a certain size a man should form in a day.

WALL FOOTINGS

These examples in Table 15-B are for different types of wall footings. No specific building is intended. The widths and depths are listed to the closest inch and the lengths to the closest foot. The area is listed as the exact concrete area to help in getting the concrete volume by multiplying the area by the depth. The cubic feet of concrete in the last column are listed to the closest foot above any fraction. The keys are listed to the closest linear foot. Place an "x" in front of the quantity listing for the keys because this item is to be kept separate from the form items that will be added up on the recapitulation sheet for "stripping and cleaning" by laborers. Check each item in the last column as it is brought over from the takeoff sheet to the recapitulation sheet with a double check (𝗫) to avoid omitting any of them.

The system used for figuring column footings should also be used to figure wall footings. Take off the quantities for hand excavation areas, forms, keyways, and concrete. When you take off wall footings the first thing to do is to separate all the wall footings that are the same as far as width and depth are concerned. That is, check off all 1- by 2-foot footings, and if you find 50 feet here and 60 feet there and 40 feet somewhere else, add them together.

Have your own system of check marks on the drawings. Don't mark up drawings with different colored pencils, partially because of the trouble it might create for the next man who uses them but mostly because it is a waste of takeoff time. Needless time can be spent in hunting for colored pencils, trading your black pencil for the colored one, marking the drawings and then finding your black pencil again.

When you have taken off and marked all of the wall footings on your quantity takeoff sheet, it is easy to extend the totals for hand excavation area, formwork, keys, and concrete from the one set of quantities. The

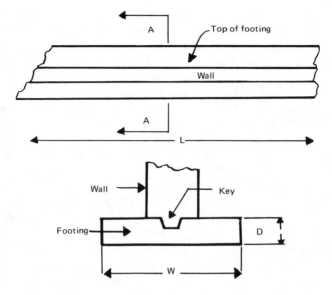

Figure 15-2 Wall footings.

Table 15-B
WALL FOOTINGS

Description	Dimensions	Square Feet Area	Square Feet Forms	Cubic Feet Concrete	Estimated Quantity
Wall					
Footings:	D x W x L	WxL	2DxL	WxLxD	
Line A	1'-0 x 2'-0 x 56'	112	112	112	
Line B	0'-10 x 1'-8 x 48'	80	80	67	
Line C	1'-0 x 3'-0 x 56'	168	112	168	
Line D	1'-2 x 3'-8 x 48'	176	112	206	
		+4	+4		
	Keys = 208	540 ⨯	420 ⨯	553 ⨯	
				27	

Concrete = 21 C.Y. ⨯
Forms = 420 S.F. ⨯
Hand Excavation Area = 540 S.F. ⨯
x Keys = 208 S.F. ⨯

hand excavation area, once again, is the exact width of the footing times the length so that you can take the area and multiply it by the depth to get the concrete required. This is a time saver. The forms, of course, will be twice the depth times the length. The keys are figured as the total length of all wall footings. These four items, hand excavation area, contact form area, keys in linear feet, and concrete in cubic yards, are extended into the last column on the right and checked ($\sqrt{}$) for future extension again to your recapitulation sheet for pricing.

The keys are not reusable. Therefore you will have to figure the material cost for new lumber at the price per linear foot plus labor for installing and removing the keys after the concrete has been placed.

The concrete quantity is always the most important item taken off. On most items, such as formwork, you can afford to be a little too high, but concrete quantities should be obtained as accurately as possible and an allowance for waste included where required. Include about 3% waste for concrete placed on the ground. Do not include waste, however, for concrete placed in forms.

All estimating is "guesstimating." Try to tie down precisely those items of cost that can be tied down. Concrete material is such an item. Depending on the specifications, concrete material will always have an exact cost, and since this cost will be about one-third to one-quarter of your total concrete bid, it is worth tying down exactly. If an item calls for 31¼ cubic feet of concrete, call it 32, picking up part of a cubic foot on each item. This is how to include waste on the job. But you should still add 3% additional waste for all concrete placed on earth or fill.

FOUNDATION WALLS ON FOOTINGS

The walls in Figure 15-3 and Table 15-C represent typical conditions. A is a typical wall below frost line. B shows that a wall is measured to the fraction of an inch for width and to the closest inch for height. The basement walls show that where a brick ledge exists, the wall must be figured the full height of 9 feet 6 inches. Then add in additional forms for the brick ledge area and subtract the concrete.

Slab seat bearing is figured in linear feet; then subtract for the concrete. Door and window box-outs have the additional forms added and, again, subtract for the concrete openings. The deductions for the concrete are in parentheses to call attention to the subtraction. The slab

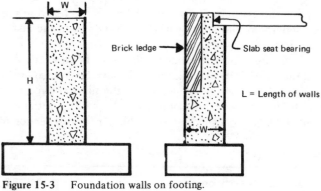

Figure 15-3 Foundation walls on footing.

Table 15-C
FOUNDATION WALLS ON FOOTINGS

Description	Dimensions	Square Feet Forms	Linear Feet Slab Seat Bearing	Cubic Feet Concrete	Estimated Quantity
Foundation Walls on Footings	W x H x L	2HxL	L	WxHxL	
Line A	1'-0 x3'-6 x 78'	1. 546	—	273	
Line B	1'-2½ x7'-10x 40'	2. 627	—	380	
Bsmt:					
(at Brick Ledge)	1'-0 x9'-6 x156'	3. 2964	—	1482	
(for Brick Ledge)	0'-4 x3'-0 x156'	1. 468	—	(− 156)	
				= Form 1 side	
Seat Bearing	0'-4 x0'-6 x156'	—	156	(− 26)	
Door Box-Out	1'-0 x7'-0 x 4'	2. 18	—	(− 28)	
Window Box-Outs 6 x 0'-8 x3'-0 x 4'		1. 60	—	(− 48)	
		−3			
		4680 ⨯	156 ⨯	1877 ⨯	
				27	

Concrete = 70 C.Y. ⨯

4"x6" Slab Seat Bearing = x 156 L.F. ⨯

1. 0' to 4' H. Forms = 1070 S.F. ⨯

2. 4' to 8' H. Forms = 650 S.F. ⨯

3. 8' to 12' H. Forms = 2960 S.F. ⨯

468 Square Feet x ½'/Square Feet + 10% = Dove-Tail Slots for Brick Ledge = 260 L.F. ⨯

seat bearing is listed in linear feet for ease in pricing and is not added into "stripping and cleaning" of forms. All walls are listed in the same takeoff and separated into 4-foot variations in height in the last column for ease in pricing when they are extended to the recapitulation sheet.

Dove-tail slots for nailing to forms (for brick anchorage) are also easily picked up at this time. If the foundation walls have pilasters on them, add in the additional formwork and concrete for the pilasters and mark that section of wall. For example, "0-foot to 4-foot-high forms with pilasters." The material and labor cost will be figured higher for this section of wall.

GRADE BEAM WALLS

All walls that bear directly on earth belong under this group. They are separated from foundation walls on footings because of the need for hand excavation and also because a 2- by 6-inch leveling plate is usually needed on the ground before the forms are set. The cost of this leveling plate is most easily put in the wall forming unit prices. Line A in Table 15-D shows a typical grade beam on caissons; B shows a wall less than 12

Table 15-D
GRADE BEAM WALLS

Description	Dimensions	2' Minimum Width Hand Excavation Area— Square Feet	Square Feet Forms	Cubic Feet Concrete	Estimated Quantity
Grade Beam					
Walls:	W x H x L	(W+1')xL	2HxL	WxHxL	
Line A	1' x3' x 50'	100	1. 300	150	
Line B	0'-8 x6' x150'	300	2. 1800	600	
Line C	1'-4½x7'-11x 83'	197	2. 1314	901	
Line D	1'-6 x8'-6 x154'	385	3. 2618	1964	
		−2	−2		
		980 ✓	6030 ✓	3615 ✓	
				27	

Concrete = 134 C.Y. ✓
Hand Excavation Area = 980 S.F. ✓
1. 0' to 4' H. Forms = 300 S.F. ✓
2. 4' to 8' H. Forms = 3110 S.F. ✓
3. 8' to 12' H. Forms = 2620 S.F. ✓

inches; *C* is shown to illustrate that the width is carried to the closest fraction of an inch, the height to the closest inch, and the length to the closest foot. Line *D* shows a wall over 8 feet high.

FOUNDATION WALLS IN GENERAL

On foundation walls take off the quantities for forms (in 4-foot heights), keys, slab seat bearing, and concrete. Separate the forms every 4 feet by height for pricing and use one price for material and labor for walls up to 4 feet high. From 4 feet to 8 feet, use a higher price. From 8 feet to 12 feet, 12 feet to 16 feet, and over, use increasingly higher unit prices per square foot of wall forms. This is because more bracing is required for higher walls and the additional cost is in both labor and material. Many contractors do not believe in breaking up their foundation wall takeoffs into different heights, but it makes a tremendous difference in cost. For example, if a 4-foot-high wall form is on a footing, you can figure a carpenter will erect about 300 square feet a day. If the wall form runs between 4 feet and 8 feet high, a carpenter can erect about 250 square feet a day; from 8 feet to 12 feet, a carpenter can erect about 160 square feet a day. This will give you some idea of the difference between a 4-foot-high and a 12-foot-high wall, as far as erection labor alone is concerned. Of course, the amount of bracing material required increases with the height of the wall. These figures are for erection labor only. They do not include removing the forms. Figure form stripping separately because laborer time only is usually used for stripping. This isn't always possible because, theoretically, the carpenters should to do the stripping if the formwork is going to be reused on the job. But most of the time laborers do it.

On form erection, figure a labor ratio of one carpenter to one-half laborer time. The cost per day for a carpenter should always include the cost of one half a day's time for a laborer. If a capenter is to set up 300 square feet of forms per day, he will need one-half day of a laborer's time to help him. The time might be spent handling forms and lumber, unloading trucks, or other ways of helping the carpenter. How to figure carpenter production per day and the carpenter-and-laborer-cost ratio and how to apply it are all discussed fully in Chapter 17.

Should you take off wall columns separately or take them off as part of the wall? This depends on how big the column is and how much extra work it causes. If a lot of work is involved, take it off separately and consider it a column to assign it to the correct price bracket. If it is only about 4 inches wider than the wall by about 1 foot long, consider it a pilaster and include the forms and concrete with the wall. If a wall has pilasters every 20 feet, do not take them off separately; they are not a large enough cost item. Figure the wall forms straight through the pilasters and then take off the pilasters separately and let the formwork double up. This simplifies the takeoff, and the extra forms will pay the difference in your pilaster cost. When pricing wall forms with pilasters, deduct 10% from the erection time used in figuring a straight wall without pilasters.

Grade beams—foundation walls without a concrete footing under them—must be treated separately. Take them off separately so you can include the cost of a mud sill in the square feet of wall area. Do this by figuring 10% more in the material cost and 10% less in the square feet erected per day. For simplicity in pricing, take off a slab seat bearing on a wall by total linear feet. These usually run about 4 inches by 6 inches. If they are not too high, brick ledges in walls should also be taken off for pricing by the linear foot. If brick ledges run 2 feet or more in height, it will most likely be necessary to place one form inside another. In other words, you will have to double-form the wall. The best way to account for this in takeoff is to double the form area. Be sure to deduct the concrete volume for this area. Try to be accurate on the concrete volume because concrete makes up a high percentage of the estimate's total cost. Take off a retaining wall with a battered face as follows: start first with the straight side of the retaining wall. This falls in the regular price for wall forming work. Then take off the battered face separately. This is a distinct problem. You can erect the straight wall and brace the other side off it, but that battered face is still a special labor cost problem. The cost of erecting the battered face can be figured closely if you reduce the production output per day on the battered face by about 20% from the estimated output on a straight wall of the same height. In wall forming takeoff work, the takeoff estimator must have notations as to the type of work that must be priced. This is needed whether he prices it himself or, more importantly, the work is to be priced by another estimator.

The form area for walls is twice the height times the length. Concrete volume in cubic feet is the height times the length times the thickness, all in feet. No deduction for reinforcing steel should ever be made from the concrete volume. In listing walls on the takeoff sheet the length should be given to the closest foot, the height to the closest inch, and the thickness to the fraction of an inch. A wall that is 12½ inches thick should be figured as a decimal—1.04 feet thick. This is necessary to get the correct concrete quantities. These same rules apply, in general, for all concrete takeoff work.

UNDERPINNING

To underpin an existing foundation wall, the work is usually done in alternate 4-foot long sections as shown in Figure 15-4A. All sections marked 1 are completed before the sections marked 2 are started. If the architect shows every third section to be done at one time for additional safety, it merely makes three separate operations of underpinning.

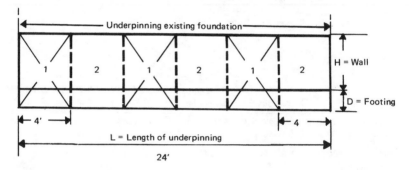

Figure 15-4A Underpinning.

The hand excavation must be figured from an estimated machine excavation line usually at a 45° angle from the bottom of the existing wall. See Figure 15-4B. This is slow work because the section of the wall is only 4 feet long and it is important not to disturb the ground beyond the inside of the existing wall. Forms are figured for one side of the wall and footing. The concrete footing is usually regular concrete and the wall is specified to be a dry-pack concrete. Sometimes the wall is regular concrete stopped about 4 inches below the existing wall and allowed to set and take its initial shrinkage.

The top 4 inches under the existing wall is then packed with a dry-pack grout which has little or no shrinkage in drying. The quantities

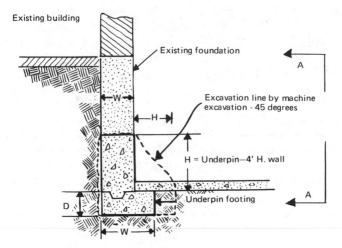

Figure 15-4B Underpinning. Elevation A-A.

Table 15-E
UNDERPINNING

Description	Dimensions	Cubic Feet Hand Excavation	Square Feet Forms	Cubic Feet Concrete	Estimated Quantity
	D				
Underpinning:	W x H x L	WxHxL	HxL	WxHxL	
Hand Excavation	4' x 4' x24'	192	—	—	
From 45° Line	2	—	—	—	
Wall	1'-6x 4' x24'	144	96	144	
Footing	2'-0x1'-6 x24'	—	36	72	
Footing Excavation	3'-0x1'-6 x24'	108	−2		
		444 ✕	130 ✕	216 ✕	

$$\frac{216}{27}$$

Concrete = 8 C.Y. ✕
Forms = 130 S.F. ✕
444 ÷ 27 = Hand Excavation = 17 C.Y. ✕
2 x 4 Keys = 24 L.F. ✕

are calculated as if the underpinning would all be done at one time. See Table 15-E.

All of these quantities must be clearly marked for "underpinning" on the recapitulation sheet. This is so that the pricing estimator can make an allowance for the slow work in his labor units, because only 4-foot lengths of wall and footing can be done at one time.

PIERS BELOW GRADE

The form material for piers below grade is usually 1-inch dimension lumber and not expensive. But the labor of forming piers is high because of the small area of forms per pier. Therefore, piers must be taken off as a separate item. The labor of placing concrete in piers is also high because of the small amount of concrete per pier. Marking the number of piers to be formed (13 in Table 15-F) is important. It tells the pricing estimator

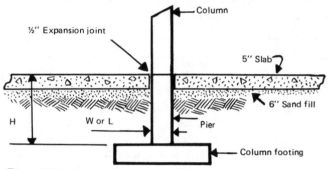

Figure 15-5   Piers below grade.

Table 15-F
PIERS BELOW GRADE

Description	Dimensions	Square Feet Forms	Cubic Feet Concrete	Estimated Quantity
Piers Below Grade:	W x L x H	(2W +2L) H	WxLxH	
F1	6 x 1' x 1' x2'-6	60	15	
F2	3 x 1' x 1' x3'-0	36	9	
F3	2 x 1'-0 x1'-4x5'-9	54	16	
F4	2 x 0'-10x1'-6x5'-9	54	16	
			−4	
	13 ✓	200 ✓	56 ✓	

$$Concrete = \frac{56}{27} = 2 \; C.Y.$$

(13) Forms = 200 S.F.

At Slab: ½"x5" Expansion Joint = 56 L.F.

the average square feet of forms per pier (here, 15 square feet). If the pricing estimator allows one hour carpenter time per pier then he can figure 8 hours times 15 square feet, or 120 square feet of pier erection per day. In fact, the number of piers to be erected sets the allowance for forming labor in square feet per day. The ½- by 5-inch expansion joint around the piers at the slab is most easily taken off under this item.

Estimating form area and concrete for piers below grade is relatively simple. See Figure 15-5. Piers below grade are piers that will have earth backfill around them or are in an unfinished space. These piers can be formed with rough lumber, and this is why they are kept separately from columns. The formwork is figured by taking twice the width plus twice the length times the height—or, once again, the contact area. The concrete volume is calculated by multiplying the width times the length times the height, as in Table 15-F.

BUILDING SLABS ON FILL

Building slabs at different elevations, such as basement floors and slabs at ground floor level, should be kept separate for estimating because of the different labor cost per cubic yard for placing these slabs. See Table 15-G.

If all slabs are of the same thickness, only the quantity for the areas should be extended, and the concrete quantity may be obtained by multiplying the total area by the slab thickness. An estimator should always look ahead for any time-saving steps he can use.

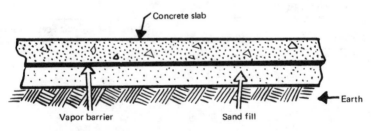

Figure 15-6   Building slabs on fill.

The hand grading of earth under fill is needed only when the ground is excavated to provide for the slab. See Figure 15-6. If the slab is to be placed on fill material furnished under the excavation specifications, then

Table 15-G
BUILDING SLABS ON FILL

Description	Dimensions	Square Feet Area	Cubic Feet Concrete	Estimated Quantity
Bldg. Slabs on Fill:	W x L x D	WxL	WxLxD	
E. Wing	40' x 63' x 0'-5	2520	1059	
Center	44' x 36' x 0'-5	1584	666	
W. Wing	40' x 63' x 0'-5	2520	1059	
Boiler Room	21' x 21' x 0'-8	441	296	
Entries	2 x 6' x 8'-6x 0'-6	102	51	
		3		
		7170 ✓	3131 ✓	
			27	

Concrete =   116 C.Y. ✓
Hand Grade Earth Under Fill = 7170 S.F. ✓
7170 S.F. x 6" = 3585 C.F. = 133 C.Y. + 25% Compaction = 6" Sand Fill = 166 C.Y. ✓
(10% Added for Lap) ≈ .006 Polyethylene Vapor Barrier = 7900 S.F. ✓
Finish Slabs = 7170 S.F. ✓
6" H. Edge Forms = 42 L.F. ✓
½" x 5" H. Expansion Joints = 672 L.F. ✓
½" x 8" H. Expansion Joints = 84 L.F. ✓

the work required of the concrete contractor would be the final hand grading of the fill material under the slab. Whenever a fill is specified (such as a 6-inch sand bed under the slab) it should be figured in cubic yards to which 25% should be added to cover compaction and loss in delivery. The polyethylene vapor barrier should have 10% added to the area required to take care of laps and waste. The necessary edge forming is easily taken off from the drawings at this time and listed to the closest linear foot. Expansion joints are obtained by referring to both drawings and specifications and should be listed to the closest linear foot.

The finish area is the total area of the slab. The concrete required may be estimated by multiplying the area of the slab times the thickness. Do this in decimals of a foot; for example, 5 inches is .42 feet. When figuring concrete for a slab on fill add 3% for waste. Also be sure to consider screed material and curing and protection items. These items are priced by the square foot of slabs. Membrane curing is priced by the exact slab area, but if paper or polyethylene materials are used add 10% for lap and waste. Most of these items are included under the last part of recapping under miscellaneous items. They can generally be picked out from the takeoff sheet without referring to the drawings. However, pick off any items you can from the drawings as you go along. This is one more timesaver.

EXTERIOR AND INTERIOR COLUMNS

All columns are figured from the floor slab below to the under-side of the beam or slab above—in other words, the contact area. The formwork is twice the width plus twice the length times the height of the column. The concrete is the width times the length times the height of the column. Note that in the quantity takeoff illustration, Figure 15-7, exterior columns are separated from interior columns even though both are tabulated on the same sheet. See Table 15-H. Both columns are figured the same way; and they might be grouped together, except that if you have too many exterior columns or too many interior columns it isn't easy to find an average price since they do not cost the same. The two types of columns cost about the same for material, but erection labor costs considerably more for exterior columns. Exterior columns must be braced on one side only, and this takes more time. Figure about 10% more labor for erecting exterior columns. Note that it is best to use a 10% differential when switching from one class of work to another. It is not difficult to vary your thinking 10% for how much work a man can or cannot do on a certain type of job. But it is difficult to say that a man can do 5% more of one type of job than he can of another. Create a separate class of work if the differential is more than about 10%. That is why on wall forms it is best to break them up into 4-foot heights; there is about a 10% differential on the forming labor. The columns are figured to the bottom of the beams so that when the beams are taken off we can figure them straight over the columns.

For column capitals or heads that vary in dimension, figure the average area times the height for concrete volume. The average area is the area at the bottom plus the top area divided by two.

Take off round columns by figuring either paper forms or steel forms separately. If paper forms are used, take off the various diameters separately and list both the number of individual forms and the total linear feet. The material must be purchased by the linear foot and, of course, may be used only once. The labor cost of erecting paper tubes is figured by the carpenter hour to erect and remove one individual column form.

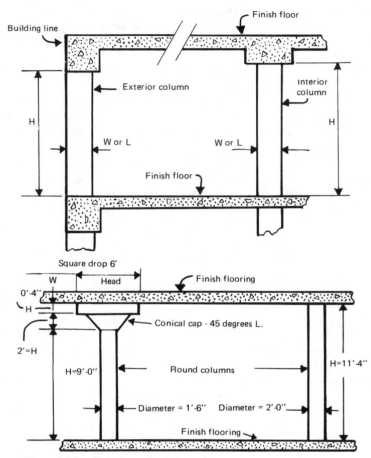

Figure 15-7   Exterior and interior columns.

If round steel column forms are used, there are companies that charge a single price for rental, erection, and removal. The concrete volume in round columns equals the area times the height. The area of a round column is 3.14 times the radius squared. The radius is one-half of the diameter. For example, if a round column has a diameter of 2 feet and is 8 feet high, the concrete volume is 3.14 times the radius squared times the column height, or 3.14 times 1 foot times 1 foot times 8.

Table 15-H
EXTERIOR AND INTERIOR COLUMNS

Description	Dimensions	Square Feet Forms	Cubic Feet Concrete	Estimated Quantity
Exterior Columns:	W x L x H	(2W+2L) H	WxLxH	
All Sides	12 x 1'-0 x 1'-6 x 8'-6	44 510	13 153	
Corners	4 x 1'-0 x 1'-0 x 8'-6	34 136	9 34	
		+4		
		650√X	187√X / 27	
				Concrete = 7 C.Y.√X
				Forms = 650 S.F.√X
Interior Columns:	W x L x H			
At Beams	6 x 1' x 1' x 8'-6	34 204	3 51	
At Slab	4 x 1' x 1' x 9'-0	34 144	9 36	
		2		
		350++	87√X / 27	
				Concrete = 4 C.Y.√X
				Forms = 350 S.F.√X

Description	Dimensions		Linear Feet 1'-6 Diam. Form	Linear Feet 2' Diam. Form	Edge Form Drop Head	Area x H Cubic Foot Concrete
Round Columns:						
Round Columns with Caps =	10 x 1'-6 Diameter	x 9'-0	90	—	—	158
Conical Caps =	10 x 1'-6 to 5'-6 Diameter	x 2'-0	—	—	—	192
Drop Heads =	10 x 6' x 6'	x 0'-4	—	—	240	120
Round Columns No Caps	10 x 2' Diameter	x 11'-4	—	114	—	360
			90√X	114√X	240√X	830√X= / 27

Concrete = 31 C.Y.√X
4" Edge Form = 240 L.F.√X
Material—2' Diameter Forms = 114 L.F.√X
Material—1'-6 Diameter Forms = 90 L.F.√X
Erect 1'-6 Diameter x 9' Forms = 10 Each√X
Erect 2'-0 Diameter x 11'-4 Forms = 10 Each√X
Erect Conical Heads = 10 Each√X
MATERIAL NOTE: Rent Metal Forms or Purchase Fiber Forms

On square column forms figure three reuses unless it is a small job where one use is all you can get and the lumber becomes scrap material.

For 20-30 story high-rise buildings, it is possible to figure about five reuses of form material. But if you plan to reuse square columns, consider that you can not reuse them just as they are because there may be a slight difference in floor height. Even if there is no height variation, they must be taken apart and put back together, and you will lose some of your bracing lumber. To figure reuse, add about 20% to the material cost to cover the cost of moving from one floor to another. Chamfer strips for exposed column corners can be added into the square foot price of column forms. Wood chamber strips will cost about 8 cents per square foot of column form for material. Plastic chamfer strips are more expensive, but should be good for more than one use. On square columns take into consideration whether they're going to be covered up with masonry walls. If they are, you can use old plywood to form them. Keep this in mind when you're pricing columns. If the finished columns are going to be exposed and have to be hand rubbed, then use a good grade of plywood and figure a little more for material cost. If they have chamfer strips on the corners, include these in the square foot price of the forming. All column forming must be analyzed in takeoff so that when pricing the formwork all necessary information is present.

EXTERIOR AND INTERIOR BEAMS

Include only contact area for beam forms. In a beam table as given on structural concrete drawings, the depth given includes the slab depth (or dimension D1 as shown in Figure 15-8). In order to obtain dimensions D2 or D, the slab depth must be subtracted from the depth given in the beam table or schedule. Exterior beams are kept separately from interior beams for pricing purposes. Both the material and the labor for exterior forms will be about 10% higher than for interior forms.

These two can be discussed together, but they must be figured separately. See Table 15-I. The formwork for a beam is the contact area, including the beam bottom and the two sides added together times the beam length. The concrete is the width times the depth below the slab times the beam length. The beam length is figured continuously over all columns. Keep exterior beams separate from interior beams because there is a considerable cost variation involved in framing beams. An exterior beam needs an outrigger knee brace to support the outside beam form. It is also more difficult for men to work on an exterior beam.

When figuring beams never base the depth on that given in the beam schedule. The depth in the beam schedule is for design purposes only,

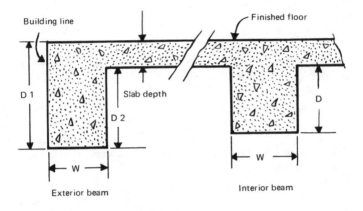

Figure 15-8   Exterior and Interior beams.

Table 15-I
EXTERIOR AND INTERIOR BEAMS

Description	Dimensions			Square Feet Forms	Cubic Feet Concrete	Estimated Quantity
				$(W+D1+D2)$ $\times L$	WxD2xL	
Exterior Beams:	W	D1/D2	L			
N. & S.	1' x	1'-6/1'	x 100'	350	100	
E. & W.	0'-6 x	1'-6/0'-6	x 100'	250	25	
				600 ✓	125 ✓	
					27	
					Concrete =	5 C.Y. ✓
					Forms =	600 S.F. ✓
				Square Feet Forms	**Cubic Feet Concrete**	
Interior Beams:	W	D	L	$(W+2D)\times L$	WxDxL	
N. & S.	3 x 1' x	1'	x 100'	900	300	
E. & W.	4 x 0'-6 x	1'	x 100'	1000	200	
Stairs	2 x 0'-6 x	0'-6	x 25'	75	13	
				1975 ✓	513 ✓	
					27	
					Concrete =	19 C.Y. ✓
					Forms =	1975 S.F. ✓

and is always figured from the top of the floor slab to the bottom of the beam. For takeoff work the estimator needs the contact area. This is figured by subtracting the thickness of the floor slab from the depth given in the beam schedule. All beams are figured to the bottom of the floor slabs, and in figuring floor slabs all slabs are figured straight over all beams. If the two sides of a beam are of different depths—and this is true of almost all exterior beams—list both depths on your takeoff sheet. This will give you all the information you need for formwork.

The easiest way to figure beams is to start the one corner of the building and work your way around the structure so when you're through you'll be sure to have all the exterior beams. Take all of the first floor exterior beams at once. Don't try to combine one floor with the one above unless it is a floor plan on which the architect has noted that the second floor is identical to the first floor. But be careful; some plans are marked "similar," but there is a difference between similar and identical. Similar means that they are alike, but not identical. If the plans are marked "similar," do not assume that the two floors are the same. Take the first floor off, and when you get to the second floor compare it to the first floor takeoff. If everything on the second floor fits everything on the first floor, then you could indicate that the second floor beams are identical to those on the first floor and just repeat your final answers. You don't have to rewrite all your calculations. In some cases "similar" means changing the depths of the beams on two sides of the building but not on the other two. You could change only these figures and obtain the answers by repeating the first floor takeoff.

When figuring beam work, remember that the form area is equal to the width plus twice the depth if the beam is the same depth on both sides. If the two depths are different for the beam, add each depth separately to get the formwork. That is the point in writing both of them down on your takeoff sheet. You always want contact area in taking off formwork.

Figure the bottom of the beam forms straight through the top of the column. In other words, a column under a beam will make a hole in the bottom of the beam forms. It is not necessary to deduct this from the beam formwork. The important thing is to always figure the right amount of concrete needed. As far as formwork is concerned, it is necessary to obtain only a reasonably accurate figure. Remember that the concrete is about 25% of the total cost of the job.

You can put anything in the exterior beams category which will cost roughly the same per unit to form as the exterior beams. If you have a canopy slab that would be 2 feet wide by 8 feet long over a doorway, include it in the exterior beam class because the forming will cost the same. Otherwise, you will have too many different categories to price.

When you finish taking off the exterior beams, mark "exterior beams plus canopies" on your takeoff sheet. This is for your own information. You could also put concrete shelves or window ledges in the exterior beam class. To form a shelf or ledge that projects outside the face of the building requires an outrigger type of forming similar to that needed for exterior beams.

If the beam is odd shaped, figure the contact area required to form it so that the beam can be cast. If there is a high percentage of this type of forming, then figure that the forming labor output per day will decrease by 10-20% or more in accordance with your experience, judgment, or plain common sense.

Beam takeoff should be separated floor by floor so that the concrete yardage in the beams can be combined with the concrete yardage in the slab for each individual floor. Knowing the total concrete yardage in any one floor system will help you figure the number of placings required in that floor and the number of days of crane time to allow for moving forms, reinforcing steel and placing concrete. That is why you should take everything off floor by floor. These figures will also help you figure hoist time.

Some contractors prefer to keep the beam bottoms separate from the beam sides for pricing. The only reason for doing this is that actually the shores are only shoring up the beam bottom. But you don't just build the beam bottom; the sides are an integral part of the beam. Therefore, by taking an average size beam (1 foot wide by 2 feet deep) and figuring a shore under the beam every 4 feet, the costs can be distributed over the beam sides and bottom in one operation. This eliminates the need to take off and price the beam bottoms separately.

There are two reasons for separating beams into categories for pricing that we have not discussed. Very large beams, say 3 feet by 6 feet and up, will increase both material and labor costs. If the height of the shores required for the beams is over about 14 feet, this will increase both material and labor costs. These form areas should be kept separate and marked plainly for the "pricing" work. Once again, this increase in cost must be based on experience, comparison with previous work, or common sense.

List beam width and depth on the quantity takeoff sheet to the fraction of an inch. This is necessary for extending the correct concrete quantity, especially if the beam is long. The length, of course, need only be to the closest foot for accuracy.

SHORED FLAT SLABS

In estimating shored flat slabs you must obtain the area of the slab and edge forms, the finish, and the concrete. The slab form area can be taken as the total area of the slabs running over the top of the beams. See Figure 15-9. In other words, we are doubling up on the area already included in the beam bottoms. This is done for simplicity in takeoff work when the beam bottom area is not too great. By doing this, the slab form area is the same as the finishing area, and the concrete for the slab form area may be found by multiplying this area by the depth. This is a common takeoff technique. The extra formwork included in the flat slabs, over the beams, is not generally a large amount. However, in pricing forms for flat slabs, keep in mind that the quantities will be a little high. All edge forms for slabs are listed by height in total linear feet. Be sure to figure edge forming at all stairwells, shafts and other openings in the slab.

In order to get the contact area of the slabs only when figuring shored slab forms, subtract the beam bottom area from the total slab finish area. This is easily obtain from the takeoff for the exterior and interior beams. Some contractors do not subtract beam bottoms from slab areas, but use the slab area for form area also. They generally tend to price the forming for slabs at a lower rate, which compensates for a too-high form area. Either method is acceptable as long as the takeoff estimator and pricing estimator understand each other. See Table 15-J.

As you take off each item, mark the floor number and the part of the building which it came in the description column on your takeoff sheet. This will save you time if you have to refer back to your takeoff. This information is particularly handy when the architect brings out addenda or when you have to add in the alternates required to make your final bid complete. Alternates, especially, can be marked on your takeoff sheet as you work so that when you finish the base bid you can quickly refer back to the alternates.

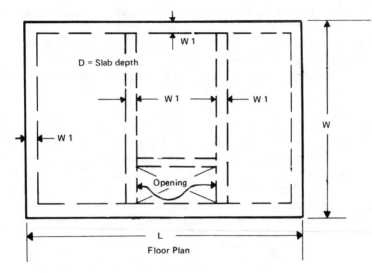

Figure 15-9 Shored flat slabs.

Table 15-J
SHORED FLAT SLABS

Description	Dimensions	Square Feet Area	Cubic Feet Concrete	Estimated Quantity
Shored Flat Slabs:	L x W x D	LxW	LxWxD	
Total Area	100′ x 40′ x 0′-9	4000	3000	
Less Opening	30′ x 8′ x 0′-9	−240	−180	
		3760 ✓	2820 ✓	
			27	
			Concrete = 105 C.Y. ✓	
			Finish Area = 3760 S.F. ✓	
Deduct Beam Bottoms:	L x W1			
Exterior Beams	2 x 100′ x 1′	200		
Exterior Beams	2 x 40′ x 1′	80		
Interior Beams	2 x 40′ x 1′	80		
Interior Beams	1 x 30′ x 0′-6	15		
Total Beam Bottoms		375		

Total Area = 3760 − 375 = Form Shored Slabs = 3385 S.F. ✓

When you take off slabs, always take the width as the horizontal dimension as you would be looking at the slab. Take the length as the vertical dimension regardless of which is longer. It is important to have a system so that everything is listed in the same manner. It saves time to know that you always list slabs in the following order: horizontal dimension times vertical dimension times thickness. Be sure to mark all dimensions clearly: (0'-7" *not* 7") to avoid errors.

There are two items that affect forming costs on shored flat slabs. One is the floor-to-floor height of the building and the other is the slab thickness. The greater the floor-to-floor height, the more bracing lumber is required. The thicker the slab, the more shoring material is required. For practical takeoff purposes, three heights are sufficient: under 12 feet, 14 feet to 16 feet, and over 16 feet from floor to floor. For practical takeoff purposes, two thickness dimensions are sufficient: 6 inches to 8 inches and 10 inches to 12 inches. Variations can be adjusted by a percentage based on experience and judgment.

Round off the slab quantities when you extend them to the extreme right-hand column of the takeoff sheet for future extension to the recap and pricing sheet. Slab forms and finish area should be rounded off to the closest 10 square feet. In other words, 8,374 square feet should be rounded off to 8,370 square feet. This is for simplicity in extension of quantities to the recap sheet. Edge forms should be marked to the closest 10 linear feet. Concrete, of course, is always marked to the cubic yard above any fraction. No allowance is made for waste other than marking each individual concrete item to the cubic foot above any fraction. Remember that all quantities extended to the right hand column of the takeoff sheet are marked with a check, thus (√). This means they have been checked and are to be moved to the recap sheet for pricing.

In marking stairwell outs or other openings in a floor slab, mark clearly "deduct stairwell" in the description column. In extending the quantities into the columns for floor area and concrete volume, put a minus sign in front of the number and brackets around it to be sure to deduct these items from the total.

Small openings in slabs need not be deducted. A 2- by 4-foot ventilation duct opening, for example, could be ignored, unless there are several such duct openings and they occur frequently enough to warrant listing them on your quantity takeoff.

This is another reason why the man who takes off a job and the man who prices it must understand each other very well. The man who prices an item must know how the takeoff was made.

On slab forms figure three reuses of material including plywood decking. This applies to both large and small jobs because you will not buy the plywood for one small job and then throw it away. You have to spread your costs over all your jobs, so assume that your slab forms will average three uses. You might get half a dozen uses or more out of some of your plywood deck forms. You'll probably only get one use out of the 2-inch lumber used for wedges and bracing. Figure an average of three uses on most jobs except for a multistory building where there is a repetition of the identical floor framing. In this case, figure on five or six uses of material as a maximum. This is a matter of judgment and experience. Reuse of material in multistory building tends to increase the erection output of forms per carpenter day. However, this additional output would be not more than 10% over that in average conditions unless previous experience on similar buildings under similar working conditions would justify figuring a higher output per man.

PAN SLABS

The slab form area is taken as the total area of the slabs running over the top of the beams. This is the same way that flat slabs are figured. Once again, we are doubling up on the area already included in beam bottoms. This simplifies the takeoff because the slab forms and finish area are identical. Beam bottoms should be deducted to get the correct slab area if they would amount to more than 10% of the slab area. The edge forms are listed by their height in total linear feet. The concrete is best taken off as follows: figure the total volume for the pan slab and subtract the pans (or voids) from it. For example, a typical pan slab would be made up with 12- by 30-inch pans with a 2-inch concrete topping over the pans. The total depth of the slab is the 12-inch pan plus the 2-inch topping, or 14 inches. The total volume including the pans is the slab area times the depth, 14 inches. To subtract the pans, list them by their size (12 inches by 30 inches) times the linear feet. This will give the total voids in the slab and can be subtracted from the total volume, and leave the concrete quantity in the slab. The size of any pan, and the cubic feet in it per linear foot, can be taken from tables furnished by the manufacturers of these pans.

The pan sizes vary both in depth and width, but the method of deducting for the pans is always the same. In one floor slab system, it is possible to have variations both in pan depth and width. In some floor framing systems there are tapered-end pans. The last 3-foot section of pans where they bear on a beam or wall is tapered to a smaller width to give more concrete width for bearing. The tables supplied by pan manufacturers will also give the correct deduction for these tapered-end pans. Most structural designers will not draw the complete floor layout of all of the pans. They will sketch in each end of the floor layout, and leave it to the takeoff man to figure out how many pans will fit across the floor. Remember to check the specifications for pan construction. Sometimes, the specifications will call for tapered-end pans, and they will not be shown on the drawings. The structural designer will specify intermediate bridging joists when the distance from bearing to bearing is too great, and these must be deducted from the pan outs.

Although the erection and removal of pans may be taken care of by the company that rents these materials, the contractor still has to furnish the means of supporting these pans.

The pan edges are generally nailed down on 2- by 6-inch or 2- by 8-inch planks which form the bottom of the joists. These joist bottoms are supported on shores—positioned about 4 feet center to center—from the floor below. This is called the centering for the pans. This centering and shoring for the pans is generally done by the concrete contractor and is referred to as slab forming for pan construction.

The company that rents the pan forms may also erect, center, shore, and remove the pans. However, the sides of the beams, a ledger to support the slabs, all vertical surfaces, all edge forming, and drop heads must be formed by the concrete contractor. You must remember that when you sublet the slab forming, you're going to send a crew out to the job to do nothing but beam framing. This increases your labor and supervision costs on this part of the work.

All of the general notes mentioned under shored flat slabs also apply to pan slabs. Some contractors prefer to use solid plywood decking under pan slab construction. This increases their forming material and labor costs, but they claim the decking provides better working conditions and that this offsets the increased costs.

SLABS ON METAL FORMS

Corrugated formwork is a corrugated metal form left in position after concrete has been placed. The corrugated metal may be galvanized or plain, standard weight or heavy duty. It may even have reinforcing rods welded across the top of the corrugations so it can be used for longer spans. Generally the thickness of floor specified is that area above the top of the corrugation, or dimension D in Figure 15-10. The depth of the

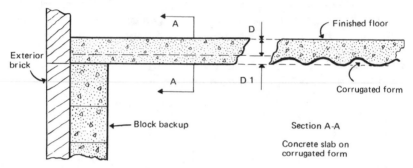

Figure 15-10 Slabs on metal forms.

corrugation is shown as D1. To figure the concrete volume, the depth of D plus D1 should be used. The concrete gained by figuring to the bottom of the corrugation will be used in the sag of the corrugated sheets.

To figure the amount of corrugated formwork needed, take the area of the slab plus 7% for lap and waste. The edge forms are listed by the height required in total linear feet for each height. The finish area will be the exact slab area to the closest 10 square feet. The concrete volume will be the slab area times the thickness of the slab, including to the bottom of the corrugations in the form. If the specifications show 2 inches of concrete above the top of the corrugations, figure ½ inch for the corrugations or figure the slab as 2½ inches thick. Include the corrugations in your figuring as if they were solid concrete. This will cover waste and possible sag in the form material.

Similar types of construction use welded wire mesh with heavy, waterproof paper backing for the slab forms. The reinforced paper is stretched tightly and clipped to steel beams or joists. This work is figured as described above except that in takeoff the concrete slab should be figured as ½ inch thicker than the drawings show. This covers waste and the sag that will occur in the paper-backed form. Forming with paper-backed forms will cost about 10% less than if corrugated formwork is used. Many contractors prefer to use corrugated formwork if they have an option as it provides better and cheaper concrete placing conditions.

STAIRS ON FILL

Shored stairs need forms on the bottom for the risers, and sometimes they need forms on both sides. The easiest way to take them off is by the total linear foot of riser with a notation as to where forms are needed. They can then be priced by the linear foot of risers with all price units included. A stair on fill would need forms for the risers only. The concrete in stairs should be figured at 1 cubic foot per linear foot of risers. This is on the high side, but it will take care of waste. Stairs must be finished on both treads and riser unless the riser is metal and part of a steel form. See Figure 15-11.

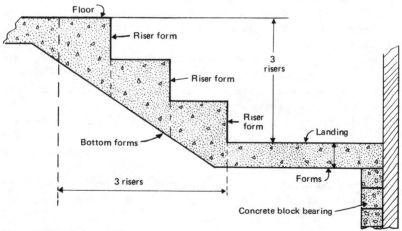

Figure 15-11 Stairs on fill.

The takeoff items to consider here are sand fill, form risers, finish treads and risers, and concrete. All stairs are taken off by the linear feet of risers. If a stair has eight risers that are 4 feet long, it has 32 linear feet of risers. You can then calculate all of your takeoff items. For the sand fill area figure 1 square foot per linear foot of riser (or 32 square feet). To form the risers list them by the linear foot (32 linear feet). List the treads and risers by the linear foot to estimate the finishing area. To take off the concrete in a stair; figure 1 cubic foot per linear foot of riser (32 cubic feet). Metal stair nosings are also equal to the linear foot of riser (32 linear feet). Abrasive grits in stair treads are figured as 1 square foot per linear foot of tread. We always have one less tread than risers, so 7 treads times 4 feet equal 28 square feet of treads or abrasive grits.

SHORED STAIRS

Form risers and stair bottoms, finish treads and risers, and concrete are taken off here. Shored stairs, as in stairs on fill, are taken off by linear feet of risers, and from this you can figure the takeoff items. If a shored stair has eight risers that are 4 feet long, it has 32 linear feet of risers. The forming of the riser and the stair bottom is priced by the linear foot of risers (or 32 linear feet). This is based on the theory that there is one square foot of stair bottom to each linear foot of riser. The finishing of treads and risers, concrete quantity and other items are taken off as listed under stairs on fill. Partial steel forms may also be included under shored stair forming.

One type of steel stair form comes to the job with the reinforcing steel welded to steel risers which are welded to channel stringers. This means that the carpenters must first set the steel stair forms by means of a light chain hoist. When pricing these items, they should be listed on the takeoff sheet by units or by flights of stairs. Next, the carpenters must form the stair bottom only, since the risers are steel. This saves one-half of the forming costs. The finishing of the stairs is of the treads only, since the risers are of steel. The difference in cost between forming stairs without steel forms and with steel forms is not too great. With all factors considered the use of steel forms should cost about 10% more than a stair formed all in wood.

STAIR LANDINGS

Shored stair landings are taken off when shored stairs are. They are listed separately on the takeoff because only a little forming is required at each landing and this increases the labor cost. The form area for these slabs and the total number of landings should be provided. It is easier to estimate the amount of time required to erect one landing than to estimate all the landings by the square foot. Seldom does a carpenter do better than 60 square feet per day on this type of work. The area to be finished is the same as the form area. The concrete is grouped with the stairs for pricing.

PAN-FILL STAIRS

Only finishing and concrete fill are taken off here. The stair treads are figured as 1 square foot per linear foot of tread. Most of the tread pans are 2 inches deep and landing pans are 3 inches deep. However, be sure to figure about 10% waste on the concrete fill. The treads are listed in total linear feet for pricing the finishing. The landings are taken off in square feet for finishing.

MISCELLANEOUS CONCRETE IN BUILDING

At this point all of the main concrete work has been taken off from both the structural and the architectural drawings. Now, search out the less obvious items of concrete. Start with the first drawing and check each section for concrete not taken off before. Some of the items you might find are: mechanical equipment bases, locker bases, cabinet bases, toilet room curbs, window sills, concrete fill placed on precast slabs, lightweight roof fill for drainage, and roof curbs for openings through the roof. These items will all have forms, finishing, and concrete to take off and recap for pricing. Generally, the formwork will be taken off and listed by the various heights times the linear feet for each height. The finishing will be listed by the linear foot or square foot, whichever is easier to price. The concrete will, of course, be in cubic yards.

The grouting of column bases is best priced by the individual column. For example: grout column bases, 12 by 12 inches by 1 inch thick. This gives the quantity of material and an idea of the labor involved.

It is best to take off and recap each of these miscellaneous concrete items separately. This makes it easier to estimate the labor time for the formwork and placing of the concrete. Each item can best be priced by estimating the labor time for the individual item itself. After you have checked all drawings for miscellaneous concrete work inside the building, take off all exterior walks, paving, and curbs from the site plan.

EXTERIOR SIDEWALKS

If sand fill is to be placed on undisturbed earth, hand grading must be figured in addition to machine excavation in order to get the subgrade exact. See Figure 15-12. This is taken off and priced as a square foot item. The sand fill will be given in cubic yards with 25% added for compaction and loss. Expansion joints are listed by the size (i.e., ½ inch by 5 inches) in linear feet with 5% added for waste. Poured sealant is listed by the size (i.e., ½ inch by 1 inch) in linear feet. We have assumed a 6-inch thick slab.

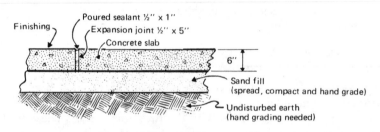

Figure 15-12 Exterior sidewalks.

Sand fill, edge forms, expansion joints, finishing, and concrete are taken off here. Exterior sidewalks are taken from—and checked off on—the site plan. The area of the sidewalks will first give you the area of subgrade hand grading required. This area times the sand fill depth will give you the sand fill volume in cubic yards. But to this you must add 25% for compaction and loss. The edge forms are taken off and listed by the height in linear feet of each height. Expansion joints are taken off by the thickness times the height in total linear feet for each size. The finishing area is the same as the subgrade area. In some localities, private walks (within property lines) must be separated from public walks, because there is a permit fee or tax on public walks only. The concrete, of course, is the area of the walks times the depth put into cubic yards. Add at least 3% to the concrete for waste.

EXTERIOR PAVING

Sand fill, edge forms, expansion joints, finishing, and concrete are taken off here. The same outline of takeoff as given for exterior sidewalks will apply to exterior paving. These two categories must be kept separate because of the difference in finishing generally specified between sidewalks and paving. The labor required for placing sidewalks is higher than for placing paving and this is another reason for separating them.

EXTERIOR STRAIGHT CURB AND GUTTER

Straight curbs and curb-and-gutter work are taken off separately because their forming and finishing are figured in two separate price brackets. The forms for straight curbs are simple and equal to twice the height. The forms for curb and gutter are more complex because the area to be estimated includes areas H1, H2 and D in Figure 15-13. The face dimension, H2, requires a curved form, but it can be grouped, for pricing purposes, with H1 and D for total form area. The finishing of straight curbs costs one-half of that required to finish a curb and gutter. Both are priced by the linear foot of either straight curb or curb and gutter. Hand grading of earth for curb work is best priced by the linear foot of straight curb or curb and gutter because it is a rather indeterminate quantity.

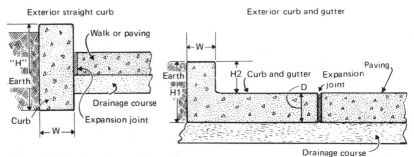

Figure 15-13 Exterior straight curb.

EXTERIOR STRAIGHT CURB

Hand excavation, forms, finishing and concrete are taken off in this section. How much hand excavation will be required for curbs is questionable. Therefore, to simplify the problem, take off the hand excavation and price it by the linear foot of curb. The cost varies with the size of the site, the amount of curbing, and the type of ground. A typical straight curb is 6 inches wide by 2 feet high with 1 foot, 6 inches below the ground. The formwork area is twice the height times the length. Since the formwork is only 2 feet high and will be buried in the ground for the most part, it belongs in the same pricing class as wall footings. The finishing of the exposed curb top should be given in linear feet. The concrete is given in cubic yards.

EXTERIOR CURB AND GUTTER

Hand excavation, forms, finishing, and concrete are taken off in this section. The same outline of takeoff as given for straight curbs will apply to curb and gutter. A typical curb and gutter is 6 inches thick with a 2-foot wide apron and a 6-inch high rise at the curb. The formwork is slightly more complicated due to the apron and is more costly than for a straight curb. The concrete is given in cubic yards.

The takeoff of concrete quantities is now completed and a discussion of material and labor form costs for each item follows.

CHAPTER 16
ESTIMATING FORMWORK

A series of labor output tables for formwork erection follow this chapter. These tables are used as the basis of the formwork prices below and in the following chapter. Tables 16-A through 16-I are included at the end of this chapter so that contractors can use them as a basis for developing their own data.

FORMS FOR COLUMN FOOTINGS (See Table 16-A, Page 506)

The material requirements for forming column footings are simple. If 2-inch material is used, figure 2 board feet of lumber per square foot for the 2-inch plank. See Figure 16-1, Section A-A. Then assume one-half board foot of lumber for the stakes and bracing. Estimate 2½ board feet of lumber per square foot of form. If you assume 50 cents per board foot for material, two and one-half times the 50 cents a board foot would total $1.25 per square foot of form if the form was used only once. When figuring form usage for column footings and wall footings, figure on five uses of the material. To determine the cost per square foot, divide 5 uses into $1.25 per square foot. The answer, of course, is 25 cents per square

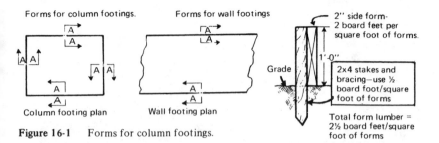

Figure 16-1   Forms for column footings.

foot of form for material cost. This applies whether the lumber is to be used on one job or six jobs. The erection labor output on column footings will vary from 120 to 200 square feet per day depending upon the size of the footings.

When forming footings under average conditions, allow one hour of carpenter's time to set up each footing. In other words, a footing that is 5 feet by 5 feet by 1 foot would be 4 sides by 5 feet, or 20 square feet per footing times 8 hours, or 160 square feet of forming per day. When there are various footing sizes on a job, base your production output on the average size footing. This production output per day (160) is entered in the labor per day column (L per D) for future conversion to a unit price based on the use of a carpenter and one-half of a laborer's time. Allow 600 square feet per day for laborer time to strip, clean, stack, or move column footing forms. Stripping time is figured as an average for the entire job and is listed as one item.

If the total quantity for column footings was less than a day's work—say, 100 square feet—don't bother to look at how many column footings there are in the job. Call it 100 square feet per day for erection.

FORMS FOR WALL FOOTINGS (See Table 16-A, Page 506)

The material requirements for wall footings are developed as for column footings. That is, figure on 2½ board feet of lumber per square foot of contact area. For erection, figure an average of 350 square feet per day. For stripping and cleaning, figure 600 square feet per day. You should also list 2-by-4 keys under wall footing forms. The 2-by-4 material has two-thirds of a board foot per linear foot. Since it is scrap material when dug out of the footing, figure on only one use. The labor of installing and removing the 2-by-4 keys should be figured at 500 linear feet per day.

FOUNDATION WALL FORMS (See Table 16-A, Page 506)

Assuming an average of 50 cents per square foot per month rental on wall forms and 50 cents per board foot for lumber, we have the following material costs:

Material form costs for walls under 4 feet high on footings will be rental of 50 cents per square foot per month plus 25 cents for ½ board foot of bracing lumber or 75 cents for three uses. This is 25 cents per use plus 4 cents for ties, nails, and oil, or 29 cents per square foot of wall form per use.

Material form costs for walls 4 feet to 8 feet high on footings are figured at a form rental rate of 50 cents per month plus 25 cents for ½ board foot of bracing lumber or 75 cents. For three uses, this is 25 cents per use plus 8 cents for ties, nails, and oil, or 33 cents per square foot of wall forms.

Material form costs for walls 8 feet to 12 feet high on footings are: a form rental rate of 50 cents per month plus 38 cents for ¾ of a board foot of bracing lumber, or 88 cents. For three uses, this is 29 cents per use plus 8 cents for ties, nails and oils, or 37 cents per square foot of wall forms.

For walls 12 feet to 16 feet high on footings, figure the materials cost the same as 8- to 12-foot-high walls but add 12 cents for ties, nails and oil, or 41 cents per square foot of wall forms.

Material form costs for walls over 16 feet high on footings will be a form rental of 50 cents per month plus 50 cents for 1 board foot of bracing lumber, or $1.00. For three uses, this is 33 cents per use plus 12 cents for ties, nails, and oil, or 45 cents per square foot of wall forms.

The erection labor for wall forms on footings should average as follows:

$$\begin{aligned}
&\text{0 feet to} \quad \text{4 feet high—300 square feet per day} \\
&\text{4 feet to} \quad \text{8 feet high—250 square feet per day} \\
&\text{8 feet to 12 feet high—160 square feet per day} \\
&\text{12 feet to 16 feet high—120 square feet per day} \\
&\text{Over 16 feet high—100 square feet per day}
\end{aligned}$$

The stripping and cleaning labor on wall forms should average as follows:

0 feet to 4 feet high—500 square feet per day
4 feet to 8 feet high—450 square feet per day
8 feet to 12 feet high—400 square feet per day
12 feet to 16 feet high—300 square feet per day
Over 16 feet high—200 square feet per day

The above figures are averages for standard conditions. They will vary with each job, depending upon its special problems.

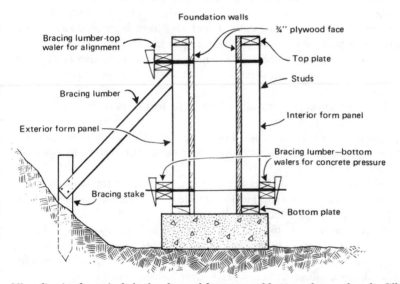

When figuring forms include the plywood face, top and bottom plate, and studs. When estimating bracing lumber include walers, diagonal bracing and stakes. Figure the stakes to support two sides of the wall forms, and then divide by two to get the correct average figure for one side of the forms. The bracing lumber varies with the height of the wall as follows:

    Under 8 feet high = .5 board foot per square foot of forms
    8 feet to 16 feet high = .75 board foot per square foot of forms
    Over 16 feet high = 1.0 board foot per square foot of forms

Figure 16-2   Foundation wall forms.

Vary these costs for wall forming as follows: If a wall has pilasters in it spaced about 10 feet to 20 feet center to center, include the forming area of the pilasters with the wall form totals. Figure that it will cost about 10% more to form these pilasters in with the wall forms than it would cost to form a straight wall form. Therefore, increase the material and labor cost of wall forms with pilasters by 10% above the cost of straight wall forming.

If a wall has no concrete footing under it, the wall forms must be set on a 2-by-6 inch mud sill. This type of wall is called a grade beam. The cost of this mud sill must be included. It can be done on a linear foot basis for the material and labor required. However, it is easier to include the cost of the 2-by-6-inch mud sill in the square foot price of the wall forming. For wall forms on mud sills (or grade beams), increase the material and labor cost of wall forms on footings by 10%.

For seat bearing box-outs in the wall for slabs, price the linear feet of slab bearing on the wall. The material cost is about 40 cents per linear foot and the labor output is about 200 linear feet per day. No stripping on seat bearing is figured since this comes down with the wall stripping.

Figure a carpenter and one-half of a laborer's rate for wall forming on all jobs except residential foundation work. Because of the small size of most residential foundations, eliminate the one-half laborer's rate. In other words, figure the same labor output per day as given previously but use the rate for straight carpenter time when setting up the labor price for erection. This assumes that carpenters will erect residential foundations with the same production per day and not need laborers to handle forms and bracing lumber for them.

PIERS BELOW GRADE (See Table 16-B, Page 508)

On piers below grade (see Figure 16-3), figure regular lumber for forms—1-by-6 or 1-by-8—because they will not be exposed. The material, then, works out to be 2½ board feet of lumber. This at 50 cents per board foot would be $1.25, divided by three uses, or 42 cents for material plus 7 cents each use for nails and ties, or 49 cents per square foot of forms. Forms for piers below grade should be erected at the rate of between 100 and 150 square feet per day. The important thing is the square feet of forming in each pier. A man will only average about one pier per day. Piers have to be set up on an individual basis. Small stub piers may have only 10 square feet of forms in each one. Ten square feet times eight hours is only 80 square feet per day. Larger piers, 1 foot

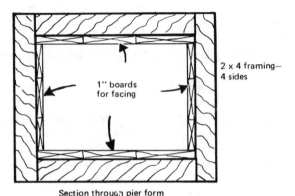

Section through pier form

Use 1 board foot per square foot for facing.
Use 1.5 board feet per square foot for framing and bracing.
Total material = 2.5 board feet per square foot of forms.

Figure 16-3   Pier form below grade.

square by 5 feet high, have 20 square feet per pier. At one of these per hour, a man could make and erect 20 square feet times 8 hours, or 160 square feet per day.

The point made here is: there is a proper time to figure square footage and a proper time to look at the individual item. On piers below grade figure 500 square feet of stripping per day.

EXTERIOR AND INTERIOR COLUMNS (See Table 16-B, Page 508)

Always figure a plywood facing for columns as in Figure 16-4. In addition, 1¼ board feet of lumber per square foot of form is used for bracing. We shall figure plywood at $1.00 per square foot. Add 10 cents per square foot for adjustable clamps. The total is $1.10 for plywood and clamps, 65 cents for lumber for a total of $1.75 a square foot per use. On columns figure an average of three uses for material. But check the columns out carefully. If you are bidding on a small building with only one story and perhaps only 10 or 12 columns in it, it may be necessary to form all 10 or 12 columns at once. When the job is over the lumber may have only scrap value. In this case figure $1.75 a square foot for material and only one use for the forms.

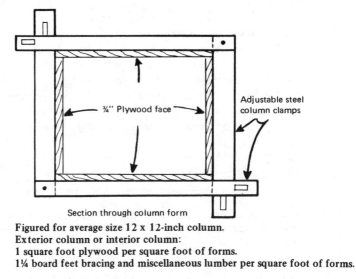

Figured for average size 12 x 12-inch column.
Exterior column or interior column:
1 square foot plywood per square foot of forms.
1¼ board feet bracing and miscellaneous lumber per square foot of forms.

Figure 16-4 Formwork for exterior and interior columns.

Nails, oil, and chamfer strips must also be figured into this estimate. Nails and oil will cost about 8 cents per square foot with no reuses. If there are chamfers at each corner of the column, the cost will go up another 8 cents per square foot of form. The material cost for columns including nails, oil, and chamfer strips for corners would therefore be 74 cents per square foot of form based on three uses for material ($1.75 for form materials divided by three uses equals 58 cents plus 8 cents for nails and oil plus 8 cents for chamfer strips). If the material can be stored and reused sometime in the future on another job, figure three reuses. Do not try to charge everything off on one job.

When estimating labor costs on exterior columns, figure a man can make and erect 80 square feet of column forms per day for the first use. If the forms can be reused, figure a man can rework and re-erect the same column forms at a rate of 110 square feet per day. Now, if we are going to use the forms a third time, add up 80 for the first use, 110 for the second and 110 for the third use when they are re-erecting the same columns. This gives a total of 300 square feet of column forms for three uses. Divide this by three uses and you'll come up with 100 square feet per day for each use.

On many high-rise structures—particularly apartment buildings where the height often remains constant from floor to floor—it is possible to figure many reuses. If the columns can be used without remaking them, they'll be reused a lot more than the normal three times.

For interior columns, assume that a man can make and erect 90 square feet per day the first time because it is easier to work around interior columns of a building than around exterior columns on the outside edge of the building. Assume that a man can rework and re-erect the same columns at 120 square feet per day. For three uses, the average is 110 square feet per day.

EXTERIOR AND INTERIOR BEAMS (See Table 16-C, Page 510)

The cost of forming exterior beams is considerably greater than it is for forming interior beams. See Figure 16-5. When estimating exterior

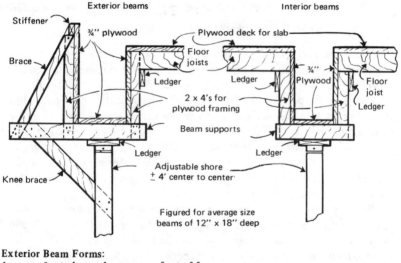

Exterior Beam Forms:
1 square foot plywood per square foot of forms.
2 board feet lumber per square foot of forms.
1 shore per 16 square feet of forms.
(4 feet center-to-center by 4 feet beam girth)

Figure 16-5 Formwork for exterior and interior beams.

Interior Beam Forms:
1 square foot plywood per square foot of forms.
1 board foot lumber per square foot of forms.
1 shore per 16 square feet of forms.
(4 feet center to center by 4 feet beam girth)

Figure 16-5   Formwork for exterior and interior beams. (Continued)

beams, figure 1 square foot of plywood per square foot of beams. The cost then would run $1.00 per square foot for the plywood. The forming and bracing lumber will run 2 board feet per square foot, or, again, $1.00. The adjustable shore rental ($1.50 divided by 16 square feet) is worth 9 cents per square foot of forms, and all of this totals $2.09 per square foot of forms for one use. For three uses this is about 70 cents plus the cost of nails, oil, and possibly chamfer strips. Including all of these the cost would run about 82 cents per square foot of forms. Figure a man can make and erect 75 square feet of exterior beam forming per day. If the forms can be reused, figure, for three uses, a man can make and erect an average of 90 square feet per day.

Since exterior beams are more difficult to form than interior beams, it will take a man 20% longer to form an exterior beam. Remember that an exterior beam will require outriggers and knee-braces as supports.

On interior beams, figure that a man can make and erect these the first time at a rate of 90 square feet per day. For three uses a man should be able to rework and erect an average of 110 square feet per day.

When stripping and cleaning beam forms, a man ought to be able to handle 230 square feet of exterior and 300 square feet of interior beam forms per day.

SHORED FLAT SLABS (Table 16-D, Page 512)

When figuring forms for flat slabs, break the flat slabs into groups of 10 to 12 feet, floor to floor; 14 to 16 feet, floor to floor; and 18 to 20 feet, floor to floor. See Figure 16-6. Very few jobs go over 12 feet, floor to floor.

Break the forming down to show whether a slab is 6 inches, 8 inches, 10 inches, or 12 inches deep. Shoring and bracing has to be a lot closer together for a 10- or 12-inch slab than it does for a 6-inch slab. The cost of material for a 6-inch slab—figuring 1 square foot of plywood, 1.5 board feet of lumber, and shores—is $1.75 a square foot per use. A 12-inch slab will run $2.00 per use.

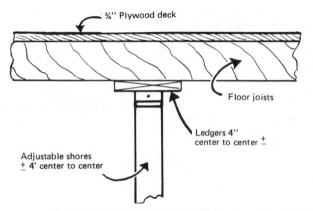

All slabs = 1 square foot plywood per square foot of forms.
All slabs = 1 shore per 16 square feet of forms (4 feet by 4 feet).
Joists—Ledgers and miscellaneous bracing lumber:
10 to 12 feet floor to floor, under 8-inch slabs: 1.5 board feet per square foot of forms
10 to 12 feet floor to floor, 10-inch to 12-inch slabs: 2.0 board feet per square foot of forms
14 to 16 feet floor to floor, under 8-inch slabs: 2.0 board feet per square foot of forms
14 to 16 feet floor to floor, 10-inch to 12-inch slabs: 2.25 board feet per square foot of forms.

Figure 16-6   Formwork for shored flat slabs.

On a 6-inch slab, figure a man can erect 140 square feet of formwork a day. On a 12-inch slab, cut that down to 120 square feet a day. Figure 350 square feet a day for stripping either slab. For unusually thick slabs (over 12 inches), the increase in cost is caused by additional bracing lumber and placing of the shores closer together. These must be analyzed carefully to figure out the additional cost.

SHORED PAN SLABS (See Table 16-E, Page 514)

On pan joist construction (see Figure 16-7), figure 2- by 8-inch centering supports for pans. Solid plywood decking is fine but the cost of materials gets to be too expensive. Figure the lumber for centering at 1½ board feet of lumber per square foot. And figure 25 square feet of floor per shore. The shore rental is about $2.50 per month divided by 25 square

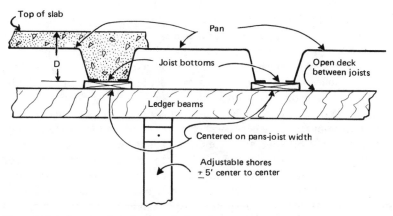

Open deck = shores and centering only for pans.

All slabs = 1 shore per 25 square feet of forms (5 feet by 5 feet).

All slabs = Forms figured as total slab area.

Joist bottoms, ledger beams and miscellaneous bracing lumber:

10 feet to 12 feet (floor to floor) D = 8½ inches to 16½ inches = 1.5 board feet per square foot of forms

14 feet to 16 feet (floor to floor) D = 8½ inches to 16½ inches = 1.75 board feet per square foot of forms

This method of forming for pans is cheaper than solid plywood decking, both for material and labor. However, some contractors always make a solid deck of plywood when using pan forms. The reasons are they feel that a solid deck is safer, and that the pans can be set in place more quickly on a solid deck than on an open deck.

Figure 16-7   Formwork for shored pan slabs.

feet or about 10 cents per square foot of floor forms. One and one-half board feet of lumber costs 75 cents plus the shores at 10 cents comes to 85 cents per square foot of floor for one use. Three reuses would come to 28 cents per use plus 4 cents for nails makes 32 cents per square foot of forms. Don't add anything for form oil. Figure 240 square feet of pan forms can be erected in a day, and 600 square feet can be stripped in a day. This will be less for heavier slabs and greater floor-to-floor heights.

The erection of formwork for thicker slabs with 16-foot floor-to-floor heights should be accomplished at the rate of 220 square feet per day. Stripping should proceed at the rate of 550 square feet per day.

STAIRS ON FILL AND SHORED STAIRS (See Table 16-B, Page 508)

For shored stairs, the lumber will run 7 board feet per linear foot of tread and riser, including the stair bottom forms and necessary shoring. Seven board feet of lumber at 50 cents per board foot is $3.50 per linear foot per use. Do not figure reuses unless it is a high building with great possibilities for reuses. In most cases, the shores and risers are all scrap anyway, so that the material for shored stairs will cost $3.50 per square foot. The stair bottoms are the only items that might be salvaged for reuse. On labor, figure stair erection at 20 linear feet of risers per day. This includes the stair bottom and all necessary shoring and bracing for the stair forms. Stripping should be done at a rate of 100 square feet per day.

Stairs on fill require only about half of the material needed for shored stairs. Stairs on fill, therefore, should cost $1.75 for material. Erection of stairs on fill should proceed twice as quickly as shored stair erection, so figure 40 linear feet of risers per day. If steel risers are used, it is necessary only to form the stair bottom and this would cost $1.75 for material. It should be possible to erect these stairs at the rate of 40 linear feet per day.

Stair landings should be priced by the item. If the stair landing measures 6 feet by 8 feet, it has 48 square feet in it and two carpenters can handle the job in half a day. Therefore, allow one day or 48 square feet of landing per day per carpenter for the job. On large buildings stair landings may total 1,300 to 1,400 square feet, and it would be difficult to figure these landings on a per-landing basis. In this case, figure one carpenter can form 50 square feet per day. This figure, however, is a matter of judgment.

SIDEWALKS, PAVING, AND CURBS

Sidewalks and paving have edge forms in which the material will cost about 30 cents per square foot, and about 400 linear feet can be erected in a day. See Figure 16-8. For straight curb or curb and gutter work, the material costs about 30 cents per square foot, and erection labor should average about 250 square feet per day. This labor production will vary considerably depending on whether the curb is broken into short lengths or in long lengths. In short lengths, production may drop to 200 square feet per day while in long lengths production may rise to 300 square feet per day.

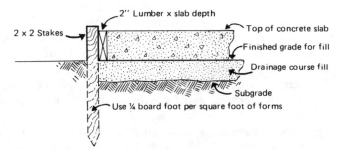

Total form lumber = 2¼ board feet per square foot of forms.
This item is more easily priced by setting it up for item required per linear foot.
Example: 2 x 6 edge forms = 100 linear feet.

Figure 16-8   Formwork for sidewalks, paving, and curbs.

DETERMINE UNIT OUTPUT

When pricing formwork it is most important to establish the number of material reuses required for each job. For labor, the important thing is to determine a unit output. But base the unit output figures on actual fact. Check out the figures with men in the field, but don't say, "I think a man can do 200 square feet a day. What do you think?" Nine times out of 10, they will agree with you. The best way to get a good estimate of unit output is to have two or more people make their own evaluations of the unit output of a particular item. Jot these figures down on separate pieces of paper. Perhaps one person will evaluate a job at 200 square feet per day and another person will evaluate it at 300 square feet per day. If the people are knowledgeable, the best thing to do is to average the two figures and use 250 square feet per day.

STRIPPING AND CLEANING

Stripping and cleaning of forms and stacking them on the job is figured as straight labor time with no carpenter time included. All formwork items of erection are added together for one item of "stripping and cleaning." This is priced as an average labor output per day for all types of formwork. Stripping and cleaning labor will vary from 40 cents to $1.20

per square foot and this is why it can all be grouped under one item. By figuring the percentage of footings, walls, columns, beams, and slabs in a job, an average labor output for stripping all the work can easily be found. If a job is all footings and walls, the stripping labor should average about 500 square feet per day. If a job has about equal quantities of foundation work as compared to columns, beams and slabs, the stripping labor should average about 400 square feet per day. If a job has a high percentage of columns, beams, and slabs in it, the stripping should average about 300 square feet per day.

If a laborer costs $200 per day and can strip 500 square feet per day, stripping labor will run 40 cents per square foot ($200 divided by 500 square feet equals 40 cents per square foot). If only 300 square feet can be stripped, the cost will run 67 cents per square foot. When a job has a high percentage of pan slabs (which require only shoring and centering the pans), the stripping labor per day should be increased to average about 400 square feet per day, or 50 cents a square foot. Some contractors price stripping and cleaning with the erection, and while this is strictly a matter of personal preference, pricing the two jobs together does make the operation a bit more complex. If an estimate has a page with forty items on it and the contractor has to go through every one of those items to get the stripping priced, he adds a lot of unnecessary extra work to his estimate. Then, once the two figures are added in, he doesn't know exactly how much of it was for erection and how much of it was for stripping. Pricing the stripping separately is a much simpler method and makes keeping field costs easier and more accurate.

Conversion Factors For Concrete Forms

To convert from	to	multiply by
	Length	
	Pressure or Stress (Force per Area)	
kilogram-force per square meter	newton per square meter (N/m²)	9.807
kip per square inch (ksi)	kilogram-force per square centimeter (kgf/cm²)	70.31
pound-force per square foot	kilogram-force per square meter (kgf/m²)	4.882
pound-force per square foot	newton per square meter (N/m²)	47.88
pound-force/ square inch (psi)	kilogram-force per square centimeter (kgf/cm²)	0.07031
pound-force per square inch (psi)	newton per square meter (N/m²)	.6895
	Bending Moment or Torque	
inch-pound-force	meter-kilogram-force (m-kgf)	0.01152
inch-pound-force	newton-meter (NM)	0.1130
foot-pound-force	meter-kilogram-force (m-kgf)	0.1383
foot-pound-force	newton-meter (NM)	1.356
meter-kilogram-force	newton-meter (NM)	9.807
	Mass	
ounce-mass (avdp)	gram (g)	28.35
pound-mass (avdp)	kilogram (kg)	0.4536
ton (metric)	kilogram (kg)	1000E
ton (short, 2000 lbm)	kilogram (kg)	907.2
	Mass per Volume	
pound-mass per cubic foot	kilogram per cubic meter (kg/m³)	16.02
pound-mass per cubic yard	kilogram per cubic meter (kg/m³)	0.5933
pound-mass per gallon (U.S.)[b]	kilogram per cubic meter (kg/m³)	119.8
pound-mass per gallon (Can.)[b]	kilogram per cubic meter (kg/m³)	99.78
	Temperature[e]	
degrees Celsius (C)	deg Celsius (C)	$t_K = (t_C + 273.15)$
degrees Fahrenheit (F)	kelvin (K)	$t_K = (t_F + 459.69)/1.8$
degrees Fahrenheit (F)	kelvin (K)	$T_C = (t_F - 32)/1.8$

[a] This selected list gives practical conversion factors of units found in concrete technology. The reference source for information on International System of Measurement units and more exact conversion factors is "Metric Practice Guide" ASTM E380.

[b] Symbols of metric units are given in parentheses.

[c] One U.S. gallon equals 0.8327 Canadian gallon.

[d] One liter (cubic decimeter) equals 0.001m³ or 1000 cm³.

[e] These factors convert one temperature reading to another and include the necessary scale corrections. To convert a difference in temperature from Fahrenheit degrees to Celsius or Kelvin degrees, divide by 1.8 only, that is, change from 70 to 88° For 18/1.8=10°C. To convert °C to °F, use $t_F = 1.8 t_C + 32$.

TABLE 16-A

Use Table 16-A to estimate footing forms and wall forms. The second column gives the material required per square foot of forms, or the contact area. The third column gives the average number of uses for material and the fourth column gives the cost per square foot for each use. Remember that nails, ties, and oil must be applied for each use. The fifth column gives an average erection figure per carpenter including ½ laborer's time helping him. The sixth column for stripping and cleaning forms is set up for straight laborer time. The item for 2- by 4-inch keys includes placing and removing. The item for 2- by 6-inch mud sills is for figuring them separately, in which case the wall forming is figured as if it had a footing under it. The table gives labor output for straight wall forming and underneath are variations for walls with pilasters and walls with the 2- by 6-inch mud sill included in the square foot of wall forming per day. The last item, "brick ledge or seat bearing," provides some information for estimating the forming of wall box-outs up to 4 inches wide by 8 inches high. Note that the material costs are based on 50 cents per board foot for lumber and 50 cents per square foot per month for form rental. These should be adjusted for your own costs if the variation is over 10%.

Table 16-A
FOOTING AND WALL FORMS

| Type | MATERIAL | | | LABOR | |
	Material per S.F.	Uses	Cost per Use	C+1/2L Erection per Day	Laborer Strip & Clean per Day
Ind'l Col. Ftg's	2½ B.F. Lumber	5	.25	120-200	600 S.F.
Wall Ftg's	2½ B.F. Lumber	5	.25	350 S.F.	600 S.F.
Walls 0' to 4' High	1 S.F. Forms = .50 .5 B.F. Lumber = .25 Oil, nails, ties = .04	3	.29	300 S.F.	500 S.F.
Walls 4' to 8' High	1 S.F. Forms = .50 .5 B.F. Lumber = .25 Oil, nails, ties = .08	3	.33	250 S.F.	450 S.F.
Walls 8' to 12' High	1 S.F. Forms = .50 .75 B.F. Lumber = .38 Oil, nails, ties = .08	3	.37	160 S.F.	400 S.F.
Walls 12' to 16'	1 S.F. Forms = .50 .75 B.F. Lumber = .38 Oil, nails, ties = .12	3	.41	120 S.F.	300 S.F.
Walls over 16'	1 S.F. Forms = .50 1 B.F. Lumber = .50 Oil, nails, ties = .12	3	.45	100 S.F.	200 S.F.
2" x 4" Keyways In & Out	2/3 B.F. Lumber	1	.33	500 L.F./Day	
2" x 6" Mud Sills	1 B.F. Lumber/L.F.	4	.13	350 L.F./Day	

Walls/Pilasters 10' to 20' O/C = 10% less production
Walls on mud sills or grade beams = 10% less production
Brick ledge or seat bearings = 200 L.F./day & (.50 material)
Walk edge: 400 L.F./day (.25 S.F.-material)
Curb forms: 300 S.F./day (.25 S.F.-material)

(Lumber .50/B.F.; Form Rental .50/S.F./Mo.)

TABLE 16-B

Use Table 16-B to estimate wall pilasters, piers, and columns. The third column gives the number of uses and the fourth column gives the cost per square foot per use. The number of times a column can be reused will vary and must be adjusted to each job. The fifth column gives the square feet per day for making and erecting columns for the first use. The sixth column gives the square feet per day for reworking and re-erecting these columns for all subsequent uses. The quantity in the brackets is the average time for three uses of the columns. For example: make and erect equals 80 square feet per day plus rework and erect equals 110 times 2 uses equals 300 square feet per day divided by 3 uses equals 100 square feet per day average for three uses.

Use the "foundation wall pilasters" line to figure, pilasters too large or too complicated to be included in with the wall forming costs. Use the line "piers below grade" to estimate all piers that can be formed with 1-inch lumber including piers in crawl spaces. The lines for exteror columns and interior columns are based on material costs of $1.00 per square foot for plywood and 50 cents per board foot for lumber. The line "add chamfers" is for wood chamfer strips at all four corners of the columns and is based on 8 cents per linear foot of chamfer, which is 8 cents per square foot of column forms, figuring 12- by 12-inch columns. Of course the cost to add chamfers is only if required or specified for a particular job and must be for each use. Adding nails and oil is also for each use. Some cost data on stair forming is provided below the table. For most jobs, only one use should be figured for stair-forming material.

Table 16-B
WALL PILASTERS, PIERS AND COLUMNS, STAIR FORMING

		MATERIAL			LABOR	
					C+1/2L	
Type	Material per S.F.	Uses	Cost per Use	Erection per S.F. per Day		Laborer Strip & Clean per Day
				Make & Erect	Rework & Erect	
Fdn. Wall Pilasters	2½ B.F. Lumber	3	.42	100 to 150	130	500
Piers Below grade	2½ B.F. Lumber	3	.42	100 to 150	130	500
Ext. Col's & Adj. Clamps No Chamfers	1 S.F. Plywood = $1.00/S.F. 1¼ B.F. Lumber = .63/S.F. Clamps = .12/S.F.	1	1.75	80	110	(3 uses = 100) 300
Int. Col's & Adj. Clamps No Chamfers	1 S.F. Plywood = $1.00/S.F. 1¼ B.F. Lumber = .63/S.F. Clamps = .12/S.F.	1	1.75	90	120	(3 uses = 110) 350
Add Chamfers	1 L.F./S.F. of Col.	1	.08	.12	.12	—
Add Nails— Oils	Per/S.F. of Col.	1	.08	—	—	—

Stairs-Shored Riser + Bot.   Material 1 Use = 7 B.F. x .50 = $1.75/S.F.   Labor   = 20 S.F./Day

Stairs on Fill Risers or Shored Stair Bottom   Material 1 Use = 3½ B.F. = $1.75/S.F.   Labor   = 40 L.F./Day

Set: Stair Builder Forms: 1 Flight = 2C. + L. ½ Day = C. + ½L = 1 Day
   Risers + Stringers
   W/Reinforcing = 1 Flight

(Lumber = .50/B.F.; Plywood = $1.00/S.F.)

TABLE 16-C

Use Table 16-C for estimating beam forming. The top table is for 1-inch sheathing boards and is not used much. Most beam forming is with plywood as given in the lower table. The fourth column gives the material cost for one use per square foot and must be divided by the number of uses figured. The fifth column gives the square feet per day to make and erect beams on the first use. The sixth column gives the square feet per day to rework and re-erect these same beams. All beams are figured for contact area for both sides and the bottom with an average beam size of 12- by 18-inches deep. This includes adjustable shores at 4 feet center to center, with the cost distributed per square foot of forms. Chamfers (if required) and nails and oil must be added for each use. For example, assume you have an exterior beam with plywood forms that requires chamfers and you figure three uses. The material square foot cost is $2.16 per use divided by three uses equals 72 cents per use plus 4 cents for chamfers plus 8 cents for nails and oil or 84 cents per square foot per use. The labor per day equals 75 square feet for the first use plus 95 square feet times two uses equals 265 square feet divided by three uses equals 90 square feet per day for a carpenter and ½ laborer's time. If a carpenter and ½ laborer's rate per day equals $340, then $340 divided by 90 square feet equals $3.78 plus 8 cents for chamfer strips equals $3.86 per square foot per use.

Table 16-C
BEAM FORMING

				Sheathing		
	MATERIAL			**LABOR**		
Type	Material per S.F.	Uses	Cost per Use	C+1/2L Erection per S.F. per Day		Labor—S.F. Strip & Clean per Day
				Make & Erect	Rework & Erect	
Ext. Beams + Adj. Shores No Chamfers	3 B.F. Lumber = $1.50/S.F. Head & Shores .16	1	1.66	(3U=90) 75	95	230
Int. Beams + Adj. Shores No Chamfers	2 B.F. Lumber = $1.00/S.F. Head & Shores = .16	1	1.16	(3U=110) 90	115	300
Add Chamfers	½ L.F./S.F.	1	.04	.08	.08	—
Add Nails & Oil	Per S.F.	1	.04	—	—	—
	Plywood					
Ext. Beams + Adj. Shores No Chamfers	1 S.F. Plywood = $1.00/S.F. 2 B.F. Lumber = $1.00/S.F. Shores .16	1	2.16	(3U=90) 75	95	230
Int. Beams + Adj. Shores No Chamfers	1 S.F. Plywood = $1.00/S.F. 1 B.F. Lumber = .50/S.F. Shores .16	1	1.66	(3U=110) 90	115	300
Add Chamfers	½ L.F./S.F.	1	.04	.08	.08	—
Add Nails—Oil	Per S.F.	1	.08	—	—	—

(Bottom+ 2 Sides; Lumber = .50/B.F.; Plywood = $1.00/S.F.)

TABLE 16-D

Use Table 16-D for shored flat slabs. The first column gives the thickness of the concrete slab. The third column shows that the cost per square foot in the fourth and fifth columns is based on one use. The height from floor to floor is given at the left side of the table to allow for additional shoring and bracing lumber for increased heights. This is shown by the fact that the bracing lumber for a 6-inch or 8-inch slab is 1.5 board feet in the first line, but is 2 board feet in the third line. For slabs higher than 18 feet it is necessary to double-stack the shores. This doubles the cost and also increases the amount of bracing lumber required. Shores are figured to be 4 by 5 feet on centers; that is, each shore supports 20 square feet of forms, at $1.50 per month rental. Labor output per day figures and material costs for shored slab edge forms are given below the table.

For example assume a 12-inch-thick concrete slab with 18 feet from floor to floor. Figure three uses of material. Now check the bottom line in the table. The material cost per square foot is $1.00 for plywood, $1.25 for lumber and for adjustable shores at $1.50 per month rental divided by 20 square feet equals 8 cents per square foot times two (stack double) equals 16 cents per square foot. This is a total cost of $2.41 per square foot for one use divided by three uses equals 80 cents per square foot per use plus 8 cents for nails and oil equals 88 cents per square foot per use. The cost of edge forming, if required, must be added as a separate item.

Table 16-D
SHORED FLAT SLABS

	Slab	Per S.F. of Forms	Uses	Cost	Cost	Erect C+1/2L	Strip & Clean Labor
		MATERIAL/S.F.				**LABOR S.F./Day**	
	6"	1 S.F. Plywood	1	1.00			
		1.5 B.F. Lumber	1	.75			
	&	20 S.F./$1.50/Mo.			1.83	140	350
10'	8"	Adj. Shores	1	.08			
to							
12'	10"	1 S.F. Plywood	1	1.00			
high		2 B.F. Lumber	1	1.00			
	&	20 S.F./$1.50/Mo.			2.08	120	350
		Adj. Shores	1	.08			
	12"						
	6"	1 S.F. Plywood	1	1.00			
		2 B.F. Lumber	1	1.00			
	&	20 S.F./.60 Mo.			2.08	120	350
14'	8"	Adj. Shores	1	.08			
to							
16'	10"	1 S.F. Plywood	1	1.00			
high		2¼ B.F. Lumber	1	1.13			
	&	20 S.F./$1.50/Mo.			2.21	110	300
		Adj. Shores	1	.08			
	12"						
	6"	1 S.F. Plywood	1	1.00			
		2¼ B.F. Lumber	1	1.13			
	&	20 S.F. Stack			2.29	100	250
18'		Double $1.50/Mo.					
to	8"	Adj. Shores	1	.16			
20'	10"	1 S.F. Plywood	1	1.00			
high		2½ B.F. Lumber	1	1.25			
	&	20 S.F. Stack			2.41	90	200
		Double/$1.50/Mo.					
	12"	Adj. Shores	1	.16			
	Any	Oil & Nail ea. use	1	.08	.08	–	–

Slab Edge Forms = 120 S.F./Day or 200 L.F./Day
Edge Material = .50/S.F. for 2U=.25/S.F. per use

(Lumber = .50/B.F.; Plywood = $1.00/S.F.; Shores = $1.50/Mo. Rental)

TABLE 16-E

Use Table 16-E for shored pan and joist slabs. These costs are for shoring and centering only of the joists for the pans. The cost of the pans (metal, plastic, or fiber) for rental, erection, and removal must be added separately, and this is calculated on the basis of total square feet of slab needed. It is possible in many areas to get a total price for the rental, erection, and removal of these pans from a subcontractor.

The table is laid out much like Table 16-D. For comparison of costs, assume a similar slab to that taken for Table 16-D. However, in Table 16-D you need figure only joint bottoms and shores. There is no plywood deck material.

Assume you have a 12½-inch overall depth of joist plus the slab over the pans with 18 feet from floor to floor. Assume also that the material will be used three times. The lumber cost for 2 board feet is $1.00 per square foot and the adjustable shores will cost $1.50 per square foot rental divided by 25 square feet equals 6 cents per square foot times two (double-stacked) or 12 cents per square foot. This is a cost of $1.12 per square foot of slab forms divided by three uses or 37 cents per square foot per use plus 4 cents for nails (oil not needed) or 41 cents per square foot of slab area per use for material. The cost of rental, erection, and removal of the pans must be added to this figure in order to compare it to the cost of flat slab forming (88 cents per square foot per use for material) obtained in Table 16-D. Pan form slabs are cheaper than flat form slabs because there is a savings in concrete material and concrete placing labor.

Table 16-E
SHORED PAN AND JOIST SLABS

| | Slab | | Shores & Centering Only | | | | LABOR S.F./Day | |
| | | | MATERIAL/S.F. | | | | Erect | Strip & Clean |
		Per S.F. of Forms	Uses	Cost	Cost		C+1/2L	Labor
10'	8½"	1.4 B.F. Lumber	1	.70			240	600
to	&	25 S.F. per shore at			.76			
12'	10½"	$1.50 per mo. rental	1	.06				
High	12½"	1.5 B.F. Lumber	1	.75			240	600
	14½"	25 S.F. per shore at			.81			
	&	$1.50 per mo. rental	1	.06				
	16½'							
12'	8½"	1.5 B.F. Lumber	1	.75			220	550
to	&	25 S.F. per shore at			.81			
16'	10½"	$1.50 per mo. rental	1	.06				
High	12½"	1.6 B.F. Lumber	1	.80			220	550
	14½"	25 S.F. per shore at			.86			
	&	$1.50 per mo. rental	1	.06				
	16½'							
	8½"	1.8 B.F. Lumber	1	.90			170	450
		25 S.F. per shore at			1.02			
	&	$1.50 per mo. rental	1	.12				
18'		Double Stack						
to	10½"							
20'	12½"	2 B.F. Lumber	1	1.00			150	400
High	14½"	Double Stack			1.12			
	&	25 S.F. per shore at	1	.12				
	16½"	$1.50 per mo. rental						
	Any	Nails each use		.04	.04		—	—

(Lumber = .50/B.F.; Adj. Shores = $1.50/per month rental)

TABLE 16-F

Use Table 16-F for estimating formwork for clay tile and concrete joist construction of slabs. The table is laid out similar to Table 16-D. This type of slab forming is not much used, but an understanding of it should be helpful to any estimator.

Rows of shores are set up on 4-foot centers with the shores in each row also on 4-foot centers generally. Each row of shores is capped with a continuous 2-inch plank. The entire slab area is decked solid with plywood. Up to this point, this is the construction for flat slab forming.

The next step is to lay hollow clay tile on the plywood deck with openings at intervals to form concrete joists between the hollow clay tiles. When the concrete is placed it flows between the tiles, fills the joist openings and covers the hollow clay tile with about 2 inches of concrete for the finished floor. This type of floor slab is lighter weight than solid concrete, and its load-bearing capacity is not as great as the same thickness of solid concrete. In cost, this type of construction falls between shored flat slab and pan slab construction, and for this reason it is not much used.

Table 16-F
CLAY TILE AND JOIST SLABS WITH BEAM AND GIRDER CONSTRUCTION

| | | | MATERIAL/S.F. | | | | LABOR S.F./Day | |
| | | | | | | | Erect | Strip & Clean |
	Slab	Per S.F. of Forms	Uses	Cost	Cost	C+1/2L	Labor
	6½'	1 S.F. Plywood	1	1.00			
		1.1 B.F. Lumber	1	.55			
10'	&	20 S.F. per shore at					
to		$1.50 per mo. rental	1	.08			
12'	8½"	Shore			1.63	150	500
High	10½"	1 S.F. Plywood	1	1.00			
	12½"	1.3 B.F. Lumber	1	.50			
	&	20 S.F. per shore at					
	14½"	$1.50 per mo. rental	1	.08	1.58	140	450
	6½"	1 S.F. Plywood	1	1.00			
		1.2 B.F. Lumber	1	.60			
14'	&	20 S.F. per shore at					
to		$1.50 per mo. rental	1	.08			
16'	8½"				1.68	140	450
High	10½"	1 S.F. Plywood	1	1.00			
	12½"	1.4 B.F. Lumber	1	.70			
	&	20 S.F. per shore at					
	14½"	$1.50 per mo. rental	1	.08	1.78	130	400
	6½"	1 S.F. Plywood	1	1.00			
		1.4 B.F. Lumber	1	.70			
	&	Double Stack					
18'		20 S.F. per shore at	1	.16			
to	8½"	$1.50 per mo. rental			1.86	110	300
20'	10½"	1 S.F. Plywood	1	1.00			
High	12½"	1.6 B.F. Lumber	1	.80			
	&	Double Stack					
	14½"	20 S.F. per shore at	1	.16			
		$1.50 per mo. rental			1.96	100	250
	Any	Nails—Oil	1	.08	.08	—	—

(Lumber = .50/B.F.; Plywood = $1.00/S.F.; Adj. Shores = $1.50/month rental)

TABLE 16-G

Use Table 16-G to calculate labor costs for placing concrete and for finishing various concrete surfaces. These represent costs based on average production per man per cubic yard of concrete or finishing area, and must be varied to fit job conditions.

Please refer to the top table for placing concrete. The third column shows the labor time per cubic yard for a man working in a crew of 6 to 12 men placing concrete. The cost of building runways, equipment needed, and hoisting must be added separately. The next four columns show the cost of placing per cubic yard for wage scales varying from $18.50 to $25.00 per hour. These costs can be worked out to fit any wage scale. For smaller or larger crews placing less or more concrete per day, the labor time per cubic yard per man must be varied in accordance with the individual job. This variance is based on the estimator's knowledge and experience.

Now refer to the lower table for finishing concrete. The third column shows the square feet or linear feet of finishing various surfaces per hour per man. The cost of equipment must be added separately. The next four columns show the cost of finishing these surfaces for wage scales varying from $25.00 to $32.50 per hour. These costs can be worked out to fit any wage scale.

Table 16-G
PLACING AND FINISHING CONCRETE

Concrete Placing Costs — Labor Only
Add Equipment & Hoist Separately

	Type	Labor /C.Y.	Rate per Hour 18.50	20.00	22.50	25.00
			Cost per Cubic Yard			
1.	Footings—Piers	.9 Hr	16.65	18.00	20.25	22.50
2.	Foundation Walls—12"—	.8 Hr	14.80	16.00	18.00	20.00
3.	Foundation Walls 8"—	.9 Hr	16.65	18.00	20.25	22.50
4.	Ground Flr. Slabs 8"—	.8 Hr	14.80	16.00	18.00	20.00
	5"—	.9 Hr	16.65	18.00	20.25	22.50
5.	2nd Flr. & Up—Slabs	1.0 Hr	18.50	20.00	22.50	25.00
6.	Beams—All Floors	1.1 Hr.	20.35	22.00	24.75	27.50
7.	Ground floor—2" Slabs	1.1 Hr	20.35	22.00	24.75	27.50
8.	2nd Flr. & Up—2" Slabs	1.2 Hr	22.20	24.00	27.00	30.00
9.	Columns	1.2 Hr	22.20	24.00	27.00	30.00
10.	Stairs & Landings	2 Hr	37.00	40.00	45.00	50.00
11.	8" Paving/or Bsmt. Flrs.	.8 Hr	14.80	16.00	18.00	20.00
12.	Curbs	1.2 Hr	22.20	24.00	27.00	30.00
13.	Sidewalks	1 Hr	18.50	20.00	22.50	25.00
14.	Locker Base & Small Pours 1 C.Y. or less	4 Hr.	74.00	80.00	90.00	100.00
15.	Steel Stairs—Panfill	8 Hr	148.00	160.00	180.00	200.00

Set Screeds & Finish

	Type	Labor Per Hour	Rate per Hour 25.00	27.50	30.00	32.50
			Cost per Unit			
1.	Steel Trowel—Bldg.	50 S.F.	.50	.55	.60	.65
2.	Float Finish	125 S.F.	.20	.22	.24	.26
3.	Steel Trowel Walks	40 S.F.	.63	.69	.75	.81
4.	Curb & Gutter or Treads & Risers	9 L.F.	2.78	3.06	3.33	3.61
5.	Curb Tops or Treads Only	18 L.F.	1.39	1.53	1.67	1.81

TABLE 16-H

Use Table 16-H to price formwork erection costs, stripping costs, and any other item for labor output based on the number of units per day estimated per man. After the number of units per day per man are filled in for each item on the recapitulation sheet, the cost for each item per unit can quickly be picked out from the table.

The wage rates shown here should represent average wage rates for the country. The figures of $30.00 for a carpenter and $25.00 per hour for a laborer are only examples. Use your own costs. Whether the fringe benefits are included here or with the insurance and tax markup depends on the preference of the contractor. Since there is a considerable variation in wage scales throughout the country, it is necessary for each contractor to make out a table similar to this one based on his own labor rate figures. Also, since the wage scale rises every year, he must revise his table every year. Use the blank forms to create your own tables.

The first column gives the units per day per man. The second column gives the unit cost for a carpenter and the third column gives the unit cost for a laborer. The fourth column gives the unit cost for a carpenter and one-half of a laborer's rate added together. This is principally for figuring formwork erection costs based on using a crew with twice as many carpenters as laborers. The variation in units from 100 to 950 is in approximately 10% differences and will fit all conditions by adjusting the decimal point or interpolating between numbers.

Table 16-H
UNIT LABOR COSTS FOR FORMWORK OR OTHER ITEMS

Ratio		Carpenter	Laborer	C + 1/2L.
Rate/Hour		**$30.00**	**$25.00**	**$42.50**
Rate/Day		**240.00**	**200.00**	**340.00**
	100	2.40	2.00	3.40
	110	2.18	1.82	3.09
	120	2.00	1.67	2.83
	130	1.85	1.54	2.62
	140	1.71	1.43	2.43
	150	1.60	1.33	2.27
	160	1.50	1.25	2.13
	170	1.41	1.18	2.00
	180	1.33	1.11	1.89
	200	1.20	1.00	1.70
	220	1.09	.91	1.55
UNITS	240	1.00	.83	1.42
	250	.96	.80	1.36
PER	275	.87	.73	1.24
	300	.80	.67	1.13
DAY	350	.69	.57	.97
	400	.60	.50	.85
	450	.53	.44	.76
	500	.48	.40	.68
	550	.44	.36	.62
	600	.40	.33	.57
	650	.37	.31	.52
	700	.34	.29	.49
	750	.32	.27	.45
	800	.30	.25	.43
	850	.28	.24	.40
	900	.27	.22	.38
	950	.25	.21	.36

Wage Rates: Incl. Fringes—No I. & T.

Table 16-H
UNIT LABOR COSTS FOR FORMWORK OR OTHER ITEMS

Ratio		Carpenter	Laborer	C + 1/2L.
Rate/Hour				
Rate/Day				
	100			
	110			
	120			
	130			
	140			
	150			
	160			
	170			
	180			
	200			
	220			
UNITS	240			
	250			
PER	275			
	300			
DAY	350			
	400			
	450			
	500			
	550			
	600			
	650			
	700			
	750			
	800			
	850			
	900			
	950			

Wage Rates: Incl. Fringes—No I. & T.

Table 16-H
UNIT LABOR COSTS FOR FORMWORK OR OTHER ITEMS

	Ratio		Carpenter	Laborer	C + 1/2L.
	Rate/Hour				
	Rate/Day				
	100				
	110				
	120				
	130				
	140				
	150				
	160				
	170				
	180				
	200				
	220				
UNITS	240				
	250				
PER	275				
	300				
DAY	350				
	400				
	450				
	500				
	550				
	600				
	650				
	700				
	750				
	800				
	850				
	900				
	950				

Wage Rates: Incl. Fringes—No I. & T.

Table 16-I
CONVERSION FACTORS FOR CONCRETE MIX DESIGN

Cement Content per Cubic Yard			Cement Content		Water-Cement Ratio		Water-Cement Ratio	
Barrels	Bags	Pounds	Pounds	Bags	Gal. per Bag	Weight Ratio	Weight Ratio	Gal. per Bag
1	4	376	350	3.72	4	0.36	0.35	3.94
	4.5	423	400	4.25	4.5	0.40	0.40	4.50
1.25	5	470	450	4.79	5	0.44	0.45	5.07
	5.5	517	500	5.32	5.5	0.49	0.50	5.63
1.5	6	564	550	5.86	6	0.53	0.55	6.20
	6.5	611	600	6.39	6.5	0.58	0.60	6.76
1.75	7	658	650	6.92	7	0.62	0.65	7.32
	7.5	705	700	7.45				
2	8	752	750	7.98				
	8.5	799	800	8.52				
2.25	9	846						

Specify portland cement content and water-cement ratio by weight.
1 ton = 5.32 bbl (376 lb/bbl)
    = 21.28 bags (94 lb/bag)

METRIC EQUIVALENTS

Cement Content		Cement Content	
lb per cu yd	kg per m³	kg per m³	lb per cu yd
376	223	200	337
423	251	225	379
470	279	250	421
517	307	275	464
564	335	300	506
611	362	325	548
658	390	350	590
705	418	375	632
752	446	400	674
799	474	425	716
846	502	450	758
		475	801
		500	843

For:	*Multiply by:*
Cement content	
pound per cubic yard to kilogram per cubic meter	0.5933
Water-cement ratio	
gallon per bag to weight ratio	0.0888

CHAPTER 17
RECAPPING AND PRICING

Setting up the recapitulation sheet seems simple, but too many contractors make it too simple. It is not enough to put the name of the job, the architect and the date at the top of the sheet. The drawing numbers and dates, the specifications and date, and any addenda that were seen should also be noted. This information may become valuable if you are awarded the contract to do the work, and then, later, revisions are made to the drawings. At that point it becomes very important to be able to prove what was included in your original proposal.

The daily labor rate on which your daily production output for formwork is based should be marked over the top of the right-hand column, the labor totals column, on your recap sheet. In most cases this daily rate will include the pay for one carpenter per day and one-half of a laborer's rate per day. This is based on most formwork erection being done by a crew with a ratio of two carpenters to one laborer. When pricing the carpenter output per day, include all laboring time. For example, if a carpenter's rate is $240.00 per day and a laborer's rate is $200.00 per day, the daily labor rate would be marked for production output as: C. + ½ L. = $340.00 per day. This is marked at the top of the labor column. These rates may or may not include the fringe benefits. Insurance and taxes will vary from about 20% to 30% depending on where the fringe benefits are included. Insurance, taxes, and fringe benefits vary with the contractor and locality and must be individually calculated.

We are now ready to move the items for pricing from the quantity takeoff sheets to the recapitulation sheet. All items to be priced are in the extreme right-hand column of the takeoff sheets and they are marked with a check (√). After an item is moved to the recap sheet, the right-hand column of the takeoff sheet should be marked thus (X). When all items in this column on all takeoff sheets are marked thus (X), we know that we have not forgotten to forward any of them to th recapitulation sheet for pricing. This is a small detail but it helps preven errors. A completed row of double-check marks in the right-hand column of the takeoff sheets is easy to spot when a quick check is made. And a quick check is important because of the recapping system. Recapitulate items in the following order: formwork, concrete, finishing, hand-excavation, and miscellaneous items. Number the item as it is taken off and use the same number on the recapitulation sheet. This helps if you have to refer back to the "takeoff" sheets.

When you have recapped all of the formwork, stop and include a line for stripping and cleaning forms. This is priced separately because laborer time only should be figured on stripping and cleaning forms. The total stripping and cleaning area may be obtained by adding up the formwork totals that have just been recapped. Be careful to include only the square feet of forms that will be stripped. Do not include permanent forms such as corrugated formwork, edge forms, or other similar items.

When you have finished recapping the concrete, add up the total amount and include a line for concrete testing. This will be priced by the total cubic yardage of concrete used.

To recapitulate the finishing, first find the screed material. This will be a total of all slab areas for the building added up from the takeoff sheets. This includes sub-slabs, finish slabs, and concrete fill slabs. Next separate all float finish slabs. These are found from the room schedule on the drawings or from the specifications. They include all sub-slabs, most roof slabs, and slabs that will have a finish floor of terrazzo, marble, slate, ceramic tile, or similar materials. If a ceramic tile floor is to be set by the "thin-set" method in mastic, it is necessary to figure a steel-trowel finish on the concrete under it.

The remainder of the floor slabs with a steel-trowel finish can be found by subtracting the float finish areas from the total screed material area. Other finishing items such as stair treads and curbs can be obtained from the takeoff sheet. Hand rubbing of exposed surfaces will be done by the finishers and must be listed here. Basement walls and ceilings should be kept under one heading. Exposed columns and beams should be grouped together, and the area can be obtained generally from the takeoff for formwork. Walls and ceilings for finished rooms should be grouped together for pricing.

The recapitulation of the hand excavation items can be taken directly from the takeoff sheets. To set them up for pricing they should be grouped as follows: column footings, wall footings, building slabs, sidewalks, and curbs. The spreading and compaction of sand or stone fill should be listed in cubic yards. Separate the sand or stone fill at walls and footings from the same items under building slabs or sidewalks. These separations are necessary for properly pricing these items.

The recapitulation of the miscellaneous items comes from several sources. First check the takeoff sheets for any items that have not been double-checked. By now, the takeoff sheet items in the last column should all be on the recapitulation sheet. Next, check the drawings for any items that may have been missed. Next, check the specifications for any items called for that are not shown on the drawings. Then add on all items of a direct job-cost nature that you know should be on the estimate. A partial listing of these items would include: curing of slabs, hoist and crane time, equipment rental, heat protection, job cleanup, hauling debris, truck driver's time, sawman's time, and foreman's time.

Next we will discuss the takeoff and the recapping of the various miscellaneous items in the order in which they might occur.

Vapor barrier under slabs-on-grade can be obtained from your takeoff sheets, but be sure to add 10% to the slab area for lap and waste.

Perimeter insulation at walls and slabs-on-grade must be taken from the drawings in square feet and entered directly on the recapitulation sheet. Add 5% for waste here.

Waterstops—rubber, copper, or other—must be taken off the drawings in linear feet, adding 5% for waste.

Expansion joints can be obtained from the takeoff sheets in linear feet, but add 5% for waste.

The setting of column anchor bolts and base plates must be taken off the drawings. These are most easily priced by the column unit, so they should be listed on the recapitulation sheet as so many column-base units to set. If the base plates are of average size with two anchor bolts, allow one hour of carpenter time per column base unit.

Steel angles set in slabs should be listed by the linear foot to price the labor of setting them.

Dove-tail slots set in walls for anchoring masonry must be taken off the drawings in linear feet with 5% added for waste.

Steel inserts or bolts to be set in concrete beams for shelf angles to support masonry must be taken off the drawings by the unit, again adding 5% for waste.

If a floor hardener is specified on exposed concrete floors, the area can be taken off the drawings by referring to the room finish schedule.

Curing of slabs must be figured even if it is not specified. If no method of curing is specified, figure on using a liquid membrane because it costs less and is the simplest method of curing. To price the spray-cure method, use the exact floor areas for figuring. If a paper, polyethylene, plastic or similar method of curing is specified, it is necessary to add 10% to the areas for lap and waste. The material costs more, the labor of placing it is higher, and it must be removed after the curing has been completed.

Hoist or crane time can be taken off by the length of time they will be used. For high buildings, where a crane cannot be used, a hoist must be figured and hoisting engineer's time included. Many companies rent out hoists to contractors and will give a price for furnishing, erecting, dismantling, and rental based on the number of weeks rental is required. To this must be added the hoist engineer's cost for the time needed. For low buildings a crane should be figured whenever it is possible, because a crane costs less and can be moved to the exact spot needed. Crane rental time should be figured by the days required on the job, allowing time for each operation as it is needed. Other equipment rental items could include concrete pumps and piping, vibrators, and small tools.

Heat protection, if required, can best be priced by the cubic yard of concrete that needs heat protection. Analyze the job as to what time of year the work will be done and what kind of concrete work needs heat protection. January and February are the most costly months, and shored slabs above the ground are the most costly to protect. Therefore, decide how many weeks of placing will require heat protection and how many cubic yards of concrete will be placed on the average per week. This gives the number of cubic yards that need heat protection. As for material costs, remember that the ready mix producer will usually charge extra for heated concrete.

The cost of protection on the job varies depending upon materials used, type of enclosure, type of heat, and whether it is a footing or a slab above ground that must be protected. The best way to price protection is on a basis of the total cubic yardage needing protection. Labor required for heat protection can be analyzed from the number of placings required, the length of time for protecting each placing, and the number of laborers required to protect a placing. Labor for heat protection of concrete based on cubic yards can vary—from one-quarter of an hour to one hour laborer's time per cubic yard.

General cleanup of the job is dependent on its size and the number of weeks the concrete work will last. Therefore, using the recapitulation sheet for concrete, estimate how many weeks it will take to do the individual parts of a building such as footings, walls, slabs-on-ground and shored slabs. Once the total number of weeks on the job are estimated, the recapitulation sheet can be completed. On jobs up to about $100,000, allow one day of laborer's time per week for general cleanup. On jobs up to $250,000 allow 2 days of laborer's time per week, and so forth. This, however, is a matter of judgment and could vary with the type of the job and, especially, with knowledge of the architect's and owner's requirements.

The quantity of debris to be hauled away is also dependent on the size of the job and the number of weeks it will last. On jobs up to about $100,000, allow one-half truckload of debris per week of work. On jobs up to $250,000, allow one truckload of debris per week of work, and so forth. It is a rough rule of thumb to use for estimating, but it is a guide based on as much knowledge as the author has been able to obtain.

The truck driver's time is also dependent on the number of weeks the job lasts, and can be estimated better than the two previous items. For any individual job, we can estimate the number of trips per week the truck driver will make to the job and the hours per trip. On average jobs up to $100,000, allow one day of truck driver's time per week of the job. On jobs up to $250,000, allow two days of truck driver's time per week, and so forth.

The saw man's time for the job must be estimated only for the time he will be used. A saw man is a carpenter who spends all of his time running a power saw to cut materials for columns, beams, and shored slabs. His time is not included elsewhere in the estimate and must be added to the recapitulation sheet. On the average job, do not figure saw man time for footings, foundation walls, and slabs on the ground. However, when columns, beams, and shored slabs are involved, do allow full time for a saw man.

The foreman's time, of course, must be allowed for the full length of the job. On small jobs, with a "working foreman" or where the owner is his own foreman, supervision must still be taken into account. There is always layout and planning time that must be priced and placed into the estimate. A good rule to follow in small jobs is to allow the full number of weeks times a percentage of the foreman's salary. An average would be to figure 50% of his salary on supervision, and charge it in the estimate as foreman's time.

On average size jobs, when the general contractor does his own concrete work, the concrete supervision is done by the job superintendent. In this case, the supervision of the concrete work is priced as part of the job superintendent's time. On large jobs, a general contractor might decide that he needs both a concrete foreman and a job superintendent, and then time for the concrete foreman should be charged into the concrete work.

PRICING MATERIALS AND LABOR

When the recapitulation sheet is filled in with all of the required quantities, begin pricing materials and labor. The total labor costs for average concrete jobs are roughly double the material costs. Material prices for lumber, plywood, ties, and nails rise an average of 8% per year. This cost is divided between from three to five uses for lumber and plywood, so the price rises for these items per use are lessened somewhat on many jobs. And if labor output per day remains a constant unit, only daily wages rates must be changed each time carpenters and laborers get raises. Think in terms of how many square feet of a certain type of work a carpenter can do in one day. His rate of production hardly changes from year to year.

Mark down on the recapitulation sheet the square feet or units per day that you estimate the carpenter will spend on each item of formwork. When this is marked for all form erection items, pick out the cost per square foot of any form erection item from the current wage table for a carpenter and one-half of a laborer's time. For example, if the rate per day is $340.00 and the labor output is 100 square feet per day, the cost is $3.40 per square foot.

Labor output tables for formwork erection are found in Chapter 16. Use them as a basis for developing your own cost data. Labor output data for finishing concrete is in Table 16-G.

On the typical job recapitulation sheet, there is one column headed L per D, or labor unit output per day. From the formwork labor output tables, the pricing estimator fills in the L per D column and the unit material prices on the recapitulation sheet. He tempers both of these with his own evaluation of the job he is pricing, taking into account the factors that might cause average material and labor units to vary. There are three items listed in the tables. These are material costs, erection labor per day, and stripping labor per day. The erection time includes a carpenter and one-half laborer per day. The stripping time includes one laborer per day.

RECAPPING THE CONCRETE (See Table 16-G)

Next, recapitulate all of the concrete from the takeoff sheets. The concrete quantities may be combined by grouping items that belong in the same pricing bracket. Column footings and wall footings, for example, can be grouped together because they have the same labor unit. Column footings should be kept separately if there are just a few isolated column footings, since these will be expensive to place individually.

When placing concrete in foundation walls, the labor unit cost difference is based upon the thickness of the wall. A 6-inch wall, for example, will use more labor per cubic yard of concrete than will a 12-inch wall. Or, the thinner the wall the higher the labor cost per cubic yard for placing concrete. Interior and exterior columns will cost the same to place and can be grouped together.

All beams will cost the same to place and can be grouped together.

Most slabs can be grouped together because the labor for placing them will be the same. However, thin slabs—3 inches thick or less—will require more labor per cubic yard of concrete to place due to the time spent in spreading the concrete out in a thin layer.

In a multistory building it may be necessary to separate the concrete by floors, particularly if the architects or engineers specify that 5,000-pound concrete is to be used up to the 10th floor and everything above the 10th floor is to be 3,500-pound concrete. Then, there will be a difference in price of perhaps $6.00 per cubic yard of material.

In most cases the concrete will cost from $50 to $75 per cubic yard delivered. Since most jobs average about $150 to $250 per cubic yard for concrete in place including concrete, formwork, finishing and miscellaneous items, it is easy to see why the concrete yardage is important. The concrete material cost will make up about one-quarter to one-third of the total cost for the job. And if the concrete yardage is figured correctly, one-quarter to one-third of the bid will be fixed. This is most important to the successful bidder.

Always figure concrete yardage for exactly the correct quantity. Waste, if added, should be figured as a separate item, perhaps 3% of the exact yardage. But when estimating waste, price the material only. Do not add anything in the labor column because there is no labor involved when you are pricing concrete as a waste material.

All labor costs for placing concrete can be broken down and priced out by the amount of concrete to be placed, the number of men it will take to place the concrete, and the amount of time required to complete the placing operation. For example, 10 men working for one full day will provide 80 hours of work and if they placed 80 cubic yards of concrete in that time, the concrete would have been placed at the rate of one cubic yard per manhour. All cases are not this simple, but the method of figuring concrete labor costs is always the same.

The table on the labor time required to place concrete was prepared after calculating many jobs and arriving at labor units per cubic yard for placing concrete.

The following information on average labor output for placing concrete must be adjusted to fit any special conditions.

For footings and piers, use 9/10 hour laborer's rate. If a laborer costs $25 per hour, the 9/10 of $25 is $22.50 per cubic yard. The 9/10 hour remains a constant.

For 12-inch thick foundation walls use 8/10 hour. For 8-inch thick foundation walls use 9/10 hour. For building slabs-on-fill use 9/10 hour for 5-inch slabs, 8/10 hour for 6-inch to 8-inch slabs, and 7/10 hour for slabs 10 inches and over in thickness.

For second floor and higher shored slabs use 1 hour per cubic yard of concrete.

For beams on all floors, figure 1-1/10 hours per cubic yard.

For thin slabs, 3 inches or less in thickness, figure 1-2/10 hours per cubic yard.

For columns, figure 1½ hours per cubic yard.

For stairs and landings, figure 2 hours per cubic yard.

For 8-inch exterior paving work in large areas, figure 6/10 hour per cubic yard.

For curbs or curb and gutter work, figure 1-2/10 hour per cubic yard.

For sidewalks, figure 1 hour per cubic yard.

For small interior building curbs, locker bases, roof curbs, and small jobs in many different locations, figure 4 hours per cubic yard.

For steel stairs, pan-filled, figure 8 hours per cubic yard.

When trying to figure a certain concrete placing operation becomes difficult, break the job down to number of cubic yards required, time, and number of men required to place the concrete.

Concrete material prices vary considerably depending on the size of the job, the location, and the concrete supplier. Concrete prices must be obtained for each individual job. An air-entraining admixture will cost very little extra. But most other admixtures will cost extra and this cost must be added into an estimate. Two things really determine the price of concrete. One is the number of bags of cement; the other is the admixture. The simplest way to get the cost of concrete material is to call up the concrete supplier, read the specifications to him and ask, "How much will you want per cubic yard for this concrete?"

Next, add up the total concrete yardage for the complete estimate. This is done for two reasons. One is to price the testing, and the other is to know the total cubic yardage so it can be divided into the total cost to find out the average cost per cubic yard for the job. This helps point out error in the estimate, if one is present.

If testing is specified, the cost of testing will average about $.50 to $1.00 per cubic yard. On small jobs a minimum of about $100 should be used to cover the design mix cost.

FINISHING

The first item to list under the finishing category is screed material. This is determined by taking the total finishing area of all building slabs for the entire job, as added up from the takeoff sheets. As the job was taken off, the finishing for all slabs should have been listed. Any area that has to be finished must have screeds set up to do the job. So from the takeoff sheets add up all of the finishing areas, and that is the area that needs screed material. Screed material will cost about 4 cents per square foot. This price includes 2- by 4-inch screeds and stakes. The price also allows for reuse of material. Nothing is charged in the labor column; that time is included in the various types of labor finishing costs.

The second item to list on the recapitulation sheet is the float finish. The roof slab in most cases will be the biggest part of the float finish area since it will be covered by roof insulation or roofing. This area can be taken directly from the takeoff sheets. There are other float-finish surfaces, in the room schedules, for example, that will have to be figured in the recapitulation. All areas shown in the room schedule to have a ceramic tile, quarry tile or terrazzo floor finish would need a float-finish concrete floor under them. If the specifications call for the ceramic tile floor to be set by the thin-set method (that is, with mastic), then a steel-trowel concrete surface is needed. The second item, therefore, is the total float finish area as required by the drawings and specifications.

The third item on the recapitulation sheet is the steel-trowel finish on slabs. This is the screed material less the float finish areas, or the first item less the second item. The remaining finishing items—such as sidewalks, stairs, curbs and so forth—must be taken directly from the takeoff sheets.

On float-finish work figure that a man can finish an average of 125 square feet per hour. At a rate of $25 per hour for a finisher, divided by 125 square feet, this is a cost of 20 cents per square foot for float finishing. These averages are figured on a basis of square feet or linear feet per hour per man, and include setting of screeds and expansion strips as the work proceeds.

On steel-trowel finish work for building slabs, figure a man can average 50 square feet per hour, and 50 into $25 per hour would be 50 cents per square foot.

On steel-trowel finish for walks, figure 40 square feet per hour because the walks are narrow and must be scored. Dividing 40 into $25 per hour would give a cost of 63 cents per square foot for sidewalks.

For a straight curb without a gutter, figure 18 linear feet per hour which, divided into $25 per hour, is about $1.40 per linear foot of curb.

Combination curb and gutter is figured at twice that much; or allow only 9 linear feet of curb and gutter per hour. This works out to $2.80 a linear foot to finish curb and gutter work.

For stair treads, figure 18 feet per hour—the same as for the curb top. This is $1.40 per linear foot for stair treads.

For stair treads and risers, double the above costs. In other words, figure 9 linear feet per hour, or $1.40 per linear foot of tread and riser.

Another item that belongs under finishing is hand rubbing of exposed walls, columns, beams, and ceilings in accordance with the specifications or notes on the drawings. In residences, the specifications usually call for only a brush coat on the interior walls. This type of job can be priced more by the items than anything else. You probably know that a couple of laborers can do the job and you also probably know about how long the job will take. Hand rubbing an entire basement, for example, might cost $150 to $200, depending on the size of the basement. But if you get a commercial job with a real hand-rubbing problem, such as exposed beams and ceilings, then break the job down into the areas for the type of work, and the degree of difficulty involved in doing the work.

When figuring hand rubbing, keep the walls, columns, beams, and ceilings as separate areas in order to price them more correctly. It is not necessary to go back over the drawings and pick out each column and beam to be hand rubbed. This can be taken on a percentage of the total areas of each item. In other words, the plans may show 2,000 square feet of columns to be formed but only about half of these columns are exposed. The other half may be encased in masonry. In this case, figure on hand rubbing 1,000 square feet. Do not spend a lot of time on figuring out which areas will require hand rubbing, because it is the unit price that is important. The unit price value of hand rubbing must be decided after considering several factors. In the first place, read the specifications. If they say the hand rubbing has to be done with an abrasive stone and the architect wants white cement used for finishing the job, then the hand-rubbing operation will be expensive and will cost about $2.00 per square foot for columns and beams.

If exposed ceilings or the bottoms of slabs in finished rooms are to be hand rubbed, the cost would probably be about $1.00 per square foot. In most buildings, the ceiling slabs would be only in the basement or in mechanical rooms. If this is the case, they will probably not cost over 50 cents per square foot. On the other hand, an apartment building with exposed-ceiling slabs and an 8-foot ceiling that must be hand rubbed before painting will be expensive. This type of job could cost $1.00 to $1.50 per square foot.

Some specifications will call for hand rubbing with an abrasive stone, using white cement, but when you actually get out on the job they don't mean anything of the kind. Someone wrote this in the specifications and it has been there for a long time, but the architect never insists on it. It pays, therefore, to know the architect and the quality he expects for hand-rubbing work. Hand-rubbing costs are really an expression of an estimator's experience, judgment, and knowledge of the requirements for each individual job.

HAND EXCAVATION

Hand excavation or hand grading of the earth before placing concrete on the ground is the next item that should be listed on the recapitulation sheet. The quantities should be taken directly from the quantity takeoff sheets, since these areas are easy to obtain when taking off the concrete and formwork quantities. What is meant by hand excavation? After the machine excavation is finished, the surface of the ground is usually rough and too high and must be cut down another 1 or 2 inches. This is hand excavation work. Figure it on a square-foot basis because you will not be able to tell how many cubic yards of earth must be excavated.

Before figuring hand excavation on a square-foot basis, however, the existing earth must be classified into three categories: sandy soil, ordinary soil, and tough clay.

If test borings show that the work is all in sandy soil, the hand excavation cost will be at a minimum. If the footings are 3 to 4 feet underground and the condition of the earth ranges from earth to clay, the hand excavation cost should be figured as a medium price. If the footings are at the bottom of a basement and the test borings show tough clay, the hand excavation cost should be figured as a maximum price.

Also consider the type of hand work to be done. Column footings, for example, are more expensive than wall footings. For hand excavation work, use a laborer's cost of $25 per hour. Hand excavation for column footings should vary from a maximum of 50 square feet per hour in sand to 25 square feet per hour in tough clay. This is a variation from 50 cents per square foot to $1.00 per square foot. Hand excavation for wall footings should vary from a maximum of 100 square feet per hour in sand to 50 square feet per hour in tough clay. This is a variation from 25 cents per square foot to 50 cents per square foot. Hand excavation for floor slabs should vary from a maximum of 200 square feet per hour in sand to 100 square feet per hour in tough clay. This is a variation from 12½ cents per square foot to 25 cents per square foot.

Hand spreading of sand fill at ground floor level should be figured at ¼ of an hour per cubic yard, or $6.25 per cubic yard (¼ times $25.00). To spread sand at a basement floor level, figure ½ hour per cubic yard, or $12.50 per cubic yard.

If the sand fill is to be compacted by hand at foundation walls the costs should be figured at about 1 hour per cubic yard, or $25 per cubic yard. This cost figure should be added to the cost of hand spreading. When figuring the sand fill (or stone fill) quantities to the "in-place" yardage, add 25% for loss and compaction.

MISCELLANEOUS ITEMS

The last items that remain to be taken off for a complete estimate are the miscellaneous items. These are probably the most important—and the most frequently forgotten—items on the average estimate. A careless estimator will not forget all of the items but he may overlook some of them in his takeoff. These miscellaneous items can amount to about 5% of the total bid. And since the overhead and profit added to a job is generally 5-10% depending on the size of the bid, the miscellaneous items, if they are omitted, will cut sharply into the profit.

The only way to be sure all of the miscellaneous items are picked up is to check them off one by one in the specifications, and then check the drawings, sheet by sheet, for all items to be furnished or installed by the concrete contractor. Here are some of the more typical miscellaneous items:

Dove-tail anchor slots in walls, columns, and beams for anchoring masonry work to concrete—these are taken off in linear feet, to which 5% must be added for waste. The labor on these should be about 500 linear feet per day by a carpenter.

Vapor barrier under slabs-on-ground. This is taken off in square feet, to which 10% must be added for lap and waste. The area for vapor barrier, of course, is taken directly from the previous takeoff sheets. The material cost of a vapor barrier depends on the specifications, and a man should be able to lay 200 square feet per hour.

Perimeter insulation on foundation walls. This must be taken off the drawings directly. Insulation is usually placed (with mastic) in 2-foot widths just below grade for slabs-on-ground to prevent heat loss through the foundation walls. Foamed plastic is commonly used as an insulating material and costs about 40 cents per inch of thickness. Add about 3 cents per square foot for mastic to this figure. A laborer should be able to lay up 400 square feet of perimeter insulation per day.

Setting of column-base plates and anchor bolts. This can be a large item if the superstructure of the building is all structural steel framing on steel columns. Although the structural steel man will furnish the anchor bolts and setting plates, the concrete contractor has to set them. The quickest and most accurate way to price this item is by the column as a unit. Count the column bases that have two anchor bolts and a setting plate, and keep them separate from column bases that have four anchor bolts and a setting plate. Check to see if the column-base plates have unusually large and heavy anchor bolts. If they do, be sure to take this into account when pricing the labor required to set them. The time for setting anchor bolts and a setting plate for one column base will vary from a minimum of 1 hour to 2 hours, depending on the number of bolts and their size.

On this work, 1 hour of time means 1 hour for a carpenter and 1 hour for a laborer helping him. If the setting plate must be grouted in place, this will have to be a separate operation and must be added in separately. If a non-shrink grout is specified, it is best to calculate this item in cubic feet and price the material by the cubic foot because of its high cost. The labor of grouting should be priced by the column base, allowing perhaps ½ hour to 1 hour per column base per man. In this work it is much easier to figure the labor by the time unit and let the cost be set by the time allowed per unit of work.

Expansion joints. These are best taken off in linear feet for the various thicknesses and depths such as the linear feet of ½- by 6-inch expansion joints. The material prices are quoted in linear feet per thickness and depth. The concrete finishers generally set the expansion joints as the placing of the slab progresses so that this labor is included in the price of finishing per square foot and is not added on as an additional cost per linear foot of expansion joint. If a joint sealer is specified after the slab has hardened, it should be figured as an additional cost. The material cost is perhaps only about 3 cents per linear foot, and a laborer should pour about 200 linear feet per hour.

Curing and protection of slabs after they have been placed. This is figured by taking the total slab area—the same area previously noted for screed material. A spray-on membrane curing compound is the least expensive method of curing and generally costs about 3 cents per square foot for material. A laborer should be able to spray about 300 square feet per hour.

If the specifications require the use of a curing paper, the work is more expensive. If paper is to be used, add 10% to the slab area for lap and waste. The cost of curing papers varies from 6 to 25 cents per square foot, depending on the material specified. The labor for placing curing papers will vary from 200 square feet per hour down to 150 square feet per hour if the joints must be sealed with tape.

When the protection period is over, the paper must be taken up and removed from the site. This requires laborer's time figured at about 300 square feet per hour. Therefore, the obvious answer to this problem is to use the spray-on membrane solution whenever the specifications or the job permit you to do it.

Sand protection of slabs. This is very much the same as using a curing paper as far as labor is concerned. A 1-inch layer of sand will cost about 3 cents per square foot for material. This makes the cost of sand protection of slabs cheaper than paper because the material will cost less. Even if the specifications do not say that the contractor must cure the slabs, a smart contractor will include it for his own protection.

There are many miscellaneous items and they are too varied to try to mention all of them. But they must *all* be included in an estimate. Most of these items are listed in the specifications and can be picked out by reading the specifications once. At the time of reading the specifications a simple black pencil mark (√) at the right side of the page will keep the place marked for future reference. These check marks should be made at each line that will cost money in the estimate. At the time the item is entered on an estimate, the check mark can be finished thus (X). When all items in the specifications are taken off, then all check marks will be completed double-checks (X). A similar check and double-check system can be used on the drawings, or on the quantity takeoff sheets when transferring quantities to the recapitulation or pricing sheet.

It is best not to mark up the specifications or the drawings too badly because someone else might have to read them after you are through with them. A good estimator simply puts a small pencil check mark at each place where he is working. These pencil check marks do not bother any other estimator using the same drawings and specifications and are very easy to erase if necessary.

HOISTING EQUIPMENT

If a hoist is called for, there are two things that can be done about it in an estimate: (1) figure it in, or (2) forget about it if you are a subcontractor figuring for a general contractor who is going to need a hoist on the job anyway. There's no sense in figuring a hoist in your estimate and forgetting to tell him about it, because this will make your bid too high.

In most cases, a subcontractor bidding to a general contractor should say, "hoist not included, to be furnished by the general contractor," If the general contractor says he wants it included in the estimate, then figure a hoist. Include the erection and dismantling labor for the hoist, the hauling cost, the rental time for the hoist, and the time for an operating engineer if required. The "operating engineer if required" is the most costly part of hoisting.

Some companies will give a price for furnishing a hoist, erected and dismantled, and charge a monthly rental for it. They also rent other equipment. Almost any item can be rented.

In estimating the cost of a hoist, be very careful that you know whether an operating engineer will be needed to run it. If an operating engineer is not needed and the hoist will be used intermittently, it is safe enough to figure that a laborer will run the hoist. In this case, no additional labor should be figured for operating the hoist. On a large job, a percentage—say 50%—of a laborer's time is sometimes added to the estimate for hoist operation.

In general, hoists over 50 feet in height require an operating engineer. Sometimes hoists that are used intermittently on alteration work in existing buildings require an operating engineer. There is no set rule to follow. The best thing to do, if in doubt, is to call the operating engineer's local union headquarters and ask them what is required for a specific job. Even they will not always give you a definite answer. If there are any doubts still remaining, include the operating engineer for the hoist. Figure him for a full day every day that the hoist is on the job, even if the hoist is not used all of the time.

All equipment directly charged to the job is included in the hoisting equipment category. This could include crane rental, buggies, vibrators, finishing machines and so forth. Any piece of equipment that is totally expendable on the job should be charged directly in the job estimate. However, if the equipment is purchased by the office and charged in as general overhead with depreciation taken as an office overhead item, then it should not be charged as a job cost. Be careful to see that each item is charged off either as a direct job cost or as office overhead. But do not charge it off in both places. It would be nice to get paid twice for the same item, but remember that you are trying to be the low bidder.

HEAT PROTECTION

Heat protection is figured and charged for by the cubic yard of concrete that must be protected. Heat protection is figured at the end of the estimate because only then will you know how much concrete needs to be protected. If heat protection will be needed from December 1 to March 1, then the heat protection season is 3 months, or 13 weeks long.

By figuring a time schedule for placing concrete, the amount of concrete that will be placed between December 1 and March 1 can be easily calculated. Add these together to obtain the total number of concrete yards that must be protected. Another way to do this is to figure the average amount of concrete placed weekly and multiply it by 13 weeks. For example, 100 cubic yards per week times a 13-week heating season would be 1,300 cubic yards to heat and protect.

As another example, assume that the concrete portion of a job is estimated to start January 1. Heat protection, therefore, would be needed for the eight weeks of January and February. If a crew averages 100 cubic yards of concrete per week for the first 8 weeks, then heat protection would be needed for only the first 800 cubic yards of concrete.

Once the number of cubic yards of concrete that need heat protection is known, the cost per cubic yard for material and labor can be worked out.

Heated ready mix costs about $2.00 to $4.00 more per cubic yard. Add another $1.50 to $3.00 per yard to this cost to pay for protecting the concrete with straw, fuel, or canvas. For each cubic yard that has to be heat-protected, use a total of $2.00 to $4.00 in the materials column.

For heat protection labor, figure from ¼ hour to ¾ hour of a laborer's rate per cubic yard to be protected. Footings, for example, should be figured at the ¼ hour rate while shored slabs should be figured at the ¾ hour rate. If a laborer costs $25 per hour, then the labor cost would vary from $6.25 per cubic yard on footings to $18.75 per cubic yard on shored slabs.

If the heat-protection problem will be unusual, the job can be analyzed more closely. This can be done by calculating the length of time required for each individual concrete placing and the number of men required for heat protection to obtain the total number of hours for protecting the concrete. However, since the severity of winter weather is so uncertain, it is not a good practice to analyze heat protection too closely. It is better to price it by the cubic yard of concrete.

TRUCK DRIVER TIME

Another item to include is the cost of a truck driver. This will be charged directly to the job and is not the depreciation on the truck or the cost of gasoline. These figures should be in the office overhead.

Truck driver time is the time a man spends driving a truck. If he is driving a truck, he can not work on the job. So his time must be charged to the job. This cost will depend on the size of the job and the time needed to complete it. As an average, figure that the driver's time on most jobs will be 1 day per week while the job lasts. The driver might handle a truck for 2 hours on Monday, 4 hours on Wednesday and 2 hours on Friday, but he will average about 1 day a week. In other words, if the concrete work on a job is expected to take 8 weeks, add in 8 days for truck driving time. This equates the cost of labor for a truck driver's time to the total length of the job.

If the job is very large, you might figure in a full-time truck driver. Do this by charging 1 week of truck driver's time for each week in which concrete will be placed.

CLEANUP

General cleanup and debris disposal must be charged in the estimate. On small jobs, the regular labor crew can often handle the cleanup so there is no need to charge it as a special item. But on most jobs, especially with architectural supervision, cleanup must be figured as a separate item.

As an average, figure cleanup at the rate of 1 day laborer's time per week while the job lasts. For larger jobs this time might be doubled or more, depending on the type of job or even the type of architectural supervision expected.

Hauling away debris is proportionate to the labor time spent cleaning up the job. Figure one truck load of debris for each week the job lasts. This will cost about $60 to $100 per load depending on the locality, whether you pay for scavenger service or haul the debris away with your own truck. On large jobs it might be necessary to double these figures. This is a matter for the estimator's experience and judgment.

SAW MAN TIME

You may also have to include an allowance for a saw man if the job is large enough to figure a man operating the saw full time. This will occur when the job involves columns, beams and shored slabs and it is necessary to set up a table saw with a full-time operator. This, of course, will be straight carpenter time, figured for the length of time required to form all of the work required for columns, beams, and slabs.

A full-time saw man will seldom be required for footings and foundation walls since these usually are formed with prefabricated panel forms. However, it is easy enough to decide how many weeks of saw man time to allow for formwork at a carpenter's rate per week. Usually, a temporary roofed-over shed, open on two sides, is built to protect the table saw. An allowance of about $300 to $500 in an estimate will cover the cost of this shed, including erecting and dismantling it.

SUPERVISION TIME

The last item in the estimate, before totaling the material and labor columns, should be the carpenter foreman or supervision time. At this point, the estimator analyzes the method of handling the job and gives considerable thought to the supervision required.

On small jobs, a working foreman should be included in the figuring. In this case, a certain part of his time will still be spent in layout and supervision. So, perhaps one or two days per week should be charged for foreman's time.

When a job requires a crew of about 6 carpenters and 3 laborers, full-time supervision by a carpenter foreman should be figured. This is needed for planning work, layout time, and keeping all of the men busy all of the time for utmost production.

When a job requires a crew of about 12 carpenters and 6 laborers, a working labor foreman, or pusher, should be added to the cost. Each job estimated should therefore be carefully analyzed to include the very important and costly item of supervision. Supervision must be included for the full length of the job. The length of the job can be obtained by analyzing the work to be done item by item.

TAXES

Now it is possible to obtain a total material cost and a total labor cost. These two totals should be as close to actual costs as can be estimated, with no "fattened items" included. If a contractor is asked how much he can take off his bid he should not have to look over his material and labor costs again. These should be enough to cover each item of cost, but not any more than is necessary, because every item should be priced as if he wanted the job very much. If you want to "fatten up" your bid, put it in the overhead and profit, where you can tell at a glance *if* you can cut your bid, and *how much* you can cut it.

Next add on state taxes, if they apply. In some states the sales tax is on material only; in other states the tax is on material and labor, and still other states have a direct income tax applied on the gross amount of the work. A contractor must understand the state taxes of the state where he is bidding the work and on what types of jobs they apply. He must be careful to add the tax only on necessary items since this is one more amount that may unnecessarily "fatten up" the bid.

Always estimate all jobs "to get them" as far as all direct costs are concerned. In this way you'll know exactly what you have and you can say, "This is the job cost without office overhead and profit."

Insurance and taxes on labor are other items that are sometimes carelessly applied. In some offices this includes welfare, pension and other items paid on direct labor, otherwise known as "fringe benefits."

If fringe benefits are already included in the unit labor rates, do not include them also in the insurance and taxes category when it is added to the total labor cost. At this point the estimator should sit down with the accountant and examine the insurance and tax package. He should find out what percentage insurance and taxes are of the total labor cost. The estimator will know if his unit prices include fringe benefits, which roughly average 10% of the labor cost. If they do, he should ask the accountant only for the percentage of labor cost for all other insurance and taxes. These should then be added to the total labor cost. These will vary from 20 to 30%, depending on the size and operation of the company, and can be checked with the accountant about every six months for possible variations.

If the estimator does not have the fringe benefits in his unit labor costs, then he must have the accountant include them in the percentage to be added to the total labor cost. Once again, be careful to include fringe benefits, insurance, and taxes on labor costs, but include them in only one place in the estimate.

OVERHEAD AND PROFIT

Once a total material and labor job cost has been determined, add office overhead and profit onto the recapitulation sheet. The job cost should include all items that are concerned with the one job only, or are "expendable" on the job being bid. Equipment rental for a job is a good example of a completely expendable item that belongs in the job cost. Equipment that is owned by a contractor and on which his accountant is charging depreciation in his office overhead is a good example of an item that does not belong in the job cost. This is strictly an office overhead item because the cost must be distributed over many jobs.

The largest and most easily controlled item of office overhead is salaries. The owner's salary, or draw account, is the largest and most flexible. In January of each year the owner should make a projected budget or overhead expense sheet for the coming year. Only the owner or the top man in the company can complete this budget. He can get help from his accountant, estimator, superintendent, or anyone else, but only the owner or top man knows the final answer.

The owner's salary should be carefully calculated in order to keep his budget low. During the year, he can take more if it is a profitable year. Only the owner knows how much and when he intends to give salary raises. These must be included in the budget if it is to be a realistically projected overhead expense.

The budget for last year is a very good guide for next year, but it is only a guide. The owner must anticipate the coming year and plan his budget overhead expense accordingly. Such items as rent, telephone, office supplies, and so forth are routine—with the exception that an increase in the volume of business planned for next year will increase most of these items on a percentage basis. An automobile that is leased for business purposes, even though used also for personal use, can be charged off wholly as a business expense. The same will apply for gasoline credit cards. All new equipment purchases, formwork material, and so forth must be anticipated and included in the budget under overhead expenses. This is not to say that he cannot buy a truck during the year if it was not included in the budget. However, the budget must be revised to properly charge in the cost of the truck or any item added to office overhead during the year.

Once all this information has been compiled, a chart projecting business volume for the forthcoming year should be prepared.

Overhead and profit, of course, are added as one item at the end of a completed estimate, but there is no profit until the overhead has been paid. Table 17-A shows how the actual profit for the year can be figured under a budget with a fixed overhead expense and varying amounts of gross business. The principle remains the same regardless of the amount budgeted for overhead. In this example we have assumed, for the sake of simplicity, an overhead expense of $50,000. A fixed amount of 10% is shown for overhead and profit. This is "fixed" by competition. If you put more than 10% overhead and profit on the job, you will not get any work. This figure of 10% will vary with each contractor, and he can make up a similar table with his own "fixed" overhead and profit as determined by his competition. Observe what happens in the table when we start with a fixed overhead and a fixed markup for overhead and profit.

In each line of the Table 17-A the fixed overhead divided by gross business will provide the percentage of overhead. By subtracting the percentage of overhead from the 10% fixed markup for overhead and profit, you can determine the percentage of profit. This percentage of profit times the gross business will give the actual profit for the year. In line 1, the percentage of overhead exceeds the overhead and profit by about 1%. This leaves a loss of $4,500 for the year. Obviously, the budget overhead is too high for the gross business that is being done. In line 2, the percentage of overhead equals the percentage of overhead and profit. This leaves no profit and is not a good condition because there is no money for business growth. In line 3, 9% of overhead leaves 1% profit, or $5,500. This is the start of a healthy business. In line 4 we see what an increase of $50,000 in gross business on the same fixed overhead can mean in profit.

Table 17-A

Fixed Budget Overhead Next Year	Gross Business Next Year	Fixed Percent Overhead & Profit	Percent of Overhead	Percent of Profit	Actual Profit for Year
$50,000	$450,000	10%	11%+	-1%	$4,500 loss
50,000	500,000	10%	10%	0%	None
50,000	550,000	10%	9%+	1%	5,500
50,000	600,000	10%	$8\frac{1}{3}\%$	$1\frac{1}{3}\%$	10,000

Line 4 dramatically points out that the maximum of gross business must be handled without increasing overhead. Since the largest part of the overhead is office salary. Office payroll must be kept to a minimum. The smaller the office the more important this fact becomes. This has no bearing on an efficiently run job in the field. A contractor can operate at a profit on his jobs but lose money because his office overhead exceeds his overhead and profit markup.

Once a budget overhead has been established for the coming year it should be watched carefully to see that operations are staying within budget. If it becomes necessary to increase the overhead during the year because of increased business, try to make the increase in business pay for the addition to the overhead. The final conclusion to be drawn is that if the business continues to operate in accord with line 1 of the table (where operating overhead is too large for the gross volume of business that can be handled), then the business should be shut down before the invested capital slowly dribbles away. At a loss of $4,500 per year it is a simple matter to calculate how many years it will take to go bankrupt with a given amount of capital.

WAGE INCREASES

All jobs should be figured on the basis of the labor unit prices for the year in which they are being figured. However, we know that some jobs will extend into the next labor increase for the men. Therefore, the labor cost must be adjusted to take care of this increase. Since most union agreements on wages are made effective June 1, this date can be used as an example.

Assume that a job is bid in January and the labor cost is $100,000. You have figured that it will take 4 months to do the job. However, the concrete work will not start until March 1. Therefore the job will last from March 1 to June 30. For the month of June you will pay the higher labor rates. This means that ¼ of the labor cost, or $25,000, will be at the new increased labor rates. If the average increase on wage rate is 10%, you should add 10% of $25,000, or $2,500 to your labor. This $2,500 labor increase should include the insurance and taxes on that amount. On large jobs that last over a year this annual wage increase is a large sum of money.

CONCRETE COST PER CUBIC YARD

All concrete estimates should end with a cost check per cubic yard of concrete in place for the entire estimate. By dividing the total estimating cost by the total number of cubic yards required, you will get an average unit price per cubic yard for the concrete in the job. This average unit price per cubic yard can tell an experienced estimator if his estimate is within reason or if he should start looking for an error.

When an estimator is taking off a job, he forms an opinion as to how much the concrete is worth per cubic yard. By the time he completes the bid, an experienced estimator will have a definite opinion as to the price per cubic yard depending on how much of the job is in footings, walls, columns, beams and slabs. If this rough price per cubic yard is not within 10% of your actual bid per cubic yard, then it is time to look for an error.

Do not give out your bid until you are satisfied with the unit cost per cubic yard. This is the last check that can be made before submitting a bid that will determine your profit or loss on what may be a future job.

REINFORCING STEEL AND MESH

The discussion of reinforcing steel and mesh in concrete work has been left until now because reinforcing is generally handled as subcontract work. As such, it does not belong in the material and labor columns with insurance, taxes, overhead, and profit added to it. The subcontract cost of reinforcing steel and wire mesh should be added only after a bid price for the concrete work, with overhead and profit included, has been reached. A percentage (somewhere between 3 to 5%) of the subcontract cost should also be added for overhead and profit for handling the subcontract work.

On some jobs, it is necessary to take off the reinforcing steel and mesh. The reinforcing bars are taken off in linear feet to the closest 3 inches for each bar size. For lengths over 20 feet, 10% should be added for lapping the bars. These lengths are multiplied by the pounds per linear foot to convert them to a total weight for reinforcing bars needed for the job. Material and labor for reinforcing bars are priced by the pound, and unit prices for material and labor should be checked locally. Reinforcing mesh for slabs is taken off in square feet. The area of the slabs with mesh in them can be taken from the finishing areas, but 10% must be added to these areas for lap and waste to get the area of reinforcing mesh needed in these slabs. Material prices per square foot vary with the size of wire and the spacing. Unit prices for both material and labor per square foot should be checked locally.

Finally, it is most important to be extremely careful when submitting a bid proposal. To protect yourself, give not only the name of the job but also the architect's name, the section of the specifications, the drawing numbers from which the estimator worked and their dates, and any addenda that may have come through. If, as frequently happens, the general contractor provides only the concrete section of the specifications, check the general conditions for the job which always apply to each section of the specifications and may affect the bid.

ITEMS INCLUDED AND NOT INCLUDED

Always list any exceptions to the specifications by noting these as items that are "included" or "not included." The fewer exceptions made in a bid, the more help it is to the person to whom the bid is submitted. If he does not have time to contact the bidder, he must make allowances as best he can for each of the exceptions so he can compare it to the other bids. This will mean that he will add in a comfortably safe allowance for each item marked "not included." This may make your bid look higher than it should be, and the general contractor will not use it. The net result is that it costs him time that he did not have to spare in preparing his bid, and it makes you look as if you did not know how to figure a job properly.

ALTERNATES

All alternates should be figured for the same reason and they should be listed by the architect's number or by some type of description. If the general contractor gives you only the concrete section of the specifications, be sure to ask if there are alternates listed anywhere else in the specifications. Knowing about alternates in advance of taking off the quantities can be most important.

UNIT PRICES

Be particularly wary about unit prices for parts of your work. These items should always be figured cautiously. Architects frequently ask for a unit price per cubic yard of concrete or square foot of formwork. Since concrete can vary up to 100% in cost depending on where it is being placed, and formwork can vary about 300% from footings to columns and beams, a unit price is difficult to give unless a number of exceptions are made to the architect's request.

Since there is no room on the bid form for exceptions, the next best thing is to figure the most costly type of work for additions and the lowest priced work for deductions. Additions for unit prices should include overhead and profit; deductions for unit prices should not include overhead and profit. This last sentence applies to all additions and deductions on a contract. An addition to any contract price should include overhead and profit because of the cost of figuring the additional work. A deduction from any contract should not include overhead and profit because it costs the contractor money to figure out the cost of omitting the work. This procedure is fairly well established and recognized by all good architects. It is also provided for in their specifications and bid documents.

If the bid form allows only one unit price for "adds" and "deducts," an average for the work must be used unless you have a good idea as to whether the actual conditions will result mostly in one or the other. For example, if you are sure that the work will be added, you could use a high unit price with some degree of safety.

CONCLUSION

Remember what your aims should be in estimating. First, you want to be correct in the estimate; second, you want to be the low bidder. With a little carelessness you can always be the low bidder. But you want to be correct, and to be correct and also the low bidder is difficult. It means figuring your quantities accurately, using the lowest material prices possible, and analyzing your labor units to keep them as low as possible. Every amount that goes into the money columns in an estimate should be reviewed for the possibility of cutting it down. This is the only possible way to compete with the consistently careless low bidder who may also carelessly put in too much money.

Do not forget that you have made an estimate when you are awarded a job. Your labor costs should go to the foreman or superintendent on the job in a form he can understand. This does not necessarily mean in dollar figures. Perhaps days of working time would be easier for your operation. Or, perhaps with a certain size crew, the foreman can be told how far the job should have progressed at the end of each week. However this is done, it must be set up by the office and checked constantly. Waiting until the job is 50% complete to find out that you have spent 75% of the money figured for labor is too late. The time to correct a problem is when the job is 10% complete and running 20% of the money figured. If this happens and the estimate is correct, the foreman or superintendent is not doing the work as estimated. The foreman or superintendent carry part of the blame for the higher costs, but it is the responsibility of management to bring costs back into line.

TYPICAL ESTIMATE

The typical estimate on the following pages can be easily understood if a little time is spent studying it. An explanation of some points is important, however.

The quantity takeoff sheets are done first, then the data is put on the recapitulation and pricing sheets. Please note that the item numbers set up on the quantity takeoff are used on the recapitulation sheets for easy reference. On the quantity takeoff sheets the last column is kept clear except for quantities that are to be moved to the recapitulation sheet for pricing. This helps avoid the mistake of taking off a quantity and not pricing it. When every quantity in the last column has been moved to the recap sheet and double-checked thus (ꭙ), we are ready to price the job.

Note the first line of the recapitulation sheet opposite the formwork heading "$340.00 C + 1/2L." This means that the job was based on a daily wage rate of a carpenter and 1/2 laborer. If a 5% wage increase had taken place prior to unit pricing, this line would be amended to read: "$340.00 C + 1/2 + 5%" and the daily wage rate increased to $357.00.

Please refer to sheet R3 of the recap. Here the cost is given without overhead and profit. This was done because the estimate was made for a general contractor who added his overhead and profit on the total bid.

PROJECT __U. S. Post Office__   ESTIMATE NO. __6104-21__

LOCATION __S. 5th Ave. & Pine St.,__   SHEET NO. __R1 of 3__

ARCHITECT __Gen. Services Admin.,__   Smith & Assoc. __Spec's: Sec. 8, Concrete__ —Drgs. 27-1-1, 1-2 & 27-7   DUE DATE __5-8-80__
ENGINEER   to 26 (6-18-79)

SUMMARY BY __EGL__   PRICES BY __EGL__   CHECKED BY __EGL__   CONCRETE

No.	Description	L.	Quantity	Unit	Unit Price	Material Cost June 340.00	Unit Price	Labor Cost Incr. c + ¼ L.
—	**FORMS:**							
1.	Column footings (12)	200	480	S.F.	.25	$ 120.00	1.79	$ 860.00
2.	Wall footings	350	1220	"	.25	305.00	1.02	1245.00
—	2 x 4 Keyways	500	650	L.F.	.33	215.00	.71	462.00
3.	Special footing "A" — at exist. ftg. —F. 1 side	120	270	S.F.	.25	68.00	2.97	802.00
4.	Foundation walls — area ways — 5'-2 H.	250	1405	S.F.	.33	464.00	1.43	2009.00
—	" " — 8' to 12' H.	160	9875	"	.37	3650.00	2.24	22120.00
5.	Boiler pit walls in existing bldg. — 2'-4" H.	200	200	"	.33	66.00	1.79	358.00
11.	Bsmt. stairs on fill — tr. & r.	40	44	L.F.	1.75	77.00	8.93	393.00
13.	Boiler pit walls — 2 x 4 keyway in slab	500	90	"	.33	30.00	.71	64.00
14.	Boiler pit stairs on fill — tr. & r.	15	6	"	2.50	15.00	23.80	143.00
17.	New steps at 5th Ave. entry — tr. & r.	40	50	"	1.75	88.00	8.93	447.00
18.	H.D. 30 + 500 Corruform—lookout slabs—postal insp. off. + gallery	100	530	S.F.	.85	450.00	3.57	1892.00
19.	Shored stairs to bsmt. — tr. & r.	20	68	L.F.	1.75	119.00	17.85	1214.00
—	" " " " — landing	100	16	S.F.	1.00	16.00	3.57	57.00
22 to 25	Curb forms — 1180 + 280 + 350 + 280 =	300	2090	"	.30	627.00	1.19	2487.00
27.	Flag-pole base-circ. form-cone 4'-9 D. x 5'-6 H.	—	1	@	—	75.00	—	90.00
—	Strip & clean forms (laborer)	400	15540	S.F.	—	—	.53	8236.00
	CONCRETE							
1. 2.	33 74 Column footings & wall footings		107	C.Y.	75.00	5350.00	22.50	2408.00
3.	Special footing "A" @ exist. ftg.	6 bag	10	"	75.00	750.00	65.00	650.00
—	" " " —grout in 4" WF — (51 holes) Dry-Pack	3000 lb.	1	"	75.00	75.00	L 2 days	400.00
4.	Foundation walls — 16"		226	"	75.00	16950.00	20.00	4520.00
5.	Boiler pit walls — in exist. bsmt.		2	"	75.00	150.00	100.00	200.00
6, 7.	2" mud coat — bsmt. floor — 2000 lb. 4.5 bag		65	"	60.00	3900.00	27.50	1788.00
7.	6" bsmt. floor on mud coat		193	"	75.00	14475.00	22.00	4246.00
8.	4" slab in areaways (4)	6 bag	4	"	75.00	300.00	100.00	400.00
9.	8" ramp slabs on fill		3	"	75.00	225.00	80.00	240.00
10.	8" dock leveler slab	3000 lb.	1	"	75.00	75.00	80.00	80.00
11.	Bsmt. stairs on fill extr.		3	"	75.00	225.00	50.00	150.00
12.	2" mud coat — boiler pit floor — 2000 lb. 4.5 bag		1	"	60.00	60.00	137.50	138.00
13.	6" boiler pit floor on mud coat	6 bag	2	"	75.00	150.00	68.75	138.00
14.	Patch 6" floor — bsmt. floor @ boiler pit		3	"	75.00	225.00	75.00	225.00
15.	Patch 6" floor — bsmt. floor @ ft. "A"	3000 lb.	5	"	75.00	375.00	75.00	375.00
16.	2" topping on precast 1st floor 2000 lb. 4.5 bag		71	"	60.00	4260.00	25.00	1775.00
17.	New platform & steps @ 5th Ave. entry 3000 lb.		4	"	75.00	300.00	50.00	200.00
18.	3" 530 S.F. Lt. weight fill lookout slabs post. insp. off.		5	"	105.00	525.00	80.00	400.00
19.	Shored stairs intr. to bsmt. 3000 lb.		3	"	75.00	225.00	45.00	135.00
21.	6" paved area — yard — private 3000 lb.		144	"	75.00	10800.00	15.00	660.00
	SUB-TOTALS FWD. TO SH. NO. R2					**$65,780.00**		**$62,007.00**

PROJECT U. S. Post Office   ESTIMATE NO. 6104-21

LOCATION   SHEET NO. R2 of 3

ARCHITECT ENGINEER   DUE DATE 5-8-80

SUMMARY BY EGL   PRICES BY EGL   CHECKED BY EGL   CONCRETE

No.	Description	L.	Quantity	Unit	Unit Price	Material Cost	Unit Price	Labor Cost
—	CONCRETE: BRT. FWD.					$65,780.00		$62,007.00
	CONCRETE: CONT'D					—		—
22.	Curb & apron		27	C.Y.	75.00	2025.00	30.00	810.00
23.	Curb & gutter		7	"	75.00	525.00	35.00	245.00
24.	6" driveway & curb		24	"	75.00	1800.00	25.00	600.00
25.	4" public sidewalks		17	"	75.00	1275.00	25.00	425.00
26.	4" private walks		1	"	75.00	75.00	75.00	75.00
27.	Flagpole base		4	"	75.00	300.00	50.00	200.00
—	CONCRETE TESTS: TOTAL YARDS =		933	"	1.00	933.00	—	—
—	FINISHING:					—		
—	Float finish under asphalt plank & c. tile $\frac{7340}{10380} + \frac{110}{110}$		7450	S.F.	—	—	.24	1788.00
—	2" mud coat bsmt + boiler pit float finish		10490	"	—	—	.24	2518.00
—	Steel trowel bldg. slabs = 22046 − 7450 S.F. =		14596	"	—	—	.60	8758.00
—	Steel trowel lookout slabs lt. wt.		530	"	—	—	1.20	636.00
21.	Float & broom paved parking area		7740	"	—	—	.48	3715.00
22.	Steel trowel 3' apron		700	"	—	—	.75	525.00
22.	Steel trowel curb face		470	L.F.	—	—	1.67	785.00
23.	Steel trowel curb & gutter		140	L.F.	—	—	3.35	469.00
24.	Float & broom driveways		1190	S.F.	—	—	.48	571.00
—	Steel trowel curb face		60	L.F.	—	—	1.67	100.00
25.	Public sidewalks float & broom		1220	S.F.	—	—	.75	915.00
"	Public sidewalks curb		130	L.F.	—	—	1.67	217.00
26.	Private walks float & broom		50	S.F.	—	—	.80	40.00
—	Stairs tr. & riser, 44 + 6 + 50 + 68 Fin.		168	L.F.	—	—	3.35	563.00
	MISC. ITEMS					—		—
1. 2.	Hand excavation grading col. ftgs. & wall ftgs. 580 2070		2650	S.F.	—	—	.67	1776.00
3.	Hand excavation special ftg. "A"		17	C.Y.	—	—	60.00	1020.00
6.	Hand escavation bsmt. floor grading for mud coat		10380	S.F.	—	—	.17	1765.00
—	Hand excavation grade misc. areas slabs etc. (8) bldg.		1255	"	—	—	.24	301.00
—	Parking area hand grade fill		7740	"	—	—	.17	1316.00
—	Curb & gutter etc. grade fill		980	"	—	—	.34	333.00
—	Driveway hand grade fill		1190	"	—	—	.17	202.00
—	Sidewalks hand grade fill 1220 + 50		1270	"	—	—	.34	432.00
3.	Ftg. "A" drill 3" x 6" dp. dowel holes		115	@	—	—	13.75	1581.00
	Ftg. "A" set a. b's & 4 wf 13 in exist. wall		51	@	—	—	4.00	204.00
	Ftg. "A" drill 6" x 6" x 20" hole in exist. wall.		51	@	—	—	30.00	1530.00
13.	Boiler pit 3/4" x 6" rubber dumbbell water stop		90	L.F.	3.00	270.00	2.00	180.00
14.	Boiler pit set 3 x 3 L.		42	"	—	—	2.50	105.00
20.	.34" x 3½" safety treads on stairs		116	"	8.00	930.00	4.00	928.00
25.	½" x 5" exp. joints @ walks		150	"	.35	53.00	1.50	225.00
						$73,966.00		$97,860.00

PROJECT __U. S. Post Office__   ESTIMATE NO. __6104-21__

LOCATION ____   SHEET NO. __R3 of 3__

ARCHITECT ____   DUE DATE __5-8-80__
ENGINEER·

SUMMARY BY __EGL__   PRICES BY __EGL__   CHECKED BY __EGL__   CONCRETE

No.	Description	L.	Quantity	Unit	Unit Price	Material Cost	Unit Price	Labor Cost
	CONCRETE: BRT. FWD					$73,966.00		$97,860.00
	Misc. Items: CONT'D					—		—
	Screed material all slabs		43436	S.F.	.04	1737.00	—	—
	Cure & protect all slabs		43436	”	.20	8687.00	.13	5647.00
	Set base pls. & anchor bolts		10	@	—	—	20.00	200.00
	Window sash reglet 1" x 1½" form 10 x 14'		140	L.F.	.50	70.00	2.50	350.00
	Dampproofing reglet form		390	”	.08	31.00	1.75	683.00
	.008 polyethylene film under slabs on ground basmt. 10380 s.f. + 5% flr.		10900	S.F.	.04	436.00	.13	1417.00
	Liquid hardener all bldg. slabs		14596	”	.10	1460.00	.26	3795.00
	Abrasive aggregate 1/4 lb./s.f. rooms 104 & 105		2670	”	.12	320.00	.15	400.00
	Rub exposed surfaces extr.		980	”	.06	59.00	.45	441.00
	4" stone backfill under curbs & gutters 2250 s.f. + walks + 20%		33	C.Y.	12.00	396.00	—	—
	8" stone backfill under drives & parking 8930 s.f. + 20%		250	”	12.00	3000.00	—	—
	3/4" x 6" exp. joints parking & drives + pour top.		1350	L.F.	.50	675.00	.20	270.00
	Set ¾"∅ x 2' dowels & sleeves 1210'/1' =		1210	@	—	—	7.00	8470.00
	½" x 6" exp. joint bsmt. flr. @ wall + 5%		520	L.F.	.50	260.00	—	—
	Set 10 (15.3 x 3") anchors @ exist. fon. wall		18	@	—	—	21.00	378.00
	Repair crack in floor		40	L.F.	2.00	80.00	20.00	800.00
	Precast conc. window stools bsmt. 10 x 5'		50	”	8.00	400.00	7.00	350.00
	Equipment — special		None	—	—	—	—	—
	Hoist		None	—	—	—	—	—
	Truck Labor		10	Wks.	—	—	200.00	2000.00
	Cleanup haul debris		10	Wks.	60.00	600.00	200.00	2000.00
	Heat protection: start Jun. 1st to Aug. 15th		None	—	—	—	—	—
	Sawman		None	—	—	—	—	—
	Foreman		10	Wks.	—	—	1240.00	12400.00
	TOTALS					$92,177.00		$137,461.00
	L. & T.		25%					$34,365.00
	MATERIAL							$92,177.00
	COST (no O. & P.)							$264,003.00
			$264,003.00 = $283.00/C.Y.					
	Included:		933 C.Y.					

Tests, hand excavation & grading, stone fill under parking area,

drives & curbs, bsmt. window stools, & foreman.

Not Included:

Rein. steel & mesh, membrane dampproofing, overhead & profit.

QUANTITY SHEET

PROJECT __U. S. Post Office__ ESTIMATOR __EGL__ ESTIMATE NO. __6104-21__

LOCATION __S 5th Ave. & Pine St.__ EXTENSIONS __EGL__ SHEET NO. __1 of 6__

ARCHITECT __General Services Admin.__ __Smith & Associates__ CHECKED __EGL__ DUE DATE __5-8-80__
ENGINEER

CLASSIFICATION __Drgs. 27-1-1, 27-1-2 & 27-7 to 27-26 (6-18-79); Spec's., Sec. 8__ CONCRETE

No.	Description	No.	Dimensions W. x	L. x	D.	(2W+2L) D Forms S.F.	Hand Exc.Area W.L. S.F.	Conc. W.L.D. C.F.		Estimated Quantity	Unit
1.	COL. FOOTINGS										
	J	7	7'6 x 7'6	x	1'6	315	394	591	(12) Forms	480	SF
	H	1	9'6 x 9'6	x	1'9	67	90	158			
	L	2	6'6 x 6'6	x	1'5	74	83	118	Hand Exc.	580	SF
	M	2	3' x 3'	x	1'	24	12	12			
	TOTALS	12				480	580	879	C.F. = Conc. =	33	CY
2.	WALL FOOTINGS		D.	W.	L.	2. D. L. S.F. Forms	W.L. Hand Exc. Area-S.F	W.D.L. C.F. Conc.			
	C		1' x	2'	x 30'	60	60	60			
	√/16		8" x	1'8	x 25'	34	42	29			
	B		1' x	4'6	x 32'	64	144	144			
	G		1' x	5'6	x 20'	40	110	110			
	F		1' x	3'4	x 27'	54	90	90			
	C		1' x	2'	x 11'	22	22	22			
	B		1' x	4'6	x 21'	42	95	95	Forms	1220	SF
	D		1' x	2'	x 33'	66	66	66			
	E		1' x	3'6	x 93'	186	326	326			
	√/45		1' x	10'9	x 8'6	39	94	94	Hand	2070	SF
	D		1' x	2'	x 33'	66	66	66	Exc.		
	B		1' x	4'6	x 12'	24	54	54			
	G		1' x	2'	x 11'	22	22	22	2 x 4 Keys	650	LF
	F		1' x	3'4	x 20'	40	67	67			
	C		1' x	2'	x 15'	30	30	30			
	B		1' x	4'6	x 6'	12	27	27			
	AREA FOOTINGS		8" x	1'8	x 108'	144	180	120			
	√/50		1' x	5'	x 102'	204	510	510			
	√/20		10' x	1'8	x 36'	72	60	50			
						-1		+5			
	TOTALS				644'	1220	2070	1988	C.F. Conc.	74	CY
3.	SPECIAL FOOTING "A" @ Exist. Ftg.					IS S.F. Forms	Hand Exc.C.F.	C.F. Conc.	IS		
	2 Sides - "A"		1'2 x	1'W	x 226'	270	452	264	Forms	270	SF
	2 Sides - "A" Grout holes		3' x	6"	x 113	—	—	4	Hand Exc.	17	CY
	TOTALS					270	452	268	Conc.	10	CY
	Drill 3" x 6" dp - dowel holes — 226'/2'									115	@
	Set A.B.'s & 4 WF 13 in existing wall									51	@
	Drill 6" x 6" x 20" hole in existing wall									51	@

PROJECT U. S. Post Office   **ESTIMATOR** EGL   **ESTIMATE NO.** 6104-21

LOCATION   **EXTENSIONS** EGL   **SHEET NO.** 2 of 6

ARCHITECT ENGINEER   **CHECKED** EGL   **DUE DATE** 5-8-80

CLASSIFICATION

No.	Description	No.	Dimensions W x H x L	2 H.L. Forms	W.H.L. Conc.			Estimated Quantity	Unit
4.	FDN. WALLS			S.F.	C.F.				
	✓ 15 − ✓ 16		1'4 x 11' x 33'	726	483				
	✓ 15 − ✓ 16 window box outs	2	x4'x4'8x1'4	50	-50				
	✓ 46		1'4 x 11' x 92'	2024	1346				
	✓ 46 seat box out		4" x 1' x 40'	52	-13				
	✓ 46 window box out	3	1'4 x 4'8 x 4'	75	-75				
	✓ 17		1'4 x 11' x 19'	418	278				
	✓ 17 seat box out		4" x 1' x 19'	26	- 6				
	✓ 19		1' x 11' x 21'	462	231				
	✓ 19		1'4 x 11' x 14'	308	205				
	✓ 18		1' x 8'2 x 4'	66	33				
	✓ 18		1' x 10' x 4'	80	40				
	✓ 47		1'4 x 10'10 x 82'	1790	1182				
	✓ 47 seat box out		4" x 1' x 82'	110	-27				
	✓ 45		1' x 8' x 5'6	88	44				
	✓ 45 - bot		1'4 x 8' x 22'	352	234				
	✓ 45 - top		1' x 3' x 22'	132	. 66				
	✓ 45 seat box out		4" x 1' x 15'	20	-5				
	Ramp @ D.		1'4 x 11' x 14'	308	205				
	Ramp @ D.		1' x 8' x 4'	64	32				
	Ramp @ D.		1' x 10' x 4'	80	40				
	Ramp @ D.		1' x 11' x 21'	462	231				
	— —		1'4 x 11' x 10'	220	147				
	— — Seat box out		4" x 1' x 10'	15	-3				
	S. wall		1'4 x 11' x 52'	1144	761				
	S. wall		1'4 x 11' x 5'	110	74				
	S. wall seat box out		4" x 1' x 5'	8	—				
	N. & S. window box outs	5	1'4 x 4'8 x 4'	125	-125				
	26' + 35' + 35' + 40' Area walls		8" x 5'2 x 136'	1405	472				
	✓ 20 stair wall		1' x 8' x 26'	416	208				
	✓ 20 stair wall		1' x 8'8 x 8'	139	70				
	TOTALS			11280	6079				
	AREA WALL FORMS		5'-2 H.					1405	SF
	WALL FORMS		8' to 2' H.	11280−1405 =				9875	SF
	CONCRETE			6079 C.F. =				226	CY

QUANTITY SHEET

PROJECT **U. S. Post Office** ESTIMATOR **EGL** ESTIMATE NO. **6104-21**

LOCATION EXTENSIONS **EGL** SHEET NO. **3 of 6**

ARCHITECT CHECKED **EGL** DUE DATE **5-8-80**

ENGINEER

CLASSIFICATION

No.	Description	No.	W	x H x	L	2.H.L. Forms	W.H.L. Conc.			Estimated Quantity	Unit
5.	New boiler pit walls (in exist. boiler room)					S.F.	C.F.				
	Walls		6"	x 2'4 H x	4'2	200	49	2'4 H	Forms	200	SF
									Conc.	2	CY
			W	L		W.L. S.F. Area					
6.	2" conc. mud coat bsmt. floor										
			89'4	x 19'8		1758		Hand Grade		10380	SF
			109'4	x 31'	70 s.f.	3320					
			71'8	x 79'	360 s.f.	5302		Float finish		10380	SF
						10380 S.F.			Conc.	65	CY
7.	6" conc. slab bsmt. floor								Finish	10380	SF
						10380 S.F.			Conc.	193	CY
			W	L	D	W.L. S.F. Area	W.L.D. C.F. Conc.				
8.	4" conc. slab in areaways										
			3'6	x 17'5	0'4	61					
			2'6	x 28'10		72		Hand Grade		320	SF
			2'6	x 28'10		72					
			3'6	x 31'8		111		Finish		320	SF
						+ 4					
						320=107 C.F.			Conc. =	4	CY
						W.L. S.F.	W.L.D C.F.				
9.	Ramp slabs - on fill		W	L	D						
		2	x 15'	x 4' x	0'8"	120	80	Finish		120	SF
								Conc.		3	CY
						W.L. S.F.	W.L.D. C.F.				
10.	Dock leveler slab		W	L	D						
			9'	x 5'6	x 0'8"	50	33	Finish		50	SF
								Conc.		1	CY
11.	Bsmt. stairs on fill + area						C.F. conc.	tr. & r.			
	Stairs - 11 Tr. & R. x 4'					44 LF.	44	Form		44	LF
	Landing - top -		4'	x 5'	x 0'9"	20 SF.	15	Finish Tr. & R.		44	LF
	Area" bot.		4'	x 12'6"	x 0'4"	50 SF.	17	Land. Finish		70	SF
							76	Conc.		3	CY

QUANTITY SHEET

PROJECT: U. S. Post Office   ESTIMATOR: EGL   ESTIMATE NO. 6104-21

LOCATION: EXTENSIONS: EGL   SHEET NO. 4 of 6

ARCHITECT ENGINEER: CHECKED: EGL   DUE DATE 5-8-80

CLASSIFICATION:

No.	Description	No.	W	x L	x D	W.L. Area S.F.	W.L.D. Conc. C.F.			Estimated Quantity	Unit
12.	New boiler pit 2" mud coat					S.F.	C.F.				
			16'	x 7'	0'2	110	19	Hand Grade		110	SF
								Float Fin.		110	SF
								Conc.		1	CY
13.	New boiler pit 6" slab										
			16'	x 7'	0'6	110	55	Finish		110	SF
								Conc.		2	CY
	2 x 4 key - top & bot.									90	LF
	¾" x 6" rubber dumb bell water stop									90	LF
14.	6" patch floor @ boiler pit + stairs					S.F.	C.F.				
	6"		2'6	x 20'		50	25	Form Stairs on fill		6	LF
			3'5	x 6'		21	11	Grade		130	SF
			2'	x 6'		12	6	Fin.		130	SF
			2'	x 20'		40	20	Tr. & R. Fin.		6	LF
	Stairs on fill - 4 Tr. & R. x 1'6					L.F. 6	6	Set 3 x 3 L =		42	LF
	TOTALS					130	68	Conc. =		3	CY
15.	Patch 6" floor @ Ftg. "A"					S.F.	C.F.				
			3'W	x 110'		330	115	Hand Grade		330	SF
								Finish		330	SF
								Conc.		3	CY
16.	2" topping on precast 1st floor					S.F.					
	Basmt. floor					+10380					
	4" bearing on wall		4"	x 171'		+ 57		Finish		10390	SF
	Less stairs		4'	x 16'		- 64					
						10390 S.F. x 2" = 1732 CF + 10% conc. 71					CY
17.	New platform & steps at 5th ave. entry						C.F.				
	Platform on fill		4' x	25'	x 6"	100	50	Hand Grade		125	SF
	Steps	2R x	25'			50 L.F. 50		Finish		100	SF
								Tr & R. Fin.		50	LF
						100 C.F. = Conc.				4	CY

PROJECT___ U. S. Post Office _____ ESTIMATOR ___EGL___ ESTIMATE NO. ___6104-21___

LOCATION_____ EXTENSIONS ___EGL___ SHEET NO. ___5 of 6___

ARCHITECT_____ CHECKED ___EGL___ DUE DATE ___5-8-80___
ENGINEER

CLASSIFICATION_____ CONCRETE

No.	Description	No.	W	L	D	W.L.	W.L.D.		Estimated Quantity	Unit
18.	Lookout slab - Postal Inspector's Office (27-10)					30 S.F.	C.F. 8			
	On corruform -		3'	x 10'	x 3"			Corruform	30	SF
								Finish	30	SF
								Lt. Wt. Conc.	1/3	CY
19.	Shored stairs & landing: to bsmt.						C.F.			
	Tr. & R. -	17 x	4'	Lg.		66 L.F.	68	Tr. & R. Form	68	LF
	Landing		4'	x 4'	x 9"	16 S.F.	12	Form landing	16	SF
								Fin. Tr. & R.	68	LF
							80	Fin. Landing	16	SF
								Conc.	3	CY
20.	¾" x 3½" steel safety treads on stairs	17 x	68 4' + 12	48 x 4' =					116	LF
						W.L. S.F. Area				
21.	Paved area - 6" - private		W	L						
			20'	x 21'		420		Hand grade	7740	SF
			99'	x 70'		6930				
			20'	x 19'4		390		Finish	7740	SF
	TOTAL					7740 S.F. = 3870 C.F. =			144	CY
22.	Curb & 3' apron - private					Form Curb - S.F.	Conc. C.F.			
	Curb		6"W	x 2'6 Av x120'		540	150	Curb forms	1180	SF
	Curb		6"W	x 2'6 Av x 20'		90	25	3' x 285' Hand grade	700	SF
	Curb		6"W	x 2'6 Av x 95'		430	119	Finish apron	700	SF
	3" apron		6"Dp.	x 3' W.x 235'		120	353	2 x 235' Face Finish curb	470	LF
	Add. conc. at paved area		6"	x 6"	x 235'	—	59			
	TOTALS					1180	706 C.F. = Conc.		27	CY
23.	& gutter Curb: S.D. 6-2-7A - private					S.F. Forms	C.F. Conc.			
	Gutter 21' + 21' + 78' + 20'		6"	x 2'	x 140'	140	140	Curb Forms	280	SF
	Curb		6"	x 6"	x 140'	140	35	Fin. C. & G.	140	LF
	TOTALS					280	175	Conc. =	7	CY

PROJECT __U. S. Post Office__ ESTIMATOR __EGL__ ESTIMATE NO. __6104-21__

LOCATION _____ EXTENSIONS __EGL__ SHEET NO. __6 of 6__

ARCHITECT _____ CHECKED __EGL__ DUE DATE __5-8-80__
ENGINEER

CLASSIFICATION _____ CONCRETE

No.	Description	No.	W	x L		Forms	W.L. Area	W.L.D. Conc.		Estimated Quantity	Unit
24.	+ Curb Conc. drives - public - det. 6 - 2-3					S.F.	S.F.	C.F.			
	6" drives: -	2 x	28'	x 17'		260	952	476	Edge Forms	350	SF
	6" drives	4 x	10'	x 6'		—	240	120	Hand grade	1190	SF
	Curb	4 x	6"W.x	2"H.x15"		90	—	30	Finish	1190	SF
									Curb Finish	60	LF
	TOTALS					350	1190	626 =	Conc. =	24	CY
			W	L		W.L.	W.L.D				
25.	Sidewalks - public - 4" - S.D. 6-2-5					S.F. Forms	S.F. Area	C.F. Conc.			
	Walk - 4"	2 x	6'	x 20'		—	240	80	Curb Forms	280	SF
	Walk - 4"		7'	x 23'		—	161	54	Grade	1220	SF
	Walk - 4"		7'	x 90'		—	630	210	Fin. Curb	130	LF
	Walk - 4"		11'	x 17'		—	187	63	Walk Finish	1220	SF
	Curb		5"	x 5"x130'		200	—	33	Exp. Jts ½" x 5"	150	LF
	Curb		6"	x 2' x 20'		80	2	20			
	TOTALS					280	1220	460 C.F. = Conc.-		17	CY
						W.L.	W.L.D.				
26.	Walks - private		W	L	D	S.F.	C.F.				
	@ flagpole		2' x	5'	0'4	10	4		Hand Gr.	50	SF
	S. stairs		4' x	10'	0'4	40	14		Finish	50	SF
						50	18		Conc.	1	CY
						3.14R.2H					
27.	Flagpole base - 6-3-1 E, Type "C"					Conc.			4'9 Dia. Forms	1	@
	4'9 D. Circ.		3.14 x 2'-4½ x 5'6 H.			98 C.F. =			Conc.	4	CY

INDEX

Other Practical References

Masonry & Concrete Construction

Every aspect of masonry construction is covered, from laying out the building with a transit to constructing chimneys and fireplaces. Explains footing construction, building foundations, laying out a block wall, reinforcing masonry, pouring slabs and sidewalks, coloring concrete, selecting and maintaining forms, using the Jahn Forming System and steel ply forms, and much more. Everything is clearly explained with dozens of photos, illustrations, charts and tables. **224 pages, 8½ x 11, $17.25**

National Construction Estimator

Current building costs in dollars and cents for residential, commercial, and industrial construction. Estimated prices for every commonly used building material. The manhours, recommended crew and labor cost for installation. Includes Estimate Writer, an electronic version of the book on computer disk - at no extra cost on 5¼" high density (1.2Mb) disk. The 1991 National Construction Estimator and Estimate Writer on 1.2Mb disk cost **$22.50** (Add $10 if you want Estimate Writer on 5¼" double density 360K disks or 3½" 720K disks.) **576 pages, 8½ x 11, $22.50. Revised annually**

Construction Estimating Reference Data

Collected in this single volume are the building estimator's 300 most useful estimating reference tables. Labor requirements for nearly every type of construction are included: site work, concrete work, masonry, steel, carpentry, thermal & moisture protection, doors and windows, finishes, mechanical and electrical. Each section explains in detail the work being estimated and gives the appropriate crew size and equipment needed. **368 pages, 8½ x 11, $26.00**

Building Layout

Shows how to use a transit to locate the building on the lot correctly, plan proper grades with minimum excavation, find utility lines and easements, establish correct elevations, lay out accurate foundations and set correct floor heights. Explains planning sewer connections, leveling a foundation out of level, using a story pole and batterboards, working on steep sites, and minimizing excavation costs. **240 pages, 5½ x 8½, $11.75**

Contractor's Guide to the Building Code Revised

This completely revised edition explains in plain English exactly what the Uniform Building Code requires, and shows how to design and construct residential and light commercial buildings that will pass inspection the first time. This revised edition is based on the 1988 code, the most recent, and covers many changes made since then. Also covers the Uniform Mechanical Code and the Uniform Plumbing Code. Suggests how to work with the inspector to minimize construction costs, what common building shortcuts are likely to be cited, and where exceptions are granted. **544 pages, 5½ x 8½, $24.25**

How to Succeed With Your Own Construction Business

Everything you need to start your own construction business: setting up the paperwork, finding the work, advertising, using contracts, dealing with lenders, estimating, scheduling, finding and keeping good employees, keeping the books, and coping with success. If you're tired of working for someone else and considering starting your own construction business, all the knowledge, tips, and blank forms you need are in this book. **336 pages, 8½ x 11, $19.50**

Masonry Estimating

Step-by-step instructions for estimating nearly any type of masonry work. Shows how to prepare material take-offs, how to figure labor and material costs, add a realistic allowance for contingency, calculate overhead correctly, and build competitive profit into your bids. **352 pages, 8½ x 11, $26.50**

Excavation & Grading Handbook Revised

Explains how to handle all excavation, grading, compaction, paving and pipeline work: setting cut and fill stakes (with both bubble and laser levels), working in rock, unsuitable material or mud, passing compaction tests, trenching around utility lines, setting grade pins and string line, removing or laying asphaltic concrete, widening roads, cutting channels, installing water, sewer and drainage pipe. This is the completely revised edition of the popular guide used by over 25,000 excavation contractors and supervisors. **384 pages, 5½ x 8½, $22.75**

Concrete and Formwork

This practical manual has all the information you need to select and pour the right mix for the job, lay out the structure, choose the right form materials, design and build the forms and finish and cure the concrete. Nearly 100 pages of step-by-step instructions cover the actual construction and erecting of nearly all site fabricated wood forms used in residential construction. **176 pages, 8½ x 11, $14.50**